Deconvolution of Images and Spectra

SECOND EDITION

Edited by
Peter A. Jansson
College of Optical Sciences
University of Arizona
Tucson, Arizona

Dover Publications, Inc.
Mineola, New York

Bibliographical Note

This Dover edition, first published in 2012, is a corrected, unabridged republication of the Second Edition of *Deconvolution of Images and Spectra,* originally published in 1997 by Academic Press, Inc., San Diego.

Any comments or questions concerning this book may be addressed to the author at pjansson@optics.arizona.edu

Library of Congress Cataloging-in-Publication Data

Deconvolution of images and spectra / edited by Peter A. Jansson. — Dover ed.
   p. cm.
   Originally published: 2nd ed. New York : Academic Press, 1997.
   Includes index.
   ISBN-13: 978-0-486-45325-5 (pbk.)
   ISBN-10: 0-486-45325-1 (pbk.)
   1. Spectrum analysis—Deconvolution. I. Jansson, Peter A.

QC451.6.D45 2012
543'.5—dc22

2006050287

Manufactured in the United States by Courier Corporation
45325102
www.doverpublications.com

Contents

7 Deconvolution Examples   236

Paul Benjamin Crilly, William E. Blass, and George W. Halsey

8 Application to Electron Spectroscopy for Chemical Analysis   264

Robert D. Davies and Peter A. Jansson

9 Deconvolution in Optical Microscopy   284

Jason R. Swedlow, John W. Sedat, and David A. Agard

Contributors

Numbers in parentheses indicate the pages on which the authors' contributions begin.

David A. Agard (284), Howard Hughes Medical Institute and Department of Biochemistry and Biophysics, The University of California, San Francisco, California 94143

William E. Blass (200, 236), Department of Physics and Astronomy, The University of Tennessee, Knoxville, Tennessee 37996

Paul Benjamin Crilly (182, 236), Department of Electrical Engineering, The University of Tennessee, Knoxville, Tennessee 37996

Robert D. Davies (264), Central Research, E. I. du Pont de Nemours and Company, Wilmington, Delaware 19898

B. Roy Frieden (361), Optical Sciences Center, The University of Arizona, Tucson, Arizona 85721

Ronald L. Gilliland (310), Space Telescope Science Institute, 3700 San Martin Drive, Baltimore, Maryland 21218

George W. Halsey (200, 236), Department of Physics and Astronomy, The University of Tennessee, Knoxville, Tennessee 37996

Robert J. Hanisch (310), Space Telescope Science Institute, 3700 San Martin Drive, Baltimore, Maryland 21218

Samuel J. Howard (397, 427), Physics Department, The Florida State University, Tallahassee, Florida 32306

Peter A. Jansson (1, 42, 76, 107, 264), Central Research, E. I. du Pont de Nemours and Company, Wilmington, Delaware 19898

Robert J. Marks II (476), Department of Electrical Engineering, The University of Washington, Seattle, Washington 98195

John W. Sedat (284), Department of Biochemistry and Biophysics, The University of California, San Francisco, California 94143

Jason R. Swedlow (284), Department of Cellular and Molecular Pharmacology, The University of California, San Francisco, California 94143

Richard L. White (310), Space Telescope Science Institute, 3700 San Martin Drive, Baltimore, Maryland 21218

ix

To my father,
the late **John H. Jansson,**
remembered for devotion to his family,
and to the advancment of science and education

Preface

The literature on deconvolution is rich with the contributions of many investigators. These contributions are, however, scattered among journals devoted to numerous specialties. Until publication of our prior work in 1984, *Deconvolution With Applications in Spectroscopy*, no single volume had provided both an overview and the detail needed by a newcomer to this field. When a specific need arose, a recent journal article or the advice of a colleague often initiated a considerable, but not always successful, effort. Indeed, the lack of a suitable volume fostered this approach. Although it was sorely needed and the means to a scientific goal, deconvolution was a significant distraction. These circumstances created the opportunity for a book to provide an understanding of the relationships among deconvolution methods, and to present practical methods and results for the researcher's own field. Furthermore, other edited collections often lacked the "glue" of a common notation, accessible symbol definitions, a good index, and a unifying tutorial discussion.

The authors of our 1984 volume worked hard to create a work that was self-contained, readable, practical, and tutorial. Since publication, its material has been adapted within spectroscopy and to other fields with gratifying results. Deconvolution is now playing a more important role in the advancement of science. As predicted in 1984, image processing has been a major beneficiary of the physical-realizability constraints we promoted. Consequently, numerous and diverse advances have accrued. Among these are the discovery of the nucleus of Halley's comet and new insights in cell biology. We believe the 1984 volume can take some of the credit for making available the information that enabled these advances.

In the decade since the 1984 volume, the deconvolution field has advanced in other ways, too. Researchers have explored and refined projections onto convex sets. Understanding of relaxation methods has improved. Artificial neural networks have enjoyed rebirth and application to deconvolution. Before the 1984 volume, mere mention of deconvolution aroused well deserved skepticism. Today, the situation is different. Ten years ago, we knew that the positivity constraint held the key to practical applications yielding reliable results. Now the promise of this constraint has largely been realized.

Continuing interest in deconvolution is timely for at least two reasons: (1) the inexorable advance of computers has given us hardware for the computation-intensive programs often needed, and (2) the advantages of modern nonlinear constrained methods have made deconvolution worth the effort. Two such advantages—reduced noise sensitivity and superresolving capability—produce results that are sometimes astonishing when judged by the standards of traditional linear methods. The clear message in this work, as in the 1984 volume, is that reexamination of dogmatic "truths" can sometimes yield surprises. For years we were led to believe that frequencies beyond the cutoff of an observing instrument were not recoverable. How could they be? The information was not apparent in the data. We have learned to look elsewhere for the information needed to restore those frequencies. We found it in seemingly unimportant physical-realizability constraints such as lower and upper bounds on the solution. All the contributors to the present volume have extensive experience with one or more of the constrained nonlinear methods.

The reader of *Deconvolution of Images and Spectra* should have a background in physical science, engineering, mathematics, or statistics. A working knowledge of calculus is assumed. Previous experience with convolutions or Fourier transforms would be helpful but is not absolutely necessary, because the required material is developed. In this respect, the present work is reasonably self-contained. Throughout, we have emphasized methods that work; methods of purely theoretical interest are described or referenced only to round out the picture. *Deconvolution of Images and Spectra* is not a book of proofs. Though we recognize the place for rigor, the detail necessary would displace, in the limited space available, material conveying insight into a broad range of methods. Because it is not possible to be totally comprehensive in a book of this size and scope, we have covered areas pivotal to the evolution of the most practical and effective methods. Where details are omitted, we refer to the literature.

The present work, like its predecessor, conveys an understanding of the field and presents under one cover a selection of the most effective and practical techniques. It draws heavily on the 1984 volume: all ten of its chapters were retained. Of these chapters, one was substantially revised by including new material on neural networks and other recent advances.

Four were revised to a lesser degree. Five chapters remain essentially unchanged. Four new chapters present material on projections onto convex sets, convergence of relaxation methods, and adaptations to image processing in microscopy and astronomy.

The fourteen chapters are organized into four sections. The first section includes four chapters (Jansson) that introduce the reader to basic concepts and progress through a survey of both traditional linear and modern nonlinear methods. This section concludes with Chapter 5 (Crilly), which examines the convergence of relaxation algorithms. In the next section, Chapters 6 (Blass and Halsey), 7 (Crilly, Blass, and Halsey), and 8 (Davies and Jansson) detail specific applications of a proven method. This method was among the first to take effective advantage of a lower bound (positivity) and was the first to use an upper bound. It remains one of the most useful. In these chapters, we describe its specific applications to the field of high-resolution infrared spectroscopy via three different types of experimental apparatus: the dispersive grating spectrometer, the interferometer, and the tunable diode laser. We also describe applications to Raman spectroscopy, gamma-ray spectroscopy, gas chromatography, and electron spectroscopy for chemical analysis (ESCA). The third section, which contains Chapters 9 (Swedlow, Sedat, and Agard) and 10 (Hanisch, White, and Gilliland), details advances made in restoration of images from cell biology and astronomy. The fourth section, Chapters 11 (Frieden), 12 and 13 (Howard), and 14 (Marks), contains material suited to use by researchers expanding the horizons of deconvolution by investigating new methods. It covers topics of maximum probable estimation, Fourier spectrum continuation, and projections onto convex sets.

Much of this volume has general applicability. Chapters 1, 3, 4, 5, and 14 are not at all confined to spectroscopy or image-processing applications. Significant portions of Chapters 11, 12, and 13 are easily generalized. The remaining chapters, though dealing with specifics of spectroscopy or image processing, can function as an application guide in new areas.

We have assisted the reader by providing a symbol list at the beginning of each chapter, a consistent notation throughout for the principal quantities, and an effective index. As a whole, the volume provides an introduction to deconvolution for the scientist who needs it but does not know how to begin. The volume should also serve as a good beginning for the scientist or student preparing for research aimed at improved deconvolution methods. It provides a guide to the literature that will enable the researcher to quickly identify many of the important contributions. Although specific examples are chosen from spectroscopy and image processing, deconvolution's general applicability should make the volume useful in diverse fields that yield both single- and multidimensional data. It should find application in signal analysis and statistics, to name but two of many other possibilities.

Our final note is a sad one. Sam Howard, author of Chapters 12 and 13, passed away on July 8, 1995. His contributions to our field will be missed.

Acknowledgments

I begin by thanking my family for their encouragement during the writing and editing, which was all done at home. Unconditional support from my wife Lihong contributed in no small way to the quality of the effort. It is deeply appreciated.

For technical criticism and suggestions related to new material, I thank P. B. Crilly, P. Larsson, A. J. Owens, and T. J. Shan. For the reading of my contributions to the 1984 volume, much of which has been little changed, I am indebted to R. H. Hunt, J. G. Yorker, B. Rubin, and B. R. Frieden.

For their collaboration during the research reported in Chapter 8, I am indebted to my colleagues in DuPont's ESCA program, who are separately acknowledged at the end of that chapter. Over the years, I have learned much from my collaboration with colleagues in DuPont's image-processing program and am thankful to them for the insights they have provided.

Valuable suggestions on style and editorial assistance were provided by D. M. Welsh. Her help has significantly improved my contribution and I am duly grateful. My work with the literature was significantly aided by the competence and willingness of DuPont's Lavoisier Library staff. Last, but not least, I wish to acknowledge the DuPont Company for providing a flexible environment and helpful information resources.

Chapter 1 | Convolution and Related Concepts

Peter A. Jansson

College of Optical Sciences, University of Arizona, Tucson

List of Symbols

a, b, g	functions $a(x), b(x), g(x)$, with no explicit dependences shown
a', b', g'	first derivatives of functions a, b, g
a_n, b_n, g_n	sampled values of $a(x), b(x), g(x)$
$a(x), b(x), g(x)$	functions used to illustrate properties of convolution
A, B, G	functions $A(\omega), B(\omega), C(\omega)$, with no explicit dependences shown
$A(\omega), B(\omega), G(\omega)$	Fourier transforms of $a(x), b(x), g(x)$
c	constant
C	constant
$\text{cas}(x)$	$\cos(x) + \sin(x)$
$f(x), F(\omega)$	function and Fourier transform, respectively
$F'(\omega), F''(\omega)$	first and second derivatives of $F(\omega)$, respectively

$G(\zeta)$	Fourier transform of $g(x)$ given by alternative convention
$H(x)$	1 when $x > 0$, 0 when $x \le 0$
$i(x), i$	"image" data that incorporate smearing by $s(x)$; sometimes includes noise
$i_p(x)$	ideal noise-free image data
$\hat{i}_M(v, x)$	model representing idealized image data
j	imaginary operator such that $j^2 = -1$
$n(x)$	additive noise
N_a, N_b, N_g	number of samples available for functions a, b, g
$o(x), o$	"object" or function sought by deconvolution, usually the true spectrum, but also the instrument function when this is sought by deconvolution
$\hat{o}(x)$	estimate of $o(x)$
$\hat{o}_M(v, x)$	model of true spectrum or true object
q	independent variable given in scaled units of Gaussian half-widths ($q = x\sqrt{\ln 2}/\Delta x_G$)
$\mathrm{rect}(x)$	rectangle function having half-width $\frac{1}{2}$
$\mathrm{sgn}(x)$	1 when $x > 0$, -1 when $x \le 0$
$\mathrm{sinc}(x)$	$(\sin \pi x)/\pi x$
$\mathrm{Si}(x)$	$\int_0^x [(\sin x')/x'] \, dx'$
$s(x), s$	spread function, usually the instrument function, but also spreading due to other causes
$s(x, x')$	general integral equation kernel; shift-variant spread function
$v(x)$	function used to illustrate Fourier transform
$V(\omega)$	Fourier transform of $v(x)$ given by third system
x, x'	generalized independent variables and arguments of various functions
$\langle x \rangle, \langle x^2 \rangle$	first and second moments of a distribution of x
$z(x), z_1(x), z_2(x)$	functions used to illustrate the Hartley transform
$Z(\omega), Z_1(\omega), Z_2(\omega)$	Hartley transforms of $z(x), z_1(x)$, and $z_2(x)$, respectively
$Z_{2e}(\omega), Z_{2o}(\omega)$	even and odd parts of $Z_2(\omega)$
α	fractional increase in instrument response-function breadth due to convolution with narrow spectral line
β	parameter specifying influence of sharpness or smoothness criteria
$\otimes$	convolution operation
$\delta(x), \delta$	Dirac δ function or impulse
$\delta'(x), \delta'$	first derivative of $\delta(x)$
Δx	half-width at half maximum (HWHM)
$\Delta x_G, \Delta x_C$	Gaussian and Cauchy half-widths at half maximum
Δx_N	Nyquist interval
ζ	conjugate of x in alternative Fourier transform system; Fourier frequency in cycles per units of x; variable of integration

$\theta(x), \Theta(\omega)$	spurious part of solution $\hat{o}(x)$ and its Fourier transform $\hat{O}(\omega)$
$\Lambda(x)$	triangle function of unit height and half-width $\frac{1}{2}$
μ	scaled ratio of Cauchy to Gaussian half-widths, $\Delta x_C \sqrt{\ln 2} / \Delta x_G$
σ	standard deviation
σ^2	variance
$\sigma_a^2, \sigma_b^2, \sigma_g^2$	variances of a, b, g
$\sigma_A^2, \sigma_B^2, \sigma_G^2$	variances of A, B, G
$\tau(\omega), \tau$	$\tau(\omega)$, Fourier transform of $s(x)$
$\boldsymbol{v}$	vector having components v_l
v_l	parameters of a model comprising multiple peaks
$\Phi(\boldsymbol{v})$	objective function to be minimized
ω	conjugate of x; Fourier frequency in radians per units of x
Ω	cutoff frequency such that $\tau(\omega) = 0$ for $|\omega| > \Omega$
$\text{II}(x)$	positive-impulse pair $\frac{1}{2}\delta(x + \frac{1}{2}) + \frac{1}{2}\delta(x - \frac{1}{2})$
$^\text{I}\text{I}(x)$	impulse pair with positive and negative components, $\frac{1}{2}\delta(x + \frac{1}{2}) - \frac{1}{2}\delta(-\frac{1}{2})$
$\text{III}(x)$	Dirac "comb" $\sum_{n=-\infty}^{\infty} \delta(x - n)$

I. Introduction

Our daily experience abounds with phenomena that can be described mathematically by convolution. Spreading, blurring, and mixing are qualitative terms frequently used to describe these phenomena. Sometimes the spreading is caused by physical occurrences unrelated to our mechanisms of perception; sometimes our sensory inputs are directly involved. The blurred visual image is an example that comes to mind. The blur may exist in the image that the eye views, or it may result from a physiological defect. Biological sensory perception has parallels in the technology of instrumentation. Like the human eye, most instruments cannot discern the finest detail. Instruments are frequently designed to determine some observable quantity while an independent parameter is varied. An otherwise isolated measurement is often corrupted by undesired contributions that should rightfully have been confined to neighboring measurements. When such contributions add up linearly in a certain way, the distortion may be described by the mathematics of convolution.

Spectroscopy is profoundly affected by these spreading and blurring phenomena. The recovery of a spectrum as it would be observed by a hypothetical, perfectly resolving instrument is an exciting goal. Recent advances have stimulated development of restoring methods that receive increasingly wider application. Imaging systems too suffer from blur, be they telescopes trained on large and distant objects, or microscopes examining small ones that are near. Whether we concern ourselves with science that requires imaging, spectroscopy, or other experimental approaches, our state of knowledge is often defined by the resolving limit of our instruments. Through modern restoring methods, the scientist has access to information that would otherwise remain unavailable. The advent of the new methods has even changed the way we regard experimental apparatus. These methods cannot fail to have a marked influence on the progress of science.

Before we can confront the problem of undoing the damage inflicted by spreading phenomena, we need to develop background material on the mathematics of convolution (the function of this chapter) and on the nature of spreading in a typical instrument, the optical spectrometer (see Chapter 2). In this chapter we introduce the fundamental concepts of convolution and review the properties of Fourier transforms, with emphasis on elements that should help the reader to develop an understanding of deconvolution basics. We go on to state the problem of deconvolution and its difficulties.

The most important symbols introduced in this chapter are used throughout the volume; that we occasionally deviate from this notation is testimony to the diversity of applications and to the limited number of clear and convenient notational possibilities. We have avoided mathematical rigor in favor of developing an intuitive grasp of the fundamentals. The reader is referred to the outstanding and readable text by Bracewell (1986a) for added depth.

This volume deals primarily with spectra and images, which may be taken generally to represent data acquired as a functions of one and two independent variables, respectively. Data from fields as diverse as radio astronomy, statistics, separation science, and communications are suitable candidates for treatment by the methods described here. Confusion arises when we discuss Fourier transforms of these quantities, which may also be called spectra. To avoid this confusion, we adopt the convention of referring to the latter spectra as Fourier spectra. When this term is used without the qualifier, the data space (nontransformed regime) is intended.

Also, application of these methods is not limited to one- and two-dimensional problems. Most of the concepts herein may readily be extended to solve higher dimensional problems. Of particular interest are the closely related tomographic techniques of reconstruction from projections that have reached widespread application in medical imaging and industry. Works by Barrett and Swindell (1981) and by Herman (1980) serve as excellent guides to these methods and applications.

II. Definition of Convolution

A. DISCRETE CASE

Convolution, in its simplest form, is usually considered to be a blurring or smoothing operation. A "rough" or "bumpy" function is convolved with a smoothing function to yield a smoother output. Typically, each output value is identified with a corresponding input value. It is obtained, however, by processing that value *and* some of its neighbors. One simple way of smoothing the fluctuations in a sequence of numbers is to perform a moving average. Below, in row *a*, we see a smoothed sequence partially computed for the noisy values in row *g*:

g	4	3	9	8	2	1	6	5	4	2	3
a			5.2	4.6	5.2	4.4	3.6				

In this particular case, we chose to average five adjacent *g* numbers to produce each value in row *a*. We wrote each element in row *a* opposite the center of the averaging interval. Giving each element of rows *a* and *g* indices *n* and *m*, respectively, we write the moving average

$$a_n = (g_{n-2} + g_{n-1} + g_n + g_{n+1} + g_{n+2})/5 = \sum_{m=n-2}^{n+2} g_m/5. \quad (1)$$

It could suit our purpose to apply a weight to each of the *g* elements before summing:

g	4	3	9	8	2	1	6	5	4	2	3
					×	×	×	×			
b					0.1	0.3	0.5	0.1	→		
					0.2	0.3	3.0	0.5			
a	···		6.6	7.2	4.4	2.5	4.0				

In this example the sum of the b weights is 1. Therefore, we may say that b is *normalized*. We see that each element of row a is a *linear combination* of a corresponding subset of the g elements. Sequential values of a are computed by sliding the set of b factors along row g. It is usually convenient to write each a value opposite the largest weighting coefficient in row b:

$$
\begin{array}{ccccccccccc}
g_1 & g_2 & g_3 & g_4 & g_5 & g_6 & g_7 & g_8 & g_9 & g_{10} & g_{11} \\
 & & & & \times & \times & \times & \times & \times & & \\
 & & & & b_{+2} & b_{+1} & b_0 & b_{-1} & b_{-2} & \rightarrow & \text{slide} \\
a_1 & a_2 & a_3 & a_4 & a_5 & a_6 & a_7 & & & &
\end{array}
$$

Thus,

$$a_7 = b_2 g_5 + b_1 g_6 + b_0 g_7 + b_{-1} g_8 + b_{-2} g_9. \tag{2}$$

We generalize this procedure in the following equation:

$$a_n = \sum_{m=n-2}^{n+2} b_{n-m} g_m. \tag{3}$$

For the present case, b vanishes everywhere except over the interval -2 to $+2$. We also can define b as an infinite sequence of numbers. In this case we may write

$$a_n = \sum_{m=-\infty}^{\infty} b_{n-m} g_m. \tag{4}$$

This is sometimes called the discrete convolution or serial product. The values of b and g, however, may just be samples of continuous functions.

Note here that because each value of a_n requires values of g_{n-2} through g_{n+2}, N values of a can never be computed from N values of g. If N_a is the number of a values desired, and N_g the number of g values available, we must always have $N_g > N_a$, except in the trivial case where b has only one non-zero value. The representation of real-world situations usually involves some accommodation of this effect, possibly by padding the ends of arrays with zeros or by treating data values as if they were cyclic and "wrapped around," as is the case in a circulant matrix (Chapter 3, Section III.A). Often, sections of spectra can be chosen so that they begin and end in regions having base-line values of zero. Alternatively, one may choose the range of interest far from the ends of the data so that end effects can be ignored.

B. CONTINUOUS CASE

We may choose the sample interval to be as fine as we please, thus extending the concept of summation into the continuous regime. With a, b, and g now written as functions of the continuous variables x and x', we may write the convolution integral

$$a(x) = \int_{-\infty}^{\infty} b(x - x')g(x')\,dx'. \qquad (5)$$

Each value of a is a weighted integral of the g function, the function b supplying the required weight and being slid along g according to the displacement specified by x'.

If the area under $b(x)$ is unity,

$$\int_{-\infty}^{\infty} b(x)\,dx = 1, \qquad (6)$$

we may say that $b(x)$ is normalized. Equation (5) then represents a moving weighted average. In the study of instrumental resolution, we often think of the convolution integral in this way.

We may denote convolution by the symbol $\otimes$, and rewrite Eq. (5):

$$a = b \otimes g. \qquad (7)$$

The presence of the minus sign in the argument of b in Eq. (5) gives the convolution integral the highly useful property of commutativity, that is,

$$b \otimes g = g \otimes b. \qquad (8)$$

This is easily shown by redefining the variable of integration. Convolution also obeys associativity,

$$(d \otimes b) \otimes g = d \otimes (b \otimes g), \qquad (9)$$

and is distributive with respect to addition,

$$d \otimes (b + g) = (d \otimes b) + (d \otimes g). \qquad (10)$$

These properties carry back to the discrete formulation. We shall use both discrete and continuous formulations in this volume, changing back and forth as needs require. The continuous regime allows us to avoid consideration of sampling effects when such consideration is not of immediate concern. Deconvolution algorithms, on the other hand, are numerically implemented on sampled data, and we find the discrete representation indispensable in such cases.

III.   Properties

In addition to the relations just presented, we shall state without proof some other properties of convolutions. Most of these are quite easily proved.

A.   *INTEGRATION AND DIFFERENTIATION*

If b and g are peaked functions (such as in a spectral line), the area under their convolution product is the product of their individual areas. Thus, if b represents instrumental spreading, the area under the spectral line is preserved through the convolution operation. In spectroscopy, we know this phenomenon as the invariance of the equivalent width of a spectral line when it is subjected to instrumental distortion. This property is again referred to in Section II.F of Chapter 2 and used in our discussion of a method to determine the instrument response function (Chapter 2, Section II.G).

Convolution has an interesting property with respect to differentiation. The first derivative of the convolution product of two functions may be given by the convolution of either function with the derivative of the other. Thus, if

$$a = b \otimes g, \tag{11}$$

then

$$a' = b' \otimes g = b \otimes g', \tag{12}$$

where the primes denote differentiation.

This property follows from the associative and commutative properties if we allow the concept of a differentiation operator δ' that performs its function by convolution. We see that

$$a' = \delta' \otimes a = (\delta' \otimes b) \otimes g = b \otimes (\delta' \otimes g). \tag{13}$$

Such an operator is indeed the first derivative of the familiar impulse or Dirac δ function. It can, like the δ function, be represented as the limiting form of a variety of functions. In the case of δ', we have adjacent positive-

and negative-going impulses. It can be shown to have the following properties:

$$\int_{-\infty}^{\infty} \delta'(x)\, dx = 0, \tag{14}$$

$$\delta'(0) = 0, \tag{15}$$

$$\int_{-\infty}^{\infty} x\delta'(x)\, dx = -1, \tag{16}$$

$$x\delta'(x) = -\delta(x). \tag{17}$$

B. CENTRAL-LIMIT THEOREM

From experience, we know that a curve obtained by integrating a function is smoother than the function itself. This characteristic also applies to the convolution integral. We know that an instrumentally smeared spectrum is smoother than the original as it would be observed by a perfect instrument.

If we convolve two single-peaked functions, the result is smoother than either component. Two rectangle functions convolved yield a triangle; two triangle functions convolved (four rectangles convolved) produce a result that is astonishingly close to a gaussian (Fig. 1).

A Gaussian function has the form

$$f(x) = \exp(-x^2). \tag{18}$$

Here the height at the center is unity and is $1/e$ when $x = 1$.

There are various ways of normalizing the gaussian, in both abscissa and ordinate. In statistics, we often deal with

$$f(x) = [1/(\sigma\sqrt{2\pi})]\exp(-x^2/2\sigma^2), \tag{19}$$

where σ is the standard deviation, and the factor before the exponential guarantees $f(x)$ to have unit area. The central ordinate is no longer 1. The variance is given by the square of σ.

A form familiar to the spectroscopist is

$$f(x) = \exp\left[-\ln 2(x/\Delta x)^2\right]. \tag{20}$$

Here the central ordinate is again unity, and Δx represents the half-width at half maximum (HWHM), so that $f(|\Delta x|) = \frac{1}{2}$.

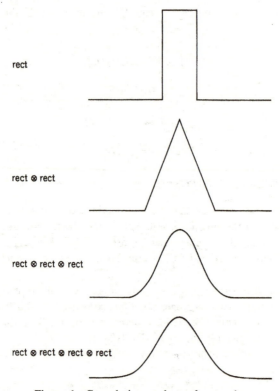

rect

rect ⊗ rect

rect ⊗ rect ⊗ rect

rect ⊗ rect ⊗ rect ⊗ rect

Figure 1   Convolution products of rectangles.

We have seen that convolving rectangles gives a gaussianlike function. A gaussian is, in fact, the exact result of an infinite number of convolutions provided that the functions convolved obey certain conditions. The rigorous statement of this property is the central-limit theorem. *A consequence of the central-limit theorem is that when a function $f(x)$ is convolved with itself n times, in the limit $n \to \infty$, the convolution product is Gaussian with variance n times the variance of $f(x)$, provided that the area, mean, and variance of $f(x)$ are finite.*

The conditions may be stated as

$$\int_{-\infty}^{\infty} f(x)\,dx < \infty, \tag{21}$$

$$\int_{-\infty}^{\infty} xf(x)\,dx < \infty, \tag{22}$$

and

$$\int_{-\infty}^{\infty} x^2 f(x)\, dx < \infty. \tag{23}$$

If nonidentical functions are convolved, the conditions are somewhat more complex (Bracewell, 1986a).

If two Gaussian functions are convolved, the result is a gaussian with variance equal to the sum of the variances of the components. Even when two functions are not Gaussian, their convolution product will have variance equal to the sum of the variances of the component functions. Furthermore, the second moment of the convolution product is given by the sum of the second moments of the components. The horizontal displacement of the centroid is given by the sum of the component centroid ' displacements. Kendall and Stuart (1963) and Martin (1971) provide helpful additional discussions of the central-limit theorem and attendant considerations.

Most peaklike functions become more gaussianlike when convolved with one another. One notable exception of interest to spectroscopists is the Cauchy function, which is the familiar Lorentzian shape assumed by lines in the spectra of gases subject to pressure broadening:

$$f(x) = \frac{1}{1 + (x/\Delta x)^2}. \tag{24}$$

If two Cauchy distributions having half-widths Δx_1 and Δx_2 are convolved, the result is a similar distribution that has a half-width of $\Delta x_1 + \Delta x_2$. If n Cauchy distributions having half-width Δx_C are convolved, the result is a Cauchy distribution of half-width $n\, \Delta x_C$.

C. VOIGT FUNCTION

One might well ask what the result would be if the Cauchy distribution were convolved with the Gaussian distribution. The result may be obtained from Eqs. (20) and (24):

$$\frac{1}{1 + (x/\Delta x_C)^2} \otimes \exp\left[-\ln 2(x/\Delta x_G)^2\right]. \tag{25}$$

Redefining the independent variable in terms of the Gaussian half-width, $q = x\sqrt{\ln 2}\,/\Delta x_G$, we may write this convolution as

$$\mu\, \Delta x_C \int_{-\infty}^{\infty} \frac{\exp(-\zeta^2)}{\mu^2 + (q - \zeta)^2}\, d\zeta, \tag{26}$$

where μ is proportional to the ratio of Cauchy to Gaussian half-widths:

$$\mu = (\Delta x_C/\Delta x_G)/\sqrt{\ln 2}. \tag{27}$$

If we require this function to have unit area when plotted as a function of q, we must modify the scale of the ordinate. The resulting expression is

$$\frac{\mu}{\pi^{3/2}} \int_{-\infty}^{\infty} \frac{\exp(-\zeta^2)}{\mu^2 + (q - \zeta)^2} d\zeta. \tag{28}$$

Multiplying this expression by π yields the Voigt function that occurs in the description of spectral-line shapes resulting from combined Doppler and pressure broadening. We elaborate on these phenomena in Section I of Chapter 2.

IV. Fourier Transforms

Although a number of effective deconvolution algorithms do not use Fourier methods, these methods shed considerable light on the performance of the algorithms. For this reason, we introduce the Fourier transform and outline some of its most useful properties. Only a brief treatment is given here. For additional detail, we again refer the reader to the excellent practical text on this subject by Bracewell (1986a).

Scale factors can be used in various ways to define Fourier transform pairs. We adopt the symmetrical convention

$$F(\omega) = \frac{1}{\sqrt{2\pi}} \int_{-\infty}^{\infty} f(x)e^{-j\omega x} dx \tag{29}$$

and

$$f(x) = \frac{1}{\sqrt{2\pi}} \int_{-\infty}^{\infty} F(\omega)e^{j\omega x} d\omega \tag{30}$$

for use in most parts of this volume. In these expressions, x is the spatial coordinate or independent variable of the measurement space (wavelength, wave number, mass, grating angle, distance, etc.) and ω the independent variable of the Fourier space given in radians per units of x. We usually denote functions of x by lowercase letters, such as the f used here, and their transforms by the corresponding capital letters.

A second symmetrical convention that has gained popularity specifies the Fourier transform pair $g(x)$ and $G(\zeta)$ to be related by

$$G(\zeta) = \int_{-\infty}^{\infty} g(x)e^{-j2\pi\zeta x}\, dx$$

and

$$g(x) = \int_{-\infty}^{\infty} G(\zeta)e^{j2\pi\zeta x}\, d\zeta,$$

where ζ is frequency in cycles per units of x. Conversion from one system to the other is easy. If x is the same for both systems, the frequency variables may be related by $\omega = 2\pi\zeta$. On the other hand, if a transform pair is known in the first system of Eqs. (29) and (30), and Fig. 2, multiplying both arguments x and ω by $\sqrt{2\pi}$ and substituting ζ for ω yields a corresponding pair in the second system. Scale changes by variable substitution and similarity theorem (Section IV.B.4) then tailor the pair to the situation at hand.

Except for symmetry, a third system bears more similarity to the first:

$$V(\omega) = \int_{-\infty}^{\infty} v(x)e^{-j\omega x}\, dx$$

and

$$v(x) = \frac{1}{2\pi}\int_{-\infty}^{\infty} V(\omega)e^{j\omega x}\, d\omega.$$

Here x and ω have the same definitions and relationship as they do in the first system. To convert a known transform pair in the first system to this system, either multiply $F(\omega)$ by $\sqrt{2\pi}$ to express $V(\omega)$, or multiply $f(x)$ by $1/\sqrt{2\pi}$ to obtain $v(x)$. As before, use the similarity theorem and scale change to adapt the pair.

A. SPECIAL SYMBOLS AND USEFUL FUNCTIONS

It is convenient to introduce some special symbols and functions. These simple representations, when combined with the various Fourier transform properties and a little practice, will enable the reader to gain a deeper understanding of, and intuition for, convolution and deconvolution.

1. Rectangle, Sinc, and Si

Consider a function, centered at the origin, defined as unity out to $\frac{1}{2}$ in both positive and negative directions and vanishing beyond $\frac{1}{2}$. More precisely, we define

$$\text{rect}(x) = \begin{cases} 0, & |x| > \frac{1}{2}, \\ \frac{1}{2}, & |x| = \frac{1}{2}, \\ 1, & |x| < \frac{1}{2}. \end{cases} \tag{31}$$

This function may be illustrated by a rectangle. A scaled version of it, $\text{rect}(x/\sqrt{2\pi})$, is shown in Fig. 2, along with its transform and a number of other functions that follow in this section. Its transform $F(\omega)$ can be computed in a straightforward way by replacing the infinite limits of integration with $\pm \sqrt{2\pi}/2$ and setting $f(x)$ equal to unity:

$$\begin{aligned} F(\omega) &= \frac{1}{\sqrt{2\pi}} \int_{-\infty}^{\infty} \text{rect}\left(\frac{x}{\sqrt{2\pi}}\right) e^{-j\omega x}\, dx \\ &= \frac{1}{\sqrt{2\pi}} \int_{-\sqrt{2\pi}/2}^{\sqrt{2\pi}/2} e^{-j\omega x}\, dx \\ &= \frac{\sin(\omega\sqrt{2\pi}/2)}{\omega\sqrt{2\pi}/2} \\ &= \text{sinc}\left(\frac{\omega}{\sqrt{2\pi}}\right), \end{aligned} \tag{32}$$

where we have used the opportunity to define the sinc function

$$\text{sinc}(x) = \frac{\sin \pi x}{\pi x}. \tag{33}$$

This function is used extensively in optics and elsewhere, as well as its integral

$$\text{Si}(x) = \pi \int_{0}^{x/\pi} \text{sinc}(x')\, dx' = \int_{0}^{x} \frac{\sin x'}{x'}\, dx'. \tag{34}$$

Some authors define sinc without the π factors, so beware of this variation when comparing works.

2. Triangle and Sinc Squared

By simple geometrical arguments and the definition of convolution, we can verify that the convolution of rect(x) with itself yields $\Lambda(x)$, which we define as

$$\Lambda(x) = \begin{cases} 0, & |x| \geq 1, \\ 1 - |x|, & |x| < 1. \end{cases} \tag{35}$$

Its transform is just $(1/\sqrt{2\pi})\mathrm{sinc}^2(\omega/2\pi)$. This may easily be shown with the aid of the convolution theorem discussed in Section IV.B.10.

3. Gaussian Function

The Gaussian function is its own Fourier transform; that is, when we define

$$f(x) = \exp(-x^2/2), \tag{36}$$

we obtain

$$F(\omega) = \exp(-\omega^2/2). \tag{37}$$

Allowing for the case of arbitrary amplitude and width, we obtain the transform pair

$$C \exp\left(-\frac{cx^2}{2}\right) \leftrightarrow \frac{C}{c}\exp\left(-\frac{\omega^2}{2c}\right). \tag{38}$$

Here and henceforth the symbol $\leftrightarrow$ indicates correspondence between the two domains. The scale factor c demonstrates the inverse relationship between the widths of the gaussians in the two domains. The wider a gaussian, the narrower is the gaussian to which it transforms.

4. The δ Function—Alone and in Pairs

The transform of $\delta(x)$ is the constant $1/\sqrt{2\pi}$. The transform of $\delta(x/\sqrt{2\pi})$ is unity owing to the scaling property $\delta(cx) = (1/|c|)\delta(x)$. Displaced from the origin, this function transforms to sinusoids having real and imaginary components. We can use two positive impulses, equally displaced from the origin in the positive and negative directions, to create two imaginary sine components in the transform that cancel each other but leave a real cosine term. If we set

$$f(x) = \delta\left(x/\sqrt{2\pi} + \tfrac{1}{2}\right) + \delta\left(x/\sqrt{2\pi} - \tfrac{1}{2}\right), \tag{39}$$

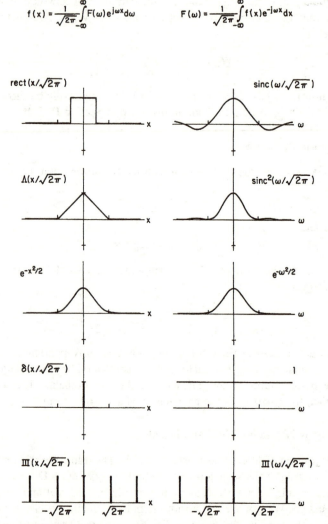

$$f(x) = \frac{1}{\sqrt{2\pi}} \int_{-\infty}^{\infty} F(\omega) e^{j\omega x} d\omega \qquad F(\omega) = \frac{1}{\sqrt{2\pi}} \int_{-\infty}^{\infty} f(x) e^{-j\omega x} dx$$

Figure 2 Fourier transform directory. The Dirac δ function having area $\sqrt{2\pi}$ is shown as a bar of unit height. Imaginary components are shown by dashed lines. The vertical axis tick mark is at 1, the horizontal axis tick mark is at $\sqrt{2\pi}$.

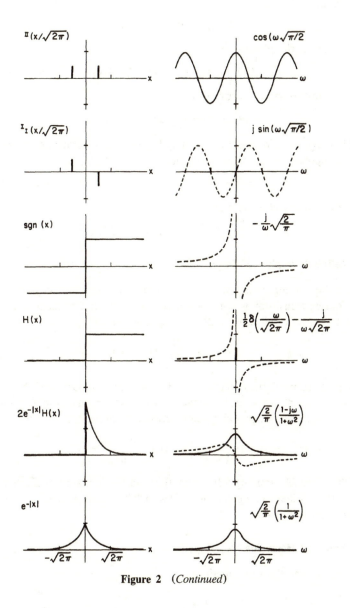

Figure 2 (*Continued*)

we obtain

$$F(\omega) = \cos\left(\omega\sqrt{\pi/2}\right). \tag{40}$$

Let us define the positive-impulse pair

$$\Pi(x) = \tfrac{1}{2}\delta(x + \tfrac{1}{2}) + \tfrac{1}{2}\delta(x - \tfrac{1}{2}). \tag{41}$$

Then we can rewrite the transform pair

$$\Pi(x/\sqrt{2\pi}) \leftrightarrow \cos\left(\omega\sqrt{\pi/2}\right). \tag{42}$$

Similarly, an odd impulse pair may be used to generate an imaginary sine component in the transform space. Let us define

$$^{\mathrm{I}}\mathrm{I}(x) = \tfrac{1}{2}\delta(x + \tfrac{1}{2}) - \tfrac{1}{2}\delta(x - \tfrac{1}{2}). \tag{43}$$

This definition yields the pair

$$^{\mathrm{I}}\mathrm{I}(x/\sqrt{2\pi}) \leftrightarrow j\sin\left(\omega\sqrt{\pi/2}\right). \tag{44}$$

5. Infinitely Replicated Impulses

The infinite string of δ functions causes certain difficulties in rigor when its Fourier transform is considered. Nevertheless, the concept is a very useful one. It is well worth defining a special symbol

$$\mathrm{III}(x) = \sum_{n=-\infty}^{\infty} \delta(x - n). \tag{45}$$

Many of its properties follow readily from the properties of $\delta(x)$. Of greatest interest, however, is the fact that its Fourier transform is a similar function that has reciprocal spacing between its impulses:

$$\mathrm{III}\left(\frac{cx}{\sqrt{2\pi}}\right) \leftrightarrow \mathrm{III}\left(\frac{\omega}{c\sqrt{2\pi}}\right). \tag{46}$$

The effect of sampling a function $f(x)$ can be simulated by multiplying it by $\mathrm{III}(x)$. Likewise, convolving $f(x)$ with $\mathrm{III}(x)$ replicates $f(x)$ infinitely in both directions.

6. Sign and Heaviside Step Functions

We may define the sign function

$$\mathrm{sgn}(x) = \begin{cases} 1, & x > 0, \\ -1, & x \le 0, \end{cases} \tag{47}$$

and note that its Fourier transform is given by

$$-\frac{j}{\omega}\sqrt{\frac{2}{\pi}}.$$ (48)

The Heaviside step, although similar, is not purely an odd function:

$$H(x) = \begin{cases} 1, & x > 0, \\ 0, & x \le 0. \end{cases}$$ (49)

We see that the $H(x)$ step is half as high as the sgn, and is raised up by $\frac{1}{2}$. The required added constant is responsible for an impulse in its Fourier transform,

$$\frac{1}{\sqrt{2\pi}}\left[\pi\delta(\omega) - \frac{j}{\omega}\right].$$ (50)

Note that we obtain a very nice ramp function through multiplying $H(x)$ by a straight line of unit slope, and that the same result can be obtained by the self-convolution of $H(x)$:

$$\text{ramp} = xH(x) = H(x) \otimes H(x).$$ (51)

One obtains an impulse by differentiating $H(x)$:

$$dH(x)/dx = \delta(x).$$ (52)

7. Truncated Exponentials and Resonance Contours

When simple electrical RC filters are treated, the truncated exponential $e^{-|x|}H(x)$ is indispensable. Its transform is given by $(2\pi)^{-1/2}(1 - j\omega)/(1 + \omega^2)$. If the truncated exponential is reflected about the origin, eliminating $H(x)$ and leaving $e^{-|x|}$, the imaginary part of the transform disappears. We obtain the transform $(2/\pi)^{1/2}/(1 + \omega^2)$. This is the resonance contour, Cauchy distribution, or Lorentzian shape encountered previously in Section III.B.

B. SOME PROPERTIES AND RELATIONSHIPS

1. Superposition

A fundamental property of the Fourier transform is that of superposition. The usefulness of the Fourier method lies in the fact that one can separate a function into additive components, treat each one separately, and then build up the full result by summing the individual results. It is a beautiful and explicit example of the stepwise refinement of complex problems. In stepwise refinement, one successfully tackles the most difficult tasks and solves problems far beyond the mind's momentary grasp by dividing the problem into its ultimately simple pieces. The full solution is then obtained by reassembling the solved pieces.

In linear superposition, the method is literally that of adding components. When treating the optics of coherent light, for example, the instantaneous values of the field vectors are superimposed. Incoherent light, on the other hand, requires us to deal with the time-averaged square of the field. In nonlinear optics, superposition breaks down as it does in other nonlinear systems. Even when it does not hold exactly, however, superposition is often useful as a first-order approximation.

We may use the concept of orthogonal functions to identify components. In the present case, our component functions are sinusoids, and we find that the Fourier transform of the sum is the sum of the Fourier transforms:

$$f_1(x) + f_2(x) \leftrightarrow F_1(\omega) + F_2(\omega). \tag{53}$$

2. Oddness and Evenness

When we say that $f(x)$ is an even function, we mean that $f(-x) = f(x)$; if it is odd then $f(-x) = -f(x)$. It may easily be verified that a real even function has a real even transform, and an imaginary even function an imaginary even transform. Real odd functions have imaginary odd transforms. Any function may be decomposed into odd and even parts. For linear operations by superposition, they may be treated separately and the results reassembled. Both real and imaginary parts of a function may have odd and even components. In this case superposition is again useful.

3. Fourier's Integral Theorem

If we transform and then inverse transform a function, we should get back to where we started. By substituting Eq. (29) into Eq. (30), we obtain

$$f(x) = \int_{-\infty}^{\infty} f(x') \left\{ \frac{1}{2\pi} \int_{-\infty}^{\infty} \exp[j\omega(x - x')] \, d\omega \right\} dx'. \tag{54}$$

The quantity in braces must have the property of a Dirac δ function, so we write

$$\delta(x - x') = \frac{1}{2\pi} \int_{-\infty}^{\infty} \exp[j\omega(x - x')] \, d\omega. \tag{55}$$

4. Sign and Scale

From the concept of linearity it follows that multiplication of a function by a constant scale factor c results in a corresponding increase in its transform:

$$cf(x) \leftrightarrow cF(\omega). \tag{56}$$

A change of scale in x or ω, however, results in a reciprocal change in the transform variable and an amplitude change as well:

$$f(cx) \leftrightarrow \frac{1}{|c|} F\left(\frac{\omega}{c}\right). \tag{57}$$

This is sometimes called the similarity theorem. The horizontal stretching of a function in one domain results in horizontal contraction and amplitude growth in the other. In fact, these changes occur in such a way that the area under the curve in the other domain remains constant.

5. Power Theorem and Rayleigh's Theorem

The previous section dealt with a horizontal scale change in one domain resulting in both horizontal and vertical changes in the other domain. Rayleigh's theorem, on the other hand, makes a statement about the area under the squared modulus. This integral, in fact, has the same value in both domains:

$$\int_{-\infty}^{\infty} |f(x)|^2 \, dx = \int_{-\infty}^{\infty} |F(\omega)|^2 \, d\omega. \tag{58}$$

In its more general form, we have the power theorem

$$\int_{-\infty}^{\infty} f(x)g^*(x) \, dx = \int_{-\infty}^{\infty} F(\omega)G^*(\omega) \, d\omega. \tag{59}$$

6. Swapping Domains

We can foresee a situation in which we already know a transform pair but require a corresponding pair with reversed domains. Specifically, suppose we know that $\alpha(x)$ has a transform $\beta(\omega)$, but our problem presents us with $\alpha(\omega)$, and we desire its x-domain transform:

$$\alpha(x) \leftrightarrow \beta(\omega), \tag{60}$$

$$? \leftrightarrow \alpha(\omega). \tag{61}$$

By using the properties of even and odd functions, it is easy to show that $\beta(x)$ is the unknown function if $\alpha(\omega)$ is an even function. On the other hand, if $\alpha(\omega)$ is an odd function, then the unknown function is $-\beta(x)$. These relationships hold for both real and imaginary components. By superposition, any function can be broken into odd and even components and treated separately. The results may then be combined.

7. Shift Theorem

Our known transform pairs are usually centered at the origin. To apply these known transforms to real situations, we must know what happens to them as they are shifted along the abscissa. The result of such a shift is a simple change in phase:

$$f(x) \leftrightarrow F(\omega), \tag{62}$$

$$f(x - c) \leftrightarrow F(\omega) \exp(-jc\omega), \tag{63}$$

$$f(x) \exp(jcx) \leftrightarrow F(\omega - c). \tag{64}$$

No change in amplitude occurs.

8. Differentiation

By using the definition of differentiation, we can show that

$$df(x)/dx \leftrightarrow j\omega F(\omega) \tag{65}$$

and

$$-jxf(x) \leftrightarrow dF(\omega)/d\omega. \tag{66}$$

An important consequence of these relationships is that differentiation increases the amplitude of high-frequency components. It is well known that data-differentiation procedures often yield unsatisfactory results when applied to noisy data. In spectroscopy, peak positions are sometimes sought by looking for a vanishing first derivative. The spectral peaks contain predominantly low and middle frequencies, but the noise often contains high frequencies as well. A possible remedy to the problem is to fit a polynomial to the data over a limited domain centered on each point at which a derivative is required. The first derivative of the polynomial is found to be a simple linear combination of the neighboring data ordinates. A whole curve may be quickly differentiated in this way. The process is one of simple convolution. This method of differentiation contains its own low-pass filter. It has been described by Savitzky and Golay (1964). Some numerical errors that appear in their paper have been corrected by Steinier *et al.* (1972). See also Chapter 3, Section III.C.5.

9. Area, Moments, and Variances

Let us write the Fourier transform of $f(x)$:

$$F(\omega) = \frac{1}{\sqrt{2\pi}} \int_{-\infty}^{\infty} f(x) \exp(-j\omega x)\, dx. \tag{67}$$

The transform evaluated at the origin is proportional to the area under $f(x)$:

$$F(0) = \frac{1}{\sqrt{2\pi}} \int_{-\infty}^{\infty} f(x) \exp(0)\, dx = \frac{1}{\sqrt{2\pi}} \int_{-\infty}^{\infty} f(x)\, dx \tag{68}$$

and

$$\int_{-\infty}^{\infty} f(x)\, dx = \sqrt{2\pi}\, F(0). \tag{69}$$

Similarly, we have

$$\int_{-\infty}^{\infty} F(\omega)\, d\omega = \sqrt{2\pi}\, f(0). \tag{70}$$

The first moment can be shown to be proportional to the slope of the transform at the origin. Precisely, we have

$$\int_{-\infty}^{\infty} xf(x)\, dx = j\sqrt{2\pi}\, F'(0). \tag{71}$$

This may be proved by exercising the differentiation property previously discussed.

The mean value of x, where $f(x)$ is a distribution, may also be called the centroid and is defined

$$\langle x \rangle = \int_{-\infty}^{\infty} xf(x)\, dx \Big/ \int_{-\infty}^{\infty} f(x)\, dx. \tag{72}$$

Substitutions from Eqs. (69) and (71) yield

$$\langle x \rangle = jF'(0)/F(0). \tag{73}$$

The second moment is given by

$$\int_{-\infty}^{\infty} x^2 f(x)\, dx = -\sqrt{2\pi}\, F''(0). \tag{74}$$

Here we twice differentiated $F(\omega)$. The mean value of x^2 is given by

$$\langle x^2 \rangle = \int_{-\infty}^{\infty} x^2 f(x)\,dx \bigg/ \int_{-\infty}^{\infty} f(x)\,dx = -F''(0)/F(0). \qquad (75)$$

The variance of a distribution $f(x)$ is given by

$$\sigma^2 = \langle (x - \langle x \rangle)^2 \rangle = \left[\frac{F'(0)}{F(0)} \right]^2 - \frac{F''(0)}{F(0)}. \qquad (76)$$

10. Convolution Theorem

We shall now present a theorem that is fundamentally important in Fourier analysis for understanding data acquisition and the performance of spectrometers. It is essential to the thought process required for both the qualitative understanding of concepts and precise mathematical analysis. Under certain circumstances, it can substantially reduce computation.

Easily proved from the definition of the Fourier transform, this theorem states that convolving two functions is equivalent to finding the product of their Fourier transforms. Specifically, if $a(x)$, $b(x)$, and $g(x)$ have transforms $A(\omega)$, $B(\omega)$, and $G(\omega)$, then

$$a(x) = b(x) \otimes g(x) \qquad (77)$$

is equivalent to

$$A(\omega) = B(\omega)G(\omega). \qquad (78)$$

The theorem works equally well whether we are using plus or minus transforms, so we may write the pair of relationships

$$a \otimes g \leftrightarrow AG, \qquad (79)$$

$$ag \leftrightarrow A \otimes G. \qquad (80)$$

Superposition may be invoked to determine the behavior of the theorem when the functions are subjected to changes in the sign of real or imaginary, odd or even, components.

We have seen in Section III.B that when two functions are convolved, the variance of the convolution product is the sum of the variances of the individual functions. That is, if

$$a = b \otimes g \qquad (81)$$

holds true, then we may write

$$\sigma_a^2 = \sigma_b^2 + \sigma_g^2. \qquad (82)$$

We see that when two gaussians of equal breadth are convolved, the result is a gaussian $\sqrt{2}$ times broader.

We know that the variance of a gaussian is the reciprocal of the variance of its transform. We apply the convolution theorem to obtain

$$\sigma_A^2 = \frac{\sigma_B^2 \sigma_G^2}{(\sigma_B^2 + \sigma_G^2)}. \tag{83}$$

The transform of our convolved equal-breadth gaussians is narrower than either individual transformed gaussian. In the present case $1/\sqrt{2}$ is the appropriate scaling.

The convolution theorem plays a valuable role in both exact and approximate descriptions of functions useful for analyzing resolution distortion and in helping us understand the effects of these functions in Fourier space. Functions of interest and their transforms can be constructed from our directory in Fig. 2 by forming their sums, products, and convolutions. This technique adds immeasurably to our intuitive grasp of resolution limitations imposed by instrumentation.

11. Fast Fourier Transform

Normally, discrete convolution involves shifting, adding, and multiplying—a laborious and time-consuming process, even in a large digital computer. The convolution theorem presents us with an alternative. It reveals the possibility of computing in the Fourier domain. What are the trade-offs between the two methods?

Conventionally, if the numbers a_i are the N_a sampled values of the function $a(x)$ over its domain of nonvanishing values, and b_i are the N_b sampled values of the function $b(x)$ over its domain of nonvanishing values, then the discrete convolution of a and b involves computing $N_a N_b$ sums and $N_a N_b$ products, or $2N_a N_b$ arithmetic operations all together. This result is demonstrated by a visualization similar to that in Section II.A. In this example, all nonvanishing values of the product are computed.

The fast Fourier transform (FFT) requires $5N \log_2 N$ (Bergland, 1969) elementary arithmetic operations to compute the Fourier transform of an array of N samples. Computation of the convolution product requires three such transforms plus $6N$ elementary operations in the transform domain.

Precise comparison of the two methods of computing a convolution requires careful attention to details such as whether aliasing, computing the "ends" of the function, matching array lengths to powers of 2, or whatever other FFT base is employed. It is apparent, however, that when $N_a = N_b$, the FFT method is superior. When $N_a \ll N_b$, the FFT method involves considerable unnecessary computation. In instrumental resolution studies, one of the two functions typically has a considerably smaller extent than the other; that is, the response function is usually narrow relative to the extent of the data. In this case, it is usually more efficient to perform convolution directly, without transformation.

The FFT can also be applied to higher dimensionality data and other sampled functions such as images. Both column- and row-wise decomposition make it possible to compute economically the transform of a two-dimensional (2-D) function such as an image by repeated application of the one-dimensional (1-D) transform (Dudgeon and Mersereau, 1984). Separability of the 2-D function into a simple product of 1-D functions can reduce the computation much further. In this case, if the coordinates are independent, the transform of the product is the product of the transforms.

The FFT is neither more nor less than an efficient algorithm for computing the discrete Fourier transform. It would draw us too far afield to explore the algorithm's details. However, it is essential that we understand the relationship between the continuous and discrete transforms to avoid serious error due to sampling.

C. SAMPLING AND THE DISCRETE FOURIER TRANSFORM

If we wish to deal with a continuous function—call it $f(x)$—in a digital computer it is convenient to employ its equivalent discretely sampled approximation. In this case $f(x)$ is usually sampled and encoded into numerical values at evenly spaced steps of its independent variable x. These values can be obtained only to a limited precision because they are subject to roundoff or quantization effects. In modern digital computers it is usually possible to attain adequate encoding precision for $f(x)$, so we shall not continue further along this line.

Let us instead turn our attention to the consequences of sampling the function at evenly spaced intervals of x. Consider the Λ function and its transform, a sinc function squared, shown in Fig. 3. Suppose that we wish to compute that transform numerically. First, let us replicate the Λ by convolving it with a low-frequency III function. Now multiply it by a high-frequency III function to simulate sampling. We see a periodically replicated and sampled Λ. The value of each sample is represented as the scaled area under a Dirac δ function.

By applying the convolution theorem, we see that replication in the x domain has produced a sampling effect in the frequency domain. The wider the replication interval, the finer is the frequency sampling. Sampling in the x domain, on the other hand, appears in Fourier space as replication. Fine sampling in x produces wide spacing between cycles in ω. The area under each scaled Dirac function of ω may be taken as the numerical value of a sample.

The discrete Fourier transform gives us the ability to compute the numbers representing one cycle of $F(\omega)$ from the number representing one cycle of $f(x)$ and vice versa. We may adjust scale factors, number of samples, sample density, and so on in this representation to elucidate many features of importance in discrete-Fourier-transform applications.

For example, suppose that we had sampled $f(x)$ too coarsely. The result would be closely spaced replicas in the Fourier domain. The high-frequency lobes of the sinc-squared cycles would overlap significantly, and high frequencies from an adjacent replica would appear superimposed on the lower frequencies. This phenomenon is known as frequency aliasing. When such aliasing might cause difficulty, it is necessary to band-limit the original by filtering it before it is sampled. If this is not done and the signal is sampled, there will be no way to separate the true from the aliased components based on the information in the sampled data alone. Both intuitively, then, and explicitly via the foregoing, fine sampling is required for adequate representation of an arbitrary function $f(x)$. The function can, in fact, be perfectly reconstructed from its evenly spaced samples provided that *at least two samples are taken for each cycle of the highest frequency that it contains*. This statement comes from the famous Whittaker–Shannon sampling theorem. The minimum sampling interval required for perfect reconstruction is known as the Nyquist interval. If $F(\omega)$ is contained entirely within frequency bounds $\pm\Omega$, evenly spaced samples of $f(x)$ must be no farther apart than the Nyquist interval $\Delta x_N = \pi/\Omega$. The sinc function provides a means of performing the reconstruction for all x:

$$f(x) = \sum_{n=-\infty}^{\infty} f\left(\frac{n\pi}{\Omega}\right) \text{sinc}\left(\frac{\Omega x}{\pi} - n\right). \tag{84}$$

The reader may wish to extend the type of analysis shown in Fig. 3 by considering the effects of "boxcar" averaging. In this process, $f(x)$ is

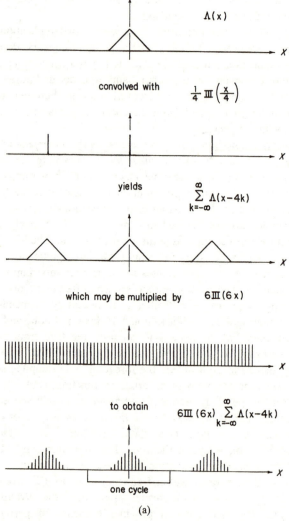

$\Lambda(x)$

convolved with $\frac{1}{4} \text{III} \left(\frac{x}{4} \right)$

yields $\sum\limits_{k=-\infty}^{\infty} \Lambda(x-4k)$

which may be multiplied by $6\text{III}(6x)$

to obtain $6\text{III}(6x) \sum\limits_{k=-\infty}^{\infty} \Lambda(x-4k)$

one cycle

(a)

Figure 3 Sampling and replication in (a) the x domain and (b) the ω domain.

convolved with a rect(x) function before sampling. Each sample then represents a uniformly weighted average. It is instructive to consider the effect of varying the number of samples used for the transform. In Chapter 12, Howard introduces the mathematics needed for practical application of the discrete Fourier transform.

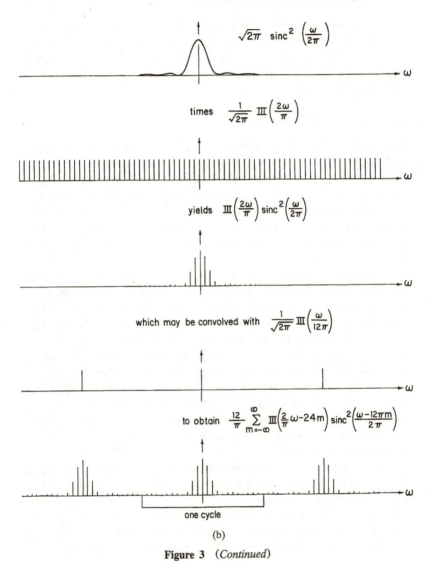

Figure 3 (*Continued*)

Development of the fast Fourier transform was a giant step that enabled the efficient computation of discrete Fourier transforms. The FFT algorithm and its variations revolutionized signal analysis and made interferometric infrared spectroscopy practical. Both NMR spectra and mass spectra are also now computed from data that are acquired in their

Fourier transform domain. The history of this algorithm goes back as far as C. F. Gauss in 1805 (Press *et al.*, 1992). Its rediscovery by Cooley and Tukey in (1965) is responsible for its current widespread use. Summaries of its properties and pitfalls are provided by Bergland (1969), Brigham (1974), and Bracewell (1986a).

D. HARTLEY TRANSFORM

Some physical laws lend themselves readily to elegant and compact expression via complex-variable mathematics. By this means we can carry, in a single expression, two related physical quantities and some of the subtlety of their relationship. There is no inherent physical significance, however, to a complex number. Complex numbers are a construct not of nature but of the mind. When we apply them, we merely take advantage of their properties in the way that we adapt numerous other concepts, such as those of vector calculus.

Expressions of many physical ideas, however, do not benefit from complex variables. Sometimes carrying a physical quantity as an imaginary component elicits pointless redundancy. In the computer, this redundancy translates to unnecessary computation and memory waste. Such redundancy is often occasioned by application of the Fourier transform to real data. The imaginary component of the transform arises from data symmetry. Only half of the transform components are needed, however, to reconstitute the data because these data bring about reflection symmetry between the positive and negative sides of the transform. It is perhaps an accident of science and mathematics history that the complex Fourier transform has, in its application, dominated an alternative formulation having many advantages.

Strictly speaking the alternative formulation, the Hartley transform (Hartley, 1942), is not a Fourier transform, though it has a Fourier kernel. This transform was thoroughly explored and advocated by Bracewell (1986b), who stated with authority that "... there is no purpose served by the complex Fourier transform that is not also served by the real Hartley transform" (Bracewell, 1986b, p. 7). The transform pair is given by

$$Z(\omega) = \frac{1}{\sqrt{2\pi}} \int_{-\infty}^{\infty} z(x) \cos(\omega x)\, dx \tag{85}$$

and

$$z(x) = \frac{1}{\sqrt{2\pi}} \int_{-\infty}^{\infty} Z(\omega) \cos(\omega x)\, d\omega. \tag{86}$$

Besides use of real variables, another immediately apparent Hartley-transform advantage is symmetry. The forward and backward relations are identical. The transform kernel cas(ωx) is defined as the sum of the sine and cosine:

$$\text{cas}(\omega x) \equiv \cos(\omega x) + \sin(\omega x). \tag{87}$$

The Hartley transform may easily be converted to a Fourier transform and vice versa by simple reflections and additions (Bracewell, 1986b). In fact, all the Fourier-transform properties and relationships described in Section IV.B have their Hartley equivalents. Among these is the Hartley version of the all-important convolution theorem. It states that $z(x)$, $z_1(x)$ and $z_2(x)$ have Hartley transforms $Z(\omega)$, $Z_1(\omega)$, and $Z_2(\omega)$, and $Z_{2e}(\omega)$ and $Z_{2o}(\omega)$ are the even and odd parts of $Z_2(\omega)$ respectively such that $Z_2(\omega) = Z_{2e}(\omega) + Z_{2o}(\omega)$, then

$$z(x) = z_1(x) \otimes z_2(x)$$

is equivalent to

$$Z(\omega) = Z_1(\omega)Z_{2e}(\omega) + Z_1(-\omega)Z_{2o}(\omega).$$

The Hartley version leads to particularly efficient convolution product computation.

Both real and imaginary curves are required to plot a Fourier transform. Attempts to simplify by plotting separate amplitude or power curves often yield little improvement because they require a phase plot for a complete description. Typical phase plots are discontinuous at $\pi/2$. This becomes awkward because such plots frequently need to be "unwrapped" for processing, improved insight, or visualization. In applications where the Hartley transform otherwise makes sense, the phase discontinuity problem simply goes away. The real Hartley transform carries all the information in its positive and negative halves.

Finally, we note that the Hartley transform, like the Fourier transform, has a discrete version (Bracewell, 1983), and a fast algorithm for its computation (Buneman, 1986; Bracewell, 1986b). This algorithm, analogous to the FFT, offers computational advantages over the FFT that commend the Hartley transform to far wider use.

V. The Problem of Deconvolution

In Section II we developed the concept of the convolution integral and its discrete approximation. No conceptual difficulty is therefore encountered in computing A from B and G:

$$A_n = \sum_{m=-\infty}^{\infty} B_{n-m} G_m. \tag{88}$$

Even considering our brief treatment of this subject thus far, we may have enough information to foresee that difficulties could arise when we attempt to compute values of G from given values of A and B. Such difficulties do indeed occur. The problem of deconvolution has therefore been the subject of a vast literature spanning the numerous special fields of science.

A. DEFINING THE PROBLEM

Let us begin our study of the problem by reformulating it in the continuous representation, with definitions of variables that have enjoyed some popularity:

$$i(x) = \int_{-\infty}^{\infty} s(x - x')o(x')\,dx'. \tag{89}$$

In optics, we seek the object function $o(x)$ when we are given the image data $i(x)$ and the known spread function $s(x)$ of the intervening optical system. This notation will be used throughout this volume. In the spectroscopy adaptation, the object o is usually the true spectrum as it would be observed by the ideal, perfectly resolving spectrometer. The spread function s may incorporate both optical blurring and electronic filtering. The image i may be the spectrum as it normally would appear on a strip-chart recorder. Regardless of which blurring effects we chose to include in s, and even in the case of nonoptical spectroscopies (e.g., electron spectroscopy for chemical analysis; see Chapter 8), in this work the quantities o, s, and i always have the same meaning in terms of the deconvolution problem: *i represents the known data; s the blurring phenomenon, the effects of which are to be eliminated; and o the function that we seek, free of spreading due to s.*

If we are observing a spectrum $o(x')$ with the aid of an instrument having a characteristic response function $s(x - x')$, then i represents the data acquired. If we have a perfectly resolving instrument, then $s(x - x')$

is a Dirac function, and our data $i(x)$ directly represent the true spectrum, that is, $o(x)$. In this case we have no need for deconvolution.

Conversely, we may observe an exceedingly narrow spectral line, so that $o(x')$ is approximated by $\delta(x')$. Now the data $i(x)$ represent the response function. This principle can, in fact, be used to determine the response function of a spectrometer. The laser, for example, is a tempting source of monochromatic radiation for measuring the response function of an optical spectrometer. Coherence effects, however, complicate the issue. We present further detail in Section II of Chapter 2.

Even if perfectly narrow spectral lines are not available, we may take a clue from this approach and measure a spectral line of known shape. Deconvolution should then yield the instrument function. This technique does indeed work (Chapter 2, Section II.G.3). Some of the information needed to generate the known line shape can even be obtained directly from the observed data.

In some fields, such as process control, determining the spread function is a special case of a more general technique called system identification. For linear systems encountered in this field, as in spectroscopy, the spread function can be determined by obtaining the system's response to a known input, then solving the associated inversion problem. In astronomical imaging, an isolated star can serve as an approximation to $\delta(x)$. Limited available light flux and shift variance of the spread function can be complicating factors. (See Chapter 10, Section III, for further detail.) In fluorescence microscopy imaging, a fluorescent bead is useful for the purpose (Chapter 9, Section II.D).

Given the difficulty of the deconvolution problem, it may seem outrageous to suggest that simultaneous extraction of *both* object and spread function should be possible. This feat, however, which is called blind deconvolution, has been accomplished in some specialized applications with increasing success. The methods involved are beyond the scope of this volume. Consult, for example, Bates and Jiang (1991) and Seldin and Fienup (1990) for pertinent discussions, and Bones *et al.* (1995) for later work.

Let us again consider the convolution integral. Equation (89) is an example of a Fredholm integral equation of the first kind. In such equations the kernel can be expressed as a more-general function of both x and x':

$$i(x) = \int_{-\infty}^{\infty} s(x, x') o(x') \, dx'. \tag{90}$$

Even in such so-called shift-variant cases, the true convolution integral may be an adequate representation if the range of x is not too large.

Some techniques for solving deconvolution problems also adapt readily to the more-general Fredholm case. The relaxation methods of, for example, Van Cittert (1931) and Jansson can be so adapted (Jansson, 1968, 1970; Jansson, Hunt, and Plyler, 1968, 1970).

Methods that have been developed for the solution of the Fredholm equation sometimes rely on continuous functional representations of $s(x - x')$ and $i(x)$. These methods are limited in usefulness for the experimentalist, who wishes to apply deconvolution techniques by computer to digitized spectra.

B. DIFFICULTIES

Difficulties in solving the deconvolution problem are revealed if we examine it in the continuous representation given by Eq. (89). Suppose that the solution for $o(x)$ is to some degree uncertain, and that it may be written as the sum of a desired solution $o(x')$ and a spurious part $\theta(x')$:

$$\hat{o}(x') = o(x') + \theta(x'). \tag{91}$$

We may then ask what the character of $\theta(x')$ must be so that Eq. (86) holds true. Certainly any $\theta(x')$ for which

$$\int_{-\infty}^{\infty} s(x - x')\theta(x') = 0 \tag{92}$$

can be summed with $o(x')$ to form a valid solution. We see the problem as one of uniqueness: how do we know that deconvolution is giving us the $o(x)$ solution that we want, free of the spurious components $\theta(x)$? The Fourier transform version of Eq. (92) gives us a clue. By the convolution theorem we have

$$\tau(\omega)\Theta(\omega) = 0, \tag{93}$$

where τ and Θ are the transforms of s and θ, respectively. Usually, instrumental distortion is characterized by the passage of low frequencies and the attenuation or complete loss of high frequencies. The transfer function $\tau(\omega)$ must be large for ω small, and small or vanishing for ω large. When ω is small, clearly $\Theta(\omega) = 0$. For ω large, when $\tau(\omega) = 0$, $\Theta(\omega)$ could take on any finite value whatsoever. The spurious component of the solution, $\Theta(\omega)$, has only high-frequency content. Filtering and smoothing have value in dealing with these high frequencies.

Beyond consideration of the spurious component $\theta(x)$, which addresses one aspect of uniqueness, far more serious problems await. The foregoing discussion has failed to contemplate the difficulty brought about by uncertainty in the data $i(x)$. The opportunity for success in solving for $o(x)$ exists only if the solution $o(x)$ itself exists. An ideal noise-free observation of $i_p(x)$ would guarantee existence in a practical (albeit not mathematical) sense to the extent that Eq. (89) correctly models physical reality. Data $i_p(x)$ did, after all, arise from a physical process that distorted a real object $o(x)$. Adding noise $n(x)$ to idealized data $i_p(x)$ such that $i(x) = i_p(x) + n(x)$, however, creates a problem that either (1) has no solution at all, (2) is ill-posed, or (3) is at best ill-conditioned.

A problem is ill-posed when at least one of three conditions obtains: either (1) its solution does not exist, (2) its solution exists but is not unique, or (3) its solution does not depend continuously on the data (Poggio *et al.*, 1985). Even if a problem passes all three tests for well-posedness, its solution may depend sensitively on small fluctuations in the data. Because all data have uncertainty, such sensitivity limits a solution's value. Problems exhibiting this behavior are said to be ill-conditioned. In deconvolution, the latter pathology is readily visualized as the large shift of intersection when small normal displacements affect near-parallel intersecting planes. Chapter 4, Section V.D.2, indeed describes the deconvolution problem in terms of intersecting hyperplanes.

C. ALTERNATIVES TO DECONVOLUTION

Much of this volume is devoted to the concept that the best solutions result from incorporating the maximum amount of prior knowledge. Assume, for example, that we know the object to have precisely four Gaussian components. Various fitting methods may be applied to determine the positions, widths, and amplitudes of these components. This approach has great value, where applicable, and is sometimes called deconvolution, although this usage is erroneous. Because these fitting methods are viable alternatives to deconvolution in many cases, we discuss them further.

Occasionally in spectroscopy, useful information may be gleaned from the observed spectrum of an isolated line without deconvolution, even though the instrument response function is wider than the line itself. We see this in the application of the method of equivalent widths to the determination of line strengths (Chapter 2, Sections II.F and II.G). When more complete knowledge is sought, we can often achieve the desired end by employing fewer degrees of freedom than a true deconvolution process utilizes.

1. Spectral Models

When we are solving for an unknown spectrum, each data point contains some information about the component spectral lines. A data point far from the center of a given line would be expected to contain very little information about that line. A hypothetical model of the spectrum could incorporate widely varying estimates of the amplitude of that line without influencing the fit to the data point in question. Three data points moderately distributed near the line center—say, spanning the interval between the half-maximum points—affix the parameters of the line more reliably. Instead of taking equally spaced samples of a trial solution as independent unknowns (the deconvolution approach), we can express the sought-after true spectrum in terms of its spectral-line parameters—amplitudes, half-widths, positions, and so on—provided that we can assume such a model with some confidence.

It is possible to estimate the influence of the spectrometer on linewidth measurements by convolving the response function with a mathematical model of the line. The resulting curve may be compared with the observed line and the true width inferred. Hunt *et al.* (1968) have employed corrections of this type supplied by the author in their determination of carbon monoxide self-broadened linewidths.

2. Objective Functions

The central-limit theorem (Section III.B) suggests that when a measurement is subject to many simultaneous error processes, the composite error is often additive and Gaussian distributed with zero mean. In this case, the least-squares criterion is an appropriate measure of goodness of fit. The least-squares criterion is even appropriate in many cases where the error is not Gaussian distributed (Kendall and Stuart, 1961). We may thus construct an objective function that can be minimized to obtain a best estimate. Suppose that our data $i(x)$ represent the measurements of a spectral segment containing spectral-line components that are specified by the N parameters $v_1, \ldots, v_N$, which may be considered as the components of a vector $\mathbf{v}$. The objective function may be written

$$\Phi(\mathbf{v}) = \int [i(x) - \hat{i}_{\mathrm{M}}(\mathbf{v}, x)]^2 \, dx, \tag{94}$$

where $\hat{\imath}_M(\mathbf{v}, x)$ is the model the parameters of which we wish to establish. We seek the parameters that minimize $\Phi(\mathbf{v})$, thereby ensuring the best fit. If we know that instrumental spreading is described by convolution of the true spectrum with a spread function $s(x)$, we may write

$$\hat{\imath}_M(\mathbf{v}, x) = \int s(x - x')\hat{o}_M(\mathbf{v}, x')\,dx', \tag{95}$$

where $\hat{o}_M(\mathbf{v}, x)$ is the model of the true spectrum based on the parameters v_l. We may then substitute $\hat{\imath}_M(\mathbf{v}, x)$ from Eq. (95) in Eq. (94). Typically the independent variable must be considered as discrete, so summations are employed instead of integrals.

Only in the simplest cases—a single Gaussian component, for example —may a conventional linear least-squares method be employed to solve for $\mathbf{v}$. More commonly, either approximate linearized methods or nonlinear methods are used.

3. Linearized Methods

One way of linearizing the problem is to use the method of least squares in an iterative linear differential correction technique (McCalla, 1967). This approach has been used by Taylor *et al.* (1980) to solve the problem of modeling images of two-dimensional electrophoresis gel separations of protein mixtures. One may also treat the components—in the present case spectral lines—one at a time, approximating each by a linear least-squares fit. Once fitted, a component may be subtracted from the data, the next component fitted, and so forth. To refine the overall fit, individual components may be added separately back to the data, refitted, and again removed. This approach is the basis of the CLEAN algorithm that is employed to remove antenna-pattern sidelobes in radio-astronomy imagery (Högbom, 1974) and is also the basis of a method that may be used to deal with other two-dimensional problems (Lutin *et al.*, 1978; Jansson *et al.*, 1983). Cohen and Sandri (1994) applied CLEAN to data compression by using it to remove nonredundant information.

4. Nonlinear Optimization

When these methods are unsuitable, nonlinear methods may be applied. The function $\Phi(\mathbf{v})$ may be considered as a surface in an $(N + 1)$-dimensional hyperspace. We seek the components of $\mathbf{v}$ that correspond to the minimum value of $\Phi(v)$, the bottom of a dishlike region of that surface. The chief difficulties in solving this problem are avoidance of trapping in local minima and overall computational efficiency. The function $\Phi(v)$ is often expensive to compute, so maximum advantage must accrue from each evaluation of it. To this end, numerous methods have been developed. Optimization is a field of ongoing research. No one single method is best for all types of problem. Where $\Phi(v)$ is a sum of squares, as we have expressed it, and where derivatives $d\Phi/dv_l$ are available, the method of Marquardt (1963) and its variants are perhaps best. Other methods may be desirable where constraints are to be applied to the v_l, or where $\Phi(v)$ cannot be formulated as a sum of squares. One method of treating the sum-of-squares problem (Lawson and Hanson, 1974) has potentially special value because it can employ constraints such as positivity (see Chapter 4). When noise is not additive but signal dependent, least squares no longer leads to a correct formulation, and $\Phi(v)$ takes on a different form.

Martin (1971) gives a brief tutorial on the subject of optimization. Fletcher (1971) and Press *et al.* (1992) give excellent reviews and present criteria for the selection of an appropriate method. Rigler and Pegis (1980) discuss the various methods from the viewpoint of optical applications. Pitha and Jones (1966) compare several methods applied to fitting infrared band envelopes.

5. A Curve-Resolving Instrument

An interesting method of fitting was presented with the introduction, some years ago, of the model 310 curve resolver by E. I. du Pont de Nemours and Company. With this equipment, the operator chose between superpositions of Gaussian and Cauchy functions electronically generated and visually superimposed on the data record. The operator had freedom to adjust the component parameters and seek a visual best match to the data. The curve resolver provided an excellent graphic demonstration of the ambiguities that can result when any method is employed to resolve curves, whether the fit is visually based or firmly rooted in rigorous least squares. The operator of the model 310 soon discovered that, when data comprise two closely spaced peaks, "acceptable" fits can be obtained with more than one choice of parameters. The closer the blended peaks, the wider was the choice of parameters. The part played by noise also became rapidly apparent. The noisy data trace allowed the operator additional freedom of choice, when he considered the "error bar" that is implicit at each data point.

6. Constraints and True Deconvolution

Although useful, and even preferable in limited circumstances, none of the aforementioned methods may properly be called deconvolution. We address ourselves in this volume to the case in which our prior knowledge of $o(x)$ is less complete and in which we seek a more general solution. In Chapter 4, we see that certain modest elements of prior knowledge of $o(x)$ may be handled within the context of deconvolution, and that this knowledge exerts a powerful beneficial effect.

More obviously, spread-function knowledge is needed. The means of determining $s(x)$ depends on the field of application. We detail specific means for both spectra and images throughout the rest of the volume in the context of each field encountered. The greatest detail appears in the following chapter, which deals with optical spectra. A text by Castleman (1996) gives a good review of spread-function considerations relevant to two-dimensional imaging. Adapting and generalizing the concepts described there and in the present volume will be useful to the reader in preparing for new applications, be they in imaging, spectroscopy, or some other field.

References

Barrett, H. H., and Swindell, W. (1981). "Radiological Imaging," 2-Vol. Ed. Academic Press, New York.

Bates, R. H. T., and Jiang, H. (1991). *In* "International Trends in Optics" (J. W. Goodman, ed.), pp. 423–437. Academic Press, New York.

Bergland, G. D. (1969). *IEEE Spectrum* **6**, 41–52.

Bones, P. J., Parker, C. R., Satherley, B. L., and Watson, R. W. (1995). *J. Opt. Soc. Am.* **A12**, 1842–1856.

Bracewell, R. N. (1983). *J. Opt. Soc. Am.* **73**, 1832–1835.

Bracewell, R. (1986a). "The Fourier Transform and Its Applications," (2nd Ed. revised). McGraw-Hill, New York.

Bracewell, R. N. (1986b). "The Hartley Transform." Oxford Univ. Press, London.

Brigham, E. O. (1974). "The Fast Fourier Transform." Prentice-Hall, Englewood Cliffs, New Jersey.

Buneman, O. (1986). *SIAM J. Sci. Stat. Comput.* **7**, 624–638.

Castleman, K. R. (1996). "Digital Image Processing." Prentice-Hall, Englewood Cliffs, New Jersey.

Cohen, N. and Sandri, G. (1994). *International Journal of Imaging Systems and Technology* **5**, 28–32.

Cooley, J. W., and Tukey, J. W. (1965). *Math. Comput.* **19**, 297–301.

Dudgeon, D. E., and Mersereau, R. M. (1984). "Multidimensional Signal Processing." Prentice-Hall, Englewood Cliffs, New Jersey.

Fletcher, R. (1971). Proceedings of the Symposium on Computer-Aided Engineering, pp. 123–155. Univ. Waterloo, Waterloo, Ontario.

Hartley, R. V. L. (1942). *Proc. IRE* **30**, 144–150.

Herman, G. T. (1980). "Image Reconstruction From Projections." Academic Press, Orlando, Florida.

Högbom, J. A. (1974). *Astron. Astrophys. Suppl. Ser.* **15**, 417–426.

Hunt, R. H., Toth, R. A., and Plyler, E. K. (1968). *J. Chem. Phys.* **49**, 3909–3912.

Jansson, P. A. (1968). Ph.D. Dissertation, Florida State Univ., Tallahassee.

Jansson, P. A. (1970). *J. Opt. Soc. Am.* **60**, 184–191.

Jansson, P. A., Hunt, R. H., and Plyler, E. K. (1968). *J. Opt. Soc. Am.* **58**, 1665–1666.

Jansson, P. A., Hunt, R. H., and Plyler, E. K. (1970). *J. Opt. Soc. Am.* **60**, 596–599.

Jansson, P. A., Grim, L. B., Elias, J. G., Bagley, E. A., and Lonberg-Holm, K. K. (1983). *Electrophoresis* **4**, 82–91.

Kendall, M. G., and Stuart, A. (1961). "The Advanced Theory of Statistics," Vol. 2, Inference and Relationship, 3-Vol. Ed. Hafner, New York.

Kendall, M. G., and Stuart, A. (1963). "The Advanced Theory of Statistics," Vol. 1, Distribution Theory, 3-Vol. Ed. Hafner, New York.

Lawson, C. L., and Hanson, R. J. (1974). "Solving Least Squares Problems." Prentice-Hall, Englewood Cliffs, New Jersey.

Lutin, W. A., Kyle, C. F., and Freeman, J. A. (1978). *In* "Electrophoresis '78" (N. Catsimpoolas, ed.), "Dev. Biochem.", Vol. 2, pp. 93–106. Elsevier, New York.

McCalla, T. R. (1967). "Introduction to Numerical Methods and FORTRAN Programming." Wiley, New York.

Marquardt, D. W. (1963). *J. Soc. Ind. Appl. Math.* **11**, 431–441.

Martin, B. R. (1971). "Statistics for Physicists." Academic Press, New York.

Pitha, J., and Jones, R. N. (1966). *Can. J. Chem.* **44**, 3031–3050.

Poggio, T., Torre, V., and Koch, C. (1985). *Nature* (*London*) **317**, 314–319.

Press, W. H., Teukolski, S. A., Vetterling, W. T., and Flannery, B. P. (1992). "Numerical Recipes in FORTRAN: The Art of Scientific Computing," 2nd Ed. Cambridge Univ. Press, New York.

Rigler, A. K., and Pegis, R. J. (1980). In *Top. Appl. Phys.: The Computer in Optical Research: Methods and Applications*, Vol. 41, (B. R. Frieden, ed.), pp. 211–268. Springer-Verlag, New York.

Savitzky, A., and Golay, M. J. E. (1964). *Anal. Chem.* **36**, 1627–1639.
Seldin, J. H., and Fienup, J. R. (1990). *J. Opt. Soc. Am.* **A7**, 428–433.
Steinier, J., Termonia, Y., and Deltour, J. (1972). *Anal. Chem.* **44**, 1906–1909.
Taylor, J., Anderson, N. L., Coulter, B. P., Scandora, A. E., Jr., and Anderson, N. G. (1980). *In* "Electrophoresis '79" (B. J. Radola, ed.), pp. 329–339. de Gruyter, Berlin.
van Cittert, P. H. (1931). *Z. Phys.* **69**, 298–308.

Chapter 2 | Distortion of Optical Spectra

Peter A. Jansson

College of Optical Sciences, University of Arizona, Tucson

List of Symbols

A_L	measured absorptance amplitude of narrow rectangular line as observed by Gaussian response function
$A_M(x)$, $A_T(x)$	measured and "true" absorptances given by $1 - T_M(x)$ and $1 - T_T(x)$, respectively
B_L	maximum absorptance amplitude of narrow-line measurement
$B_M(x)$	flux in absence of sample as measured by a spectrometer of finite resolution
c	velocity of light
C	electrical capacitance
D	linear dimension of aperature of spectrometer grating or prism

$D_M(x)$	additive offset due to stray flux or electrical errors as might be recorded by a spectrometer
$\mathscr{E}(t), \mathscr{E}_1(t), \mathscr{E}_2(t)$	quantities proportional to electric field, time dependence explicitly shown
$E(z)$	quantity proportional to net field at distance z from image center
f	collimator focal length
f'	telescope focal length
$f(x)$	function
FWHM	full width at half maximum
h	Planck's constant
HWHM	half-width at half maximum
$i(t)$	electrical current
$i(x), i$	"image" data that incorporate smearing by $s(x)$
J_N, J_P, J_D, J_V	probabilities per unit frequency of a transition between energy states for spectral lines showing natural, pressure (combined natural and collision), Doppler, and Voigt (combined) broadening
k	Boltzmann's constant
m	molecular mass
$o(x), o$	"object" or function sought by deconvolution, usually the true spectrum, but also the instrument function when this is sought by deconvolution
P'	$S\sqrt{\ln 2 / \pi} / \Delta x_D$
P_N, P_P, P_D, P_V	spectral absorption coefficients in cases of natural, pressure-broadened, Doppler-broadened, and combined Doppler- and pressure-broadened profiles, respectively
$P(x), P$	spectral absorption coefficient, where $x = v/c$ and is given in cm^{-1} (c is velocity of light)
$q(t)$	electrical charge
R	electrical resistance
$rect(x)$	rectangle function having half-width $\frac{1}{2}$
$r(x)$	normalized spectrometer response function
$r_{coh}(z), r_{inc}(z)$	unnormalized coherent and incoherent optical spectrometer response functions, respectively
S	line strength
$sinc(x)$	$(\sin \pi x) / \pi x$
$Si(x)$	$\int_0^x [(\sin x') / x'] \, dx'$
$s(x), s$	spread function, usually the instrument function, but also spreading due to other causes
t, t'	time
T	absolute temperature
$T_M(x)$	transmittance as measured $[U_M(x) - D_M(x)]/[B_M(x) - D_M(x)]$
$T_T(x)$	"true" transmittance $[U_T(x) - D_T(x)]/[B_T(x) - D_T(x)]$

U, U_1, U_2	irradiances given by time averages $\overline{[\mathscr{E}(t)]^2}$, $\overline{[\mathscr{E}_1(t)]^2}$, $\overline{[\mathscr{E}_2(t)]^2}$, respectively
U_s	scale factor accounting for entrance-slit irradiance and optical-system efficiency
$U(x), U_T(x), U$	spectral flux transmitted through sample as it would be recorded by a perfectly resolving spectrometer
$U_0(x), U_0$	spectral flux incident on sample
$U_M(x)$	flux passed through sample as measured by a spectrometer of finite resolution
$U_T(x), B_T(x), D_T(x)$	"true" versions of functions $U_M(x), B_M(x), D_M(x)$ as would be recorded by a perfectly resolving spectrometer
$U_{coh}(z), U_{inc}(z)$	irradiances in slit image plane when slit is coherently and incoherently irradiated, respectively
$U_{obs}(z)$	electrically recorded spectrum
$v(t), v_C(t)$	voltages applied to circuit and across capacitor, respectively
$V(\omega), V_C(\omega)$	Fourier transforms of voltages $v(t), v_C(t)$
w	entrance-slit width
w'	detector width
W	equivalent width
W_M	measured equivalent width
$W(SX)$	equivalent width expressed as a series
x, x'	generalized independent variables and arguments of various functions, in places given specifically in cm^{-1}
x_0	location of line center in cm^{-1}
X	pressure of sample gas multiplied by optical path length
z	distance from center of line image expressed in Rayleigh widths
z_0	half-width of geometrical image of entrance slit expressed in Rayleigh widths
Z	collision frequency
α	fractional increase in instrument response-function breadth due to convolution with narrow spectral line
γ_C	$2Z$
γ_N	sum of reciprocals of natural upper- and lower-state lifetimes
γ_P	$\gamma_N + \gamma_C$
$\delta(x), \delta$	Dirac δ function or impulse
$\Delta E, \Delta t$	energy and time uncertainties, respectively
$\Delta\lambda$	Rayleigh resolution in wavelength units
Δx_G	Gaussian half-width at half maximum
$\Delta x_L, \sigma_L^2$	half-width and variance of narrow rectangular line, respectively
$\Delta x_N, \Delta x_P, \Delta x_D, \Delta x_V$	line-profile half-widths in cm^{-1} for natural, pressure-broadened, Doppler-broadened, and combined Doppler- and pressure-broadened cases, respectively; generally $\Delta x = \Delta\nu/c$, where c is velocity of light

$\Delta x_{\mathrm{R}}, \sigma_{\mathrm{R}}^2$	half-width and variance of instrument response function, respectively
$\Delta \bar{\nu}$	Rayleigh resolution in wave-number units
$\Delta \nu_{\mathrm{C}}$	half-width $\gamma_{\mathrm{C}}/4\pi$ observed when collision broadening dominates (given in hertz)
$\Delta \nu_{\mathrm{D}}$	Doppler half-width in hertz
$\Delta \nu_{\mathrm{N}}$	natural half-width $1/4\pi\gamma_{\mathrm{N}}$
$\Delta \nu_{\mathrm{P}}$	pressure-broadened half-width $1/4\pi\gamma_{\mathrm{P}}$
ζ	variable of integration
λ	wavelength of radiant flux
$\Lambda(x)$	triangle function of unit height and half-width $\frac{1}{2}$
μ	scaled ratio $(\Delta \nu_{\mathrm{N}} + \Delta \nu_{\mathrm{C}})\sqrt{\ln 2}/\Delta \nu_{\mathrm{D}}$
ν	frequency of electromagnetic wave in hertz
$\bar{\nu}$	wave number, usually in cm^{-1}
ν_0	center frequency of transition
ξ	distance from line center in scaled units of Doppler half-width
$\xi(t)$	response function of single-stage RC circuit
$\Xi(\omega)$	transfer function of single-stage RC circuit
σ^2	variance
$\tau(\omega), \tau$	Fourier transform of $s(x)$
ψ	Gaussian line-broading function
$\psi(SX)$	$W_{\mathrm{M}} - W(SX)$
ω	conjugate of x; Fourier frequency in radians per units of x
Ω	cutoff frequency such that $\tau(\omega) = 0$ for $|\omega| > \Omega$

Because the need for deconvolution arises in many disciplines, the details of the blurring mechanisms involved are varied. It is neither possible nor desirable to cover all circumstances here. Although the content of the present chapter may diverge from the reader's own specialty, it serves to illustrate considerations needed in understanding spreading phenomena. Optical spectrum broadening originates from two principal sources. One is inherent in the phenomenon being observed; the other is caused by the spectrometer. We begin by describing inherent broadening.

I. Inherent Line Breadth

Restricting ourselves to the optical spectra of gases, we now describe phenomena that cause a spectral line to have an inherent breadth. Even observations by a perfectly resolving spectrometer would show this broadening. Can the broadening be removed? And if so, under what circumstances? To answer these questions, we must first discuss the nature of the broadening.

A. NATURAL LINE BROADENING

The Heisenberg principle tells us that the energy uncertainty of a given state is inversely proportional to the uncertainty in the time during which the corresponding energy level is occupied:

$$\Delta E \, \Delta t \approx h/2\pi, \tag{1}$$

where ΔE and Δt are the energy and time uncertainties, respectively, and h is Planck's constant. The quantity Δt may be identified with the lifetime of an energy state. There are two such states, upper and lower, associated with a simple transition involving absorption or emission of a photon. The energy uncertainties combine as an overall "fuzziness" in our knowledge of the energy converted in a transition between the two states. A distribution is therefore required to characterize the probability of seeing a transition involving light of a given frequency. For such natural line broadening, we state that the probability per unit frequency of a transition yielding frequency ν is given by

$$J_N = \frac{\gamma_N}{4\pi^2(\nu - \nu_0)^2 + \gamma_N^2/4}, \tag{2}$$

where γ_N is the sum of the reciprocals of the natural upper- and lower-state lifetimes, and ν_0 the center frequency of the emission. Equation (2) represents the Lorentzian spectral-line shape, and we recognize it as having the Cauchy mathematical form encountered in Sections III.B and IV.A.7 of Chapter 1.

B. COLLISION BROADENING

To consider gas molecules as isolated from interactions with their neighbors is often a useless approximation. When the gas has finite pressure, the molecules do in fact collide. When natural and collision broadening effects are combined, the line shape that results is also a lorentzian, but with a modified half-width at half maximum (HWHM). Twice the reciprocal of the mean time between collisions must be added to the sum of the natural life-time reciprocals to obtain the new half-width. We may summarize by writing the probability per unit frequency of a transition at a frequency ν for the combined natural and collision broadening of spectral lines for a gas under pressure:

$$J_P = \frac{\gamma_P}{4\pi^2(\nu - \nu_0)^2 + \gamma_P^2/4}. \qquad (3)$$

In this expression the line breadth is dependent on $\gamma_P = \gamma_N + \gamma_C$. The collision contribution is given by $\gamma_C = 2Z$, where Z is the collision frequency. The resulting half-width is $\Delta\nu_P = \gamma_P/4\pi$. At low pressure γ_N will dominate, and γ_C will dominate at high pressure. Half-widths for these limiting line profiles are given by $\Delta\nu_N = \gamma_N/4\pi$ and $\Delta\nu_C = \gamma_C/4\pi$, respectively.

Quantum-theoretical calculations also predict a shift in line frequency under the latter condition. References and additional detail on all of the material in this section are given by Penner (1959).

C. DOPPLER BROADENING

From statistical mechanics we know that molecular velocities obey the Maxwell distribution law. This distribution is Gaussian in form. The Doppler effect influences light emitted from the moving molecules. The light from those in motion toward the observer is shifted upwards in frequency at the observation point, and the light from those moving away is shifted downwards. When this is taken into account, it can be determined that the probability per unit frequency of observing light at a frequency ν is given by

$$J_D = \frac{1}{\Delta\nu_D}\sqrt{\frac{\ln 2}{\pi}}\exp\left[\frac{-\ln 2(\nu - \nu_0)^2}{\Delta\nu_D^2}\right], \qquad (4)$$

where the Doppler half-width is given by

$$\Delta\nu_D = \nu_0\sqrt{(2kT\ln 2)/mc^2}. \qquad (5)$$

In this expression, m is the molecular mass, k is Boltzmann's constant, T is the absolute temperature of the gas, and c is the velocity of light.

D. COMBINED EFFECTS

The effects of Doppler and collision broadening are independent. Let us consider the case of emission, which can be generalized to the case of absorption as well. Each ensemble of molecules moving with a *given specific velocity* relative to the observer emits flux according to the spectral-distribution formula for combined natural and collision broadening. The flux observed from each such ensemble appears Doppler shifted according to its velocity relative to the observer. We combine by integration the contributions from all the molecules. The same result can also be obtained by considering the individual contributions of all molecules emitting at a given frequency in the Lorentzian profile and realizing that these molecules have Gaussian-distributed velocities. We would expect the two approaches to give the same result because the convolution operation is associative and commutative.

The function that we obtain is the Voigt function encountered in Section III.C of Chapter 1. With definition of variables and scaling appropriate to the case of combined Doppler, natural, and collision broadening, the Voigt function appears as the probability per unit frequency of observing light at a frequency v:

$$J_V = \frac{1}{\Delta v_D} \sqrt{\frac{\ln 2}{\pi}} \frac{\mu}{\pi} \int_{-\infty}^{\infty} \frac{\exp(-\zeta^2)}{\mu^2 + (\xi - \zeta)^2} \, d\zeta, \tag{6}$$

where

$$\xi = \frac{v - v_0}{\Delta v_D} \sqrt{\ln 2} \tag{7}$$

and

$$\mu = \frac{\Delta v_P}{\Delta v_D} \sqrt{\ln 2}. \tag{8}$$

Although this function has been extensively investigated, it has not been evaluated in closed form. A program has been developed by Armstrong (1967) for its evaluation. References to many approximations and expansions are given by Jansson and Korb (1968). Kielkopf (1973), Puerta and Martin (1981), and others have also developed useful approximations.

When the molecular mean free path is small compared with wavelength c/v_0, the mean Doppler shift is smaller than that for a free molecule. Under these circumstances, Eq. (4) is no longer valid, even when collision and natural broadening are neglected. This phenomenon of collisional narrowing was first described by Dicke (1953). A line-shape function given by Galatry (1961) is standardized in dimensionless form by Herbert (1974), who also develops and summarizes methods of evaluating it. Rodgers (1976) describes its effect on the equivalent width and gives criteria for judging when collisional narrowing must be taken into account. Pine (1980) has tested various line-shape models by observing the fundamental vibration of hydrogen fluoride gas in the infrared with the aid of a tunable-laser difference-frequency spectrometer.

E. ABSORPTION SPECTROSCOPY

Until now we have developed equations most directly adaptable to a particular case of spectral emission. In this case, the radiant flux, once emitted, suffers no further interactions on its way to the spectrometer. The intensity of the emitted flux is directly proportional to the probability distributions J_N, J_P, J_D, and J_V.

Many cases of real spectral emission, however, are not so simple, and the blackbody law must be invoked, along with attendant considerations of the relationships between transmissivity, reflectivity, emissivity, and absorptivity. We refer the reader to Penner (1959) for development of these concepts.

1. Absorption Lines

For the present, let us turn our attention to absorption phenomena. As the flux passes through the absorbing gas, each successive lamellar slice of the gas absorbs a fixed fraction of the light incident upon it from the previous neighboring slice. The total flux absorbed by the gas is proportional to the flux incident upon it, and is given by a single integration. We recognize the result as the Bouguer–Lambert law. The transmitted spectral flux is given by

$$U = U_0 \exp(-PX), \tag{9}$$

where U_0 is the spectral flux incident upon the sample, X the pressure times the optical path length, and P the spectral absorption coefficient. The probability distribution J that we have already discussed may be adapted to this situation in the manner described by Penner (1959). At this point, we shall also begin dealing with the independent variable x in reciprocal centimeters (wave numbers), instead of the frequency ν in hertz employed until now. With this in mind, we may write

$$P(x) = cSJ(\nu), \tag{10}$$

where c is the velocity of light and S the line strength. Bearing also in mind that $\nu = cx$, we may now write the spectral absorption-coefficient profile. For the profile with purely natural broadening, we find

$$P_N(x) = \frac{S}{\pi} \frac{\Delta x_N}{(x - x_0)^2 + \Delta x_N^2}, \tag{11}$$

where $\Delta x_N = \gamma_N/4\pi c$. The quantity x_0 is the location of the line center in reciprocal centimeters. In the pressure-broadened case, where both natural and collision broadening are present, we obtain

$$P_P(x) = \frac{S}{\pi} \frac{\Delta x_P}{(x - x_0)^2 + \Delta x_P^2}. \tag{12}$$

In this profile we define $\Delta x_P = \gamma_P/4\pi c$. Doppler broadening yields

$$P_D(x) = \frac{S}{\Delta x_D} \sqrt{\frac{\ln 2}{\pi}} \exp\left[\frac{-\ln 2(x - x_0)^2}{\Delta x_D^2}\right], \tag{13}$$

where the Doppler half-width (HWHM) is given by

$$\Delta x_D = x_0\sqrt{(2kT \ln 2)/mc^2}. \tag{14}$$

Expressing the distance from the line center in terms of Doppler half-widths, we identify

$$\xi = (\nu - \nu_0)\sqrt{\ln 2}/\Delta \nu_D = (x - x_0)\sqrt{\ln 2}/\Delta x_D. \tag{15}$$

Defining $P' = P_D(0) = (S/\Delta x_D)\sqrt{(\ln 2)/\pi}$, we write the Doppler profile in a much simpler form,

$$P_D = P' \exp(-\xi^2). \tag{16}$$

By using the same definitions, the profile for combined Doppler, natural, and collision broadening may be expressed in terms of the Voigt function:

$$P_V = P'\frac{\mu}{\pi}\int_{-\infty}^{\infty} \frac{\exp(-\zeta^2)}{\mu^2 - (\xi - \zeta)^2}\, d\zeta. \tag{17}$$

2. Deconvolution of Inherent Broadening

It may have occurred to the reader that inherent line broadening could be removed by deconvolution. This appears especially attractive as a means of separating the various contributions. In the particular case of absorption spectra, however, it is not as straightforward as it at first appears. Referring to Eq. (9), we see that the various broadening-profile convolutions appear in the exponential:

$$U = U_0 \exp(-\text{natural} \otimes \text{collision} \otimes \text{Doppler} \otimes \text{spectrum}). \quad (18)$$

The data that we acquire are usually in the form of U/U_0. Only if there is no significant instrumental broadening present, or if it has already been corrected, may we remove the inherent broadening. In this case, we operate on the logarithm of the data or corrected data:

$$K(x) = \text{natural} \otimes \text{collision} \otimes \text{Doppler} \otimes \text{spectrum} = \ln[U_0(x)/U(x)]. \quad (19)$$

There is another difficulty, however. Additive instrumental noise is also logarithmically converted by the operation. This in turn may affect the validity of the noise assumptions employed in developing certain deconvolution algorithms. When absorptions are very small, we may approximate

$$XP(x) \approx 1 - U(x)/U_0(x), \quad (20)$$

and the inherent broadening convolutions are done in the same regime as those for instrumental broadening. When absorptions are very small, however, the signal-to-noise ratio (or, more properly, peak-height-to-noise ratio) is usually poor. Deconvolution requires a good signal-to-noise ratio, so that this case would not appear to be practical.

It may be surprising, then, that Pliva *et al.* (1980) have been successful in partially removing inherent broadening by operating directly on absorptance data $U(x)/U_0(x)$. Their success appears to be due in part to the stability enforced by physical-realizability constraints.

Measurements of the total integrated absorption $S = \int P(x)\,dx$ may be made by using the method of equivalent widths. Such measurements are independent of instrumental broadening and are considered further in Sections II.F and II.G.

F. INFRARED SPECTRA OF LIQUIDS

In liquids the interactions between neighboring molecules are considerably more complicated than in gases. The resultant broadening obliterates the fine line structure seen in gas spectra, leaving only broad band profiles. There are many possible contributors to this broadening. In some cases, adequate approximation is obtained by assuming that the band contour is established by collisions. Ramsay (1952) has noted that substitution of appropriate molecular density and collision diameter numbers in the collision broadening formula results in realistic band widths for certain liquid-phase systems. In such systems, the bands typically show an approximately Lorentzian profile. Approximate deconvolution of inherently broadened liquid-phase spectra may therefore be obtained on the basis of the assumption of Lorentzian shape (Kauppinen *et al.*, 1981).

It is common, however, for liquid-phase systems to include many specific absorbing species. Such species could include isotopic variations, conformational isomers, and solvent-solute interactions resulting in varied-lifetime transients associations between molecules. Distributions resulting from these effects give the Voigt profile utility in studying liquid spectra. We must understand, however, that the functions introduced here are only rough approximations when applied to the spectra of liquids because of the complexities just mentioned and others beyond the scope of this work.

II. Contribution of Dispersive Spectrometer

We now consider broadening caused by the spectrophotometric system itself. This broadening is of special interest to us in this work because it is not part of the phenomenon under observation. Instead, it represents our inability to observe this phenomenon accurately. Elimination of spectrometer broadening is the usual goal of deconvolution.

The discussion in the present section concerns only dispersive optical spectroscopy. We shall not treat the resolution characteristic of a Fourier interferometric spectrometer, which is determined by the optical path difference scanned and the apodizing function used. Nevertheless, these considerations are pertinent to Howard's work in extending Fourier interferograms. (See Chapters 12 and 13, which contain appropriate references.)

The other deconvolution methods described in this volume may, with suitable adaptation, also be applied to Fourier spectroscopy.

Broadening caused by the dispersive optical spectrometer can be divided into electrical and predominantly optical parts. When photographic detection is employed, other considerations apply as well, but we shall not discuss them here. As for optics, the idealized model of a perfect diffraction pattern integrated over the slit openings is complicated by the need to consider slit irradiation coherence. Furthermore, optical aberrations and grating imperfections can contribute significantly. These considerations might lead us to believe that empirical determination of the spectrometer response function is the best. This belief is often well justified in spite of the obstacles that also lie along this route.

A brief treatment of the major contributors to optical spreading is certainly in order because it will clarify some of the considerations in designing and adjusting the spectrometer. In some cases it will also allow us to compute a useful approximation to the optical spread function. In the same vein, we later describe the behavior of the most common, useful, and easily described form of electrical noise smoothing: the single stage *RC* filter.

A. COHERENT IRRADIATION

Consider a spectrometer used for measuring either emission or absorption spectra. The flux from the sample is typically focused onto an entrance slit. The spectral distribution of this flux is the result of an emission process, possibly also followed by absorption. It may be a line spectrum affected by broadening of the type previously discussed. In one possible optical arrangement (Fig. 1), the flux emerging from the slit is incident upon a collimating mirror, which we here take to have focal length f. The flux may then be dispersed by a prism or grating having an aperture of width D. A telescope mirror, which we take to have focal length f', is then employed to focus the dispersed flux to an image of the entrance slit.

1. Monochromatic Case

Assume first that we have closed the entrance slit to the smallest possible opening so that a perfect geometrical line is approximated. From elementary physical optics we know that monochromatic illumination of the

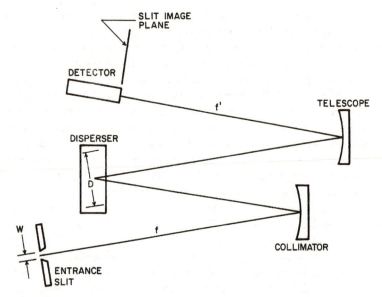

Figure 1 Optical schematic of a typical dispersive spectrometer.

idealized slit in this system gives rise to a diffraction phenomenon owing to spatial truncation of the wave fronts by the aperture of width D. Thus, an irradiance pattern is formed in the image plane having $\text{sinc}^2(z)$ dependence, where z is the distance from the center of the line image as measured in Rayleigh widths. The Rayleigh width is given by $f'\lambda/D$, where λ is the wavelength of the monochromatic light.

Let us see what happens as the entrance slit is opened to a width w. It can now be thought of as comprising a large number of narrow slits placed next to one another. If this entrance slit is uniformly filled with monochromatic light, then the image plane will show the superposition of a large number of diffraction patterns, a sinc function as shown in Fig. 2 for the field owing to each narrow elemental entrance-slit component.

Before we state this mathematically, let us consider the implication of perfectly monochromatic illumination. We shall assume that the slit elements are illuminated by light of *exactly* the same wavelength and frequency. Each given pair of points in the entrance slit will then have its own fixed phase relationship. Neither point can gain or lose phase relative to the other because the fields have *exactly* the same frequency. The fields at

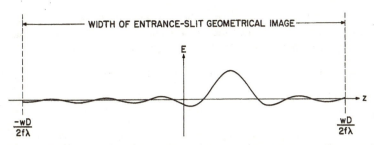

Figure 2 Diffraction image of an entrance-slit element. Abscissa x is distance in Rayleigh widths; ordinate E is proportional to electric-field amplitude.

these two points are perfectly *coherent*. Because this is true for any given pair of points in the slit, all the points have a fixed phase relationship. We call this the case of *coherent irradiation*.

From the physics of electricity and magnetism we know that electric fields obey the superposition principle. To find the net field in the image plane, we must sum the field components, not the irradiances. For simplicity, let us consider only one plane of polarization. In the limit of contributions from infinitesimal slit elements, we obtain

$$E(z) = \sqrt{U_s} \int_{-wD/2f\lambda}^{wD/2f\lambda} \operatorname{sinc}(z - z')\, dz', \tag{21}$$

where $E(z)$ is proportional to the net field at a distance z from the center of the image. We introduced the sinc function in Section IV.A.1 of Chapter 1. The constant U_s is a scale factor accounting for the entrance-slit irradiance and optical-system efficiency. The limits of integration represent the breadth of the slit image when the optical-system magnification is taken into account and are given in Rayleigh widths.

We can also express Eq. (21) as the convolution of a sinc with a rectangle,

$$E(z) = \sqrt{U_s}\, \operatorname{sinc}(z) \otimes \operatorname{rect}(fz/f'w),$$

or explicitly

$$\begin{aligned} E(z) &= \sqrt{U_s} \int_{-wD/2f\lambda}^{wD/2f\lambda} \frac{\sin[\pi(z - z')]}{\pi(z - z')}\, dz' \\ &= \sqrt{U_s} \int_{z-z_0}^{z+z_0} \frac{\sin \pi z'}{\pi z'}\, dz'. \end{aligned} \tag{22}$$

We have defined $z_0 = wD/2f\lambda$ as the half-width of the geometrical image of the entrance slit, as measured in Rayleigh widths. Referring to the Si function defined in Section IV.A.1 of Chapter 1 we now may write

$$E(z) = (1/\pi)\sqrt{U_s}\{\text{Si}[\pi(z + z_0)] - \text{Si}[\pi(z - z_0)]\}. \tag{23}$$

The irradiance in the image plane is proportional to the square of the field:

$$U_{\text{coh}}(z) = U_s\pi^{-2}\{\text{Si}[\pi(z + z_0)] - \text{Si}[\pi(z - z_0)]\}^2. \tag{24}$$

2. General Case

The irradiance of the entrance slit just described was perfectly monochromatic and coherent. If we now irradiate the entrance slit with flux having a spectrum of finite breadth, but arrange the optics so as to guarantee continued perfect spatial coherence in the entrance-slit plane, the resulting irradiance in the image plane is given by the convolution of $U_{\text{coh}}(z)$ with the spectrum of the source.

Now assume that we have placed a flat-surfaced detector of width w' in the image plane, and that each point on the surface of the detector yields an electrical-signal contribution proportional to the irradiance (not the field). Assume further that the response across the detector is uniform, and that contributions from all parts of the detector combine by superposition. The resulting signal is then proportional to the direct superposition of the irradiance contributions. If we sweep this slit detector across the spectrally dispersed irradiance, the resulting signal, as a function of time, is given by the convolution of the spectrum with $U_{\text{coh}}(z)$ and $\text{rect}(z/w')$, the rectangle function this time representing the exit slit. The unnormalized optical response function of the spectrometer is thus given by

$$\text{rect}(z/w') \otimes [\text{rect}(fz/f'w) \otimes \text{sinc}(z)]^2. \tag{25}$$

Note that the inner convolution results from the superposition of coherent fields and the outer convolution from the superposition of electrical-signal contributions.

Suppose, instead, that we had placed a lens near the plane of the exit slit, in the path of the flux emerging from that slit, and imaged the spectrometer pupil onto the detector. Then the field at each point in the exit slit would contribute equally to the field at any particular point on the detector. The net field at that point on the detector is given by linear superposition of the fields. The electrical contribution, as before, goes as the square of this field. Sweeping the spectrum again, we observe the signal given by convolving the spectrum with a different unnormalized optical response function $r_{coh}(z)$:

$$U_{obs}(z) = r_{coh}(z) \otimes U(z),\qquad(26)$$

where $U_{obs}(z)$ is the electrically recorded spectrum and $U(z)$ the spectrum as it would be recorded by a perfectly resolving spectrometer. The response function is now given by

$$r_{coh}(z) = [\text{rect}(z/w') \otimes \text{rect}(fz/f'w) \otimes \text{sinc}(z)]^2.\qquad(27)$$

In the common case where the exit-slit width is given by $w' = f'w/f$ (a typical situation is $f' = f$ and $w' = w$), we obtain

$$r_{coh}(z) = [\Lambda(fz/f'w) \otimes \text{sinc}(z)]^2,\qquad(28)$$

where, as stated earlier, f and f' are the focal lengths of the collimating and telescope elements, respectively, w and w' are the entrance- and exit-slit widths, respectively, and z is the distance in the direction of dispersion in the exit-slit plane as measured in Rayleigh widths.

B. INCOHERENT IRRADIATION

1. Monochromatic Case

Now let us assume that a monochromatic source of flux is placed in the plane of the entrance slit so that there is no constant phase relationship between the fields at any two given points in the slit. This, in itself, is a contradiction, because perfect source monochromaticity implies both spatial and temporal coherence. By definition of coherence, a constant phase relationship would result. To eliminate the possibility of such a relationship, we must require the source spectrum to have finite breadth. Let us modify the assumption accordingly but specify the source spectrum breadth narrow enough so that its spatial extent when dispersed is negligible compared with the breadth of the slits, diffraction pattern, and so on. Whenever time integrals are required to obtain observable signals from superimposed fields, we evaluate them over time periods that are long compared with the reciprocal of the frequency difference between the fields. We shall call the assumed source a quasi-monochromatic source.

Let us consider the flux from the quasi-monochromatic incoherent slit source imaged through the spectrometer as in the previous example. Slit width and focal lengths are defined as before. Again the exit-slit–plane image will consist of the superimposed fields from the sinc(z) diffraction pattern due to each vertical element. The instantaneous fields obey superposition. We may write the field resulting from two contributions, for example, as simply proportional to $\mathscr{E}_1(t) + \mathscr{E}_2(t)$. The resulting irradiance is given by the time average,

$$U = \overline{[\mathscr{E}_1(t) + \mathscr{E}_2(t)]^2} = \overline{\mathscr{E}_1^2(t)} + \overline{\mathscr{E}_2^2(t)} + \overline{2\mathscr{E}_1(t)\mathscr{E}_2(t)}. \qquad (29)$$

The irradiance U is seen to be the sum of the irradiances due to each component plus a cross term $\overline{2\mathscr{E}_1(t)\mathscr{E}_2(t)}$. But the assumption of complete incoherence and quasi-monochromaticity guarantees that no two components $\mathscr{E}_1(t)$ and $\mathscr{E}_2(t)$ could have a constant phase difference and still yield a product having a finite time integral. (Remember that the integration time must be long compared with the reciprocal of the frequency difference.) For perfectly incoherent slit irradiation, then, the resulting exit-slit–plane irradiance is the linear superposition of the diffraction-pattern irradiances, and we find that

$$U_{\text{inc}}(z) = U_s \int_{-wD/2f\lambda}^{wD/2f\lambda} \text{sinc}^2(z - z')\, dz'. \qquad (30)$$

As before, we may express an integral of this type as a convolution:

$$U_{\text{inc}}(z) = U_s \text{sinc}^2(z) \otimes \text{rect}(fz/f'w). \qquad (31)$$

This time the convolution is linear in the irradiances, not the fields. We may rewrite Eq. (30) as

$$U_{\text{inc}}(z) = U_s \int_{z-z_0}^{z+z_0} \left(\frac{\sin \pi z'}{\pi z'}\right)^2 dz'. \qquad (32)$$

Integration by parts yields

$$U_{\text{inc}}(z) = \frac{1}{\pi} U_s \left\{ \text{Si}[2\pi(z + z_0)] - \text{Si}[2\pi(z - z_0)] \right.$$

$$\left. - \frac{\sin^2[(z + z_0)/\pi]}{(z + z_0)/\pi} + \frac{\sin^2[(z - z_0)/\pi]}{(z - z_0)/\pi} \right\}, \qquad (33)$$

where z and z_0 are defined as before.

2. General Case

Let us assume now that, instead of looking at a quasi-monochromatic source, we are filling the entrance slit with a truly polychromatic source. The irradiance components in the image plane add by incoherent superposition, and we find that the dispersed spectrum in the image plane is obtained by convolving $U_{inc}(z)$ with the spectrum of the source. A detector of width w that obeys the superposition criteria established previously may be placed in the image plane and swept across the spectrum. Its contribution to the spectral broadening will be a linear superposition of irradiances over its width, resulting in convolution with $rect(z/w')$. Consequently, the electrically recorded spectrum is given by

$$U_{obs}(z) = r_{inc}(z) \otimes U(z), \tag{34}$$

where $U(z)$ is the undistorted spectrum as it would be recorded by a perfectly resolving spectrometer, and the unnormalized incoherent optical response function is given by

$$r_{inc}(z) = rect(z/w') \otimes rect(fz/f'w) \otimes sinc^2(z). \tag{35}$$

When $w' = f'w/f$, by analogy with the previous derivation, we find

$$r_{inc}(z) = \Lambda(fz/f'z) \otimes sinc^2(z). \tag{36}$$

We also obtain this result if we use a lens or mirror to image the exit slit onto a detector, provided that any blur produced by this imaging element is negligible compared with the breadth of $r_{inc}(z)$.

C. REAL SPECTROMETERS

Unlike the situation just discussed we are usually not concerned with measuring the spectrum of a highly coherent source like a laser. It is also not usual to place an incoherent source directly in the entrance slit. A more common situation than either of these extremes is one in which the source is imaged onto the entrance slit by foreoptics.

   To deal properly and completely with this general case, we must consider the effects of partial coherence. Discussion of this subject is beyond the scope of the present work. Mielenz (1967) has derived appropriate response-function formulas and gives references to the literature.

D. ELECTRICAL FILTERING

The spectrum is usually observed by passing it across the detector, thereby effectively converting it to a function of time. Wavelength or wave number may (or may not) be a linear function of time, depending on the drive mechanism employed. Over limited spectral regions, it is often adequate to assume that either wavelength or wave number is proportional to time.

Whatever its source, the electrical signal usually contains noise components having frequencies far higher than those found in the spectrum itself. Electrical filtering is then called for to remove these high-frequency components.

1. Single-Stage RC Filter

The design of optimum electrical filters is a highly developed discipline that we shall not address here. It is worthwhile, however, to discuss briefly the simple single-stage *RC* filter that is widely used. When all factors are considered, the performance advantages of elaborate designs are often outweighed by simplicity.

We shall not rely on the electrical engineer's formalism to deal with this problem. Instead, because it is instructive, we appeal to basic physics for a description of this filter's performance.

Figure 3 shows a simple single-section *RC* low-pass filter network. Kirchhoff's law tells us that

$$v(t') - i(t')R - v_C(t') = 0, \tag{37}$$

where $v(t')$ is the input signal voltage, $i(t')$ the current in the loop, R the

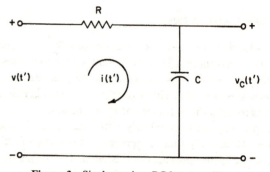

Figure 3   Single-section *RC* low-pass filter.

resistance, $v_C(t')$ the voltage across the capacitor, and t' time. Now $v_C(t') = q(t')/C$, where $q(t')$ is the charge on the capacitor with capacitance C. Furthermore, $i(t') = dq(t')/dt'$. Therefore, from Eq. (37), we find

$$\frac{d}{dt'}q(t') + \frac{1}{RC}q(t') - \frac{1}{R}v(t') = 0. \qquad (38)$$

Multiplying by the integrating factor $\exp(t'/RC)$ and integrating with respect to time from $-\infty$ to t, we obtain

$$\frac{1}{C}q(t)\exp\left(\frac{t}{RC}\right) = \frac{1}{RC}\int_{-\infty}^{t} v(t')\exp\left(\frac{t'}{RC}\right) dt'. \qquad (39)$$

Noting that $q(t)/C = v_C(t)$, and moving $\exp(t/RC)$ into the integrand, we have

$$v_C(t) = \frac{1}{RC}\int_{-\infty}^{t} v(t')\exp\left[-\frac{(t - t')}{RC}\right] dt'. \qquad (40)$$

If we let

$$\xi(t) = (1/RC)H(t)\exp(-t/RC), \qquad (41)$$

where $H(t)$ is the Heaviside step function, we obtain

$$v_C(t) = \xi(t) \otimes v(t). \qquad (42)$$

The output voltage v_C is obtained from the input voltage by a simple convolution. Referring to the directory of Fourier transforms (Chapter 1, Fig. 2), we may write

$$V_C(\omega) = \Xi(\omega)V(\omega). \qquad (43)$$

Here the frequency spectra of the output and input signals are given by $V_C(\omega)$ and $V(\omega)$, respectively, and the complex filter transfer function is given by

$$\Xi(\omega) = \frac{1}{\sqrt{2\pi}} \frac{1 - j\omega RC}{1 + (\omega RC)^2}. \qquad (44)$$

We see that the amplitude-transfer characteristic is given by $\{2\pi[1 + (\omega RC)^2]\}^{-1/2}$. The power-transfer characteristic is given by the square of this quantity. It has the form of a Cauchy function and attenuates high frequencies. Brodersen (1953) and Stewart (1967) have analyzed in detail the performance of other linear electrical filters applied in spectroscopy.

The spectrum distortion introduced by all such filters limits the spectrometer's scanning rate. The subtleties of the scan rate's dependence on electrical filtering and slit width for constant signal-to-noise ratio are explored by Blass and Halsey (1981).

2. Applications

The single-section RC filter is often used after phase-sensitive detection. In this method of observing spectra, the radiant flux is typically modulated by a mechanical chopper. The resulting signal is usually demodulated synchronously by switching. This operation can be considered as multiplication of the waveform by a bipolar square wave. Sometimes a narrow-band filter is used to reduce high-frequency noise and permit additional gain without amplifier saturation by noise before demodulation. The result of using such a narrowband filter is improved dynamic range. By such a demodulation scheme, the fundamental frequency in the signal is shifted to the origin. The modulated information is available after demodulation as a slowly varying (nearly dc) signal. We see that the effect of the RC filter is, in frequency space, equivalent to applying a square-root Cauchy amplitude-transfer characteristic centered at the modulation frequency.

The entire analysis of synchronous detection, or lock-in amplification as it is sometimes called, can be conveniently analyzed by straightforward application of the Fourier transform techniques, transform directory, and convolution theorem developed in Section IV of Chapter 1.

In concluding this section, we point out that the effect of any electrical filter composed of purely linear elements, whether they be passive like resistors, capacitors, and inductors or active like linear amplifiers, can be represented as a convolution. The various other spreading phenomena that are described by convolution in the same domain may therefore be lumped together with the electrical contribution and comprehensively called the spectrometer response function. Even inherent line broadening may be included, provided that the convolution does not appear in an exponent, as in the case of absorption spectra.

E.   MEASURING TRANSMITTANCE AND ABSORPTANCE

We have shown that the radiant flux spectrum, as recorded by the spectrometer, is given by the convolution of the "true" radiant flux spectrum (as it would be recorded by a perfect instrument) with the spectrometer response function. In absorption spectroscopy, absorption lines typically appear superimposed upon a spectral background that is determined by the emission spectrum of the source, the spectral response of the detector, and other effects. Because we are interested in the properties of the absorbing molecules, it is necessary to correct for this background, or base line as it is sometimes called. Furthermore, we shall see that the valuable physical-realizability constraints presented in Chapter 4 are easiest to apply when the data have this form.

1. Transmittance

In double-beam spectrometers, the correction is made before the spectrum is recorded by dividing:

$$T_M(x) = [U_M(x) - D_M(x)]/[B_M(x) - D_M(x)]. \qquad (45)$$

In this expression, x is the independent variable of measurement, whether it be wavelength, wave number, or other parameter. The quantity $U_M(x)$ is the flux, as measured by the spectrometer, after it has passed through the absorbing sample. The quantity $B_M(x)$ represents the flux that has passed over an identical path but in the absence of an absorbing sample. The term $D_M(x)$ represents any additive offsets that might be introduced by "stray" flux, errors in electronic amplification and digital conversion, or other causes. We call the resulting measured transmittance $T_M(x)$.

If we have a single-beam spectrometer, we may separately record spectra $B_M(x)$, $U_M(x)$, and $D_M(x)$ and apply Eq. (45) later in the computer. With special rapid-scanning spectrometers this approach may be practical, but source and detector drift usually preclude use of this procedure for high-accuracy, slow-scanning spectroscopy.

More typically, one seeks $U_M(x)$ regions of negligible absorption and connects them with a fitted curve (often a straight line is adequate over a limited region) to approximate $B_M(x)$. If $D_M(x)$ represents mainly offsets in the electronics, a shutter placed in the beam facilitates its determination.

2. Effect of Instrumental Spreading

In Eq. (45), T_M represents the "true" transmittance only if instrumental spreading is negligible. In previous sections, however, we learned that instrumental spreading may be described by the convolution product of the flux and the spectrometer response function, here called $r(x)$, that incorporates all the instrumental spreading phenomena:

$$U_M(x) = r(x) \otimes U_T(x). \tag{46}$$

Here $U_T(x)$ represents the "true" flux.

Both $B_M(x)$ and $D_M(x)$ are similarly distorted versions of true background and offset functions that we may identify, respectively, as $B_T(x)$ and $D_T(x)$. Thus we may write the measured transmittance as

$$T_M(x) = \frac{U_M(x) - D_M(x)}{B_M(x) - D_M(x)} = \frac{r(x) \otimes U_T(x) - r(x) \otimes D_T(x)}{r(x) \otimes B_T(x) - r(x) \otimes D_T(x)}$$

$$= \frac{\int r(x - x')[U_T(x') - D_T(x')]\, dx'}{\int r(x - x')B_T(x')\, dx' - \int r(x - x')D_T(x')\, dx'}, \tag{47}$$

where the limits cover a spectral region of interest.

Let us now assume that $B(x)$ and $D(x)$ are either constant or slowly varying compared with $r(x)$, so that we may take $B(x') = B(x)$ and $D(x') = D(x)$ in the denominator. The normalization of $r(x)$ then permits us to write

$$T_M(x) = \frac{1}{B_T(x) - D_T(x)} \int r(x - x')[U_T(x') - D_T(x')]\, dx'. \tag{48}$$

Again taking advantage of the slow variation of $B_T(x)$ and $D_T(x)$, we may include them in the integrand:

$$T_M(x) = \int r(x - x') \frac{U_T(x') - D_T(x')}{B_T(x') - D_T(x')}\, dx' = r(x) \otimes T_T(x). \tag{49}$$

We see that transmittance obeys the convolution relation for instrumental spreading provided that $B_T(x)$ and $D_T(x)$ are either constant or slowly varying compared with $r(x)$.

3. Absorptance

We define measured and true absorptance by the pair of relations

$$A_M(x) = 1 - T_M(x) \tag{50}$$

and

$$A_T(x) = 1 - T_T(x). \tag{51}$$

From Eqs. (49) through (51) we may obtain

$$A_M(x) = 1 - r(x) \otimes T_T(x) = r(x) \otimes [1 - T_T(x)] = r(x) \otimes A_T(x). \tag{52}$$

We see that absorptance obeys the convolution relation for instrumental spreading under the same conditions of slowly varying $B_T(x)$ and $D_T(x)$.

We note here the similarity between the terms absorp*t*ance and absor*b*ance. The latter is frequently employed in chemical infrared spectroscopy to denote $\log_{10}[1/T(x)]$, and may also be called optical density. As discussed in Section I.E.2, absor*b*ance may not be directly employed in the convolution relation for instrumental spreading.

F. MEASUREMENTS INDEPENDENT OF SPECTROMETER BROADENING

In many applications, it is convenient to define the equivalent width of a function $f(x)$ as the width of a rectangle having the same area and maximum height as $f(x)$. Following this definition for a function $f(x)$ having its maximum at $x = 0$, we have for the equivalent width

$$\left[\int_{-\infty}^{\infty} f(x)\, dx\right]\bigg/ f(0). \tag{53}$$

In absorption spectroscopy it is more useful to define the equivalent width of a spectral line as the width of an idealized, totally absorbing rectangular line having the same area (Fig. 4). The equivalent width of an isolated absorption line is therefore given by

$$W = \int A_T(x)\, dx = \int [1 - T_T(x)]\, dx, \tag{54}$$

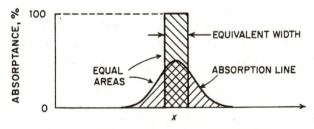

Figure 4   Equivalent width of an absorption line.

where the limits of integration are understood to extend on either side of the line until the absorption encountered is negligible. Because the area under the instrument response function is normalized to unity, we may insert that integral in the integrand of Eq. (54):

$$W = \int A_T(x') \left[\int r(x - x') \, dx \right] dx'. \tag{55}$$

Reversing the order of integration and recognizing that $A_M(x) = r(x) \otimes A_T(x)$, we obtain

$$W = \int \int r(x - x') A_T(x') \, dx' \, dx = \int A_M(x) \, dx = \int [1 - T_M(x)] \, dx. \tag{56}$$

Therefore, we have shown that

$$W = \int A_T(x) \, dx = \int A_M(x) \, dx. \tag{57}$$

The equivalent width W of an isolated absorption line is independent of spectrometer broadening effects.

Noise causes excursions above and below the desired trace when a scan is made across such an absorption line. The small positive and negative contributions made by these excursions tend to cancel in the computation of area. Such areas can thus be measured to relatively high precision. Furthermore, when lines exhibiting predominantly Doppler broadening are measured, the wings fall rapidly to zero at a moderate distance from the line center. This behavior facilitates determination of the base line and contributes to the accuracy of measurement. Accurate measurements of line strength can therefore be made by these methods (Korb et al., 1968) because the equivalent width may be expressed as a function of the line strength S times the pressure path-length product X. The quantity W is given by different series for Doppler- and pressure-broadened cases. Typically one wishes to determine S from $W(SX)/X$. A straightforward and rapidly converging application of Newton's method enables us to find the value of the product SX for which $\psi = 0$ in the expression

$$\psi(SX) = W_M - W(SX), \tag{58}$$

where W_M is the measured equivalent width.

For purely Doppler broadening, the equivalent width takes the form

$$W(SX) = SX \sum_{n=0}^{\infty} \frac{\left[-\sqrt{(\ln 2)/\pi}\, SX/\Delta x_D \right]^n}{(n+1)!\sqrt{n+1}}. \tag{59}$$

The quantity Δx_D is the Doppler half-width encountered earlier. Newton's method requires knowledge of the first derivative, which is easily provided by multiplying each term in Eq. (59) by n/SX.

When pressure broadening dominates, the situation is more complicated because the resulting Lorentzian profile contributes significant area far from the line center. A further complication in this case is that the Lorentzian half-width cannot be accurately calculated and must be measured in other experiments. If both Doppler and pressure broadening are present, however, and if the Lorentzian to Doppler half-width ratio μ is small, the correction necessitated by pressure broadening is small. In this situation an accurate value of the Lorentzian half-width may not be needed. Line strength in the case of combined Doppler and pressure broadening may be obtained from the equivalent width by the use of tables (Jansson and Korb, 1968).

G. DETERMINING THE SPECTROMETER RESPONSE

In applying the technique of deconvolution, we take as known the spectrometer response function. It seems reasonable that the more accurately we know this function, the more accurate will be the deconvolved result. Although the nonlinear methods described in Chapter 4 are more tolerant of error, they too require a knowledge of the response function.

1. Difficulties in Use of Narrow Lines

Owing to aberrations, grating defects, and so on, it may not be adequate to approximate the response function by formulas based on idealized models. If a line source could be found having the spectrum that approximates a δ function, then perhaps the measurement of such a line would adequately determine the response function. We have learned, however, that the spatial coherence of the source plays an important part in the shape of the response function. This precludes the use of a laser line source to measure the response function applicable to absorption spectroscopy. Furthermore, we would have to be sure that both the pupil and the slit are filled with the same spatial distribution of flux as that used in the observation of the spectra to be deconvolved. The latter requirement seems to be a tall order, unless we use the same continuum source employed for observation of the absorption spectra themselves. This, in turn, implies that we might consider observing a very narrow absorption line. Let us consider such an observation.

An inherent difficulty in observing a narrow absorption line is that it does not absorb very much flux. The narrower the line, the less flux absorbed. In fact, the ideal δ-function line absorbs none at all. This statement may be verified by evaluating the integrated absorptance

$$\int_{\substack{\text{entire} \\ \text{line}}} [1 - e^{-XP(x)}] \, dx \qquad (60)$$

of a narrow Gaussian absorption line $XP(x)$, and taking the limit as the $P(x)$ half-width Δx_G goes to zero, while holding the integrated absorption coefficient $X \int P(x) \, dx$ constant.

If we attempt to render a narrow absorption line observable by increasing the absorbing optical path, we succeed only in broadening the line as it is observed in the transmittance-absorptance regime. Although the shape and breadth of the absorption-coefficient curve $XP(x)$ have not been altered by this procedure, the spectrometer sees a broader spectral line because of the saturation of the absorption near the line center and increasing absorption in the wings. The term "half-width" as applied to absorption lines is almost universally used with respect to the absorption-coefficient profile $P(x)$. Used alone it is *not meaningful* as a measure of the effective breadth of isolated lines employed to determine the response function. For this measure to be meaningful, the exponential relationship between absorption coefficient and absorptance must be explicitly involved.

2. Narrow Absorption Lines

To illustrate this point, let us suppose that the half-width of a line in either the absor*b*ance or absor*pt*ance regime is truly much narrower than the half-width of the instrument response function $r(x)$. The measured absorptance then cannot be more than a few percent, at the most.

Let us establish the required relationships more precisely. Consider a narrow idealized rectangular absorption line $A_T(x) = \text{rect}(x/2 \, \Delta x_L)$ having half-width Δx_L and centered at $x = 0$. Its variance is easily found to be $\sigma_L^2 = (2 \, \Delta x_L/3)^2$. Its area is $2 \, \Delta x_L$. Now, let us assume that this line is being used to measure an instrument response function $\exp(-x^2/2\sigma_R^2)$ that has Gaussian shape and variance σ_R^2.

The resulting observed absorptance is given by the convolution of the gaussian and the narrow rectangle, and is also Gaussian to a high degree of approximation. The variance of this result is just the sum $\sigma_L^2 + \sigma_R^2$. The fractional increase in the square root of the variance is given by

$$\alpha = \frac{\sqrt{\sigma_R^2 + \sigma_L^2} - \sigma_R}{\sigma_R}. \tag{61}$$

For the observation to overestimate σ_R by a given fraction α, the linewidth is given by

$$\sigma_L = \sigma_R\sqrt{\alpha^2 + 2\alpha} \approx \sigma_R\sqrt{2\alpha}, \qquad \alpha \ll 1. \tag{62}$$

We may convert this relationship to half-width form by remembering that $\sigma_L = 2 \, \Delta x_L/3$ and expressing σ_R in terms of the Gaussian instrument function half-width Δx_R:

$$\sigma_R = \Delta x_R/\sqrt{2 \ln 2}. \tag{63}$$

Doing this by making the appropriate substitutions and solving for Δx_R, we find

$$\Delta x_R = \tfrac{2}{3}\sqrt{(\ln 2)/\alpha} \, \Delta x_L. \tag{64}$$

Because the instrument response function must have unit area, the area under the narrow line is preserved by the measurement process. Recalling that this area is $2 \, \Delta x_L$, we may write the absorptance amplitude A_L of the observation (which is Gaussian) in terms of its half-width and $2 \, \Delta x_L$:

$$A_L = \sqrt{(\ln 2)/\pi} \, 2 \, \Delta x_L/\Delta x_R. \tag{65}$$

By making the appropriate substitution for Δx_R from Eq. (64), we obtain the amplitude of the observed line as a function of α, the fractional increase in breadth:

$$B_L = 3\sqrt{\alpha/\pi}. \tag{66}$$

To hold line broadening to less than 0.1%, we find that the maximum absorptance B_L of the line must be held to under 5.4%. For a maximum broadening of 1.0%, the absorptance cannot exceed 16.9%.

Usual noise levels produce considerable error in such measurements. Slit openings wide enough to yield the high signal-to-noise ratio needed would probably be unsuitable for data acquisition runs of the spectra to be deconvolved. Whether or not we can tolerate various errors in the response-function measurement depends on the way in which it is applied. Deconvolution with an artificially wide response function yields artificially narrow deconvolved lines and probably some artifacts as well. Experimentation may be the best guide in deciding this issue.

3.   Successful Use of Absorption Lines

We may overcome the problems just described by measuring the possibly saturated absorption lines of finite breadth to which we are restricted and performing a deconvolution. In this case, we take the actual line shape as known and seek the spectrometer response function. The response function becomes the object $o(x)$, and the known line shape is identified with $s(x)$. Doppler-broadened lines with little or no Lorentzian component are best for this purpose owing to their localization of area near the line center. We may synthesize the line shape from Eqs. (9) and (16) if we know Boltzmann's constant k, the molecular mass m, the speed of light c, the absorption line frequency ν_0, the temperature T of the absorbing gas, and the product SX. The first four constants are known with more than adequate accuracy. The temperature T is easily determined, and the product SX is simply obtained from the area under the line by the equivalent-width method described in Section II.F. Only a rough estimate of pressure broadening is needed to determine its significance and account for its effect, if necessary. The optical path need not even be explicitly measured. Because only a modest narrowing is sought from the deconvolution process and because the true line shape does not impose the restriction of band limiting, a simple linear deconvolution method works adequately.

In conclusion, the observed spectrum of an isolated Doppler-broadened line, along with some well-known constants and the easily measured temperature, contains all the information needed to determine the response function. Application details for this method are available in the literature (Jansson, 1968, 1970).

III. Resolution Criteria

When we say that objects are resolved, we mean that they are rendered so that their separate and distinguishable identities are apparent. The present section briefly formalizes this intuitive idea.

A. TWO-POINT CRITERIA

In the simplest possible case, we wish to distinguish only two objects. In optics the required specification is called a two-point resolution criterion. In spectroscopy a two-point criterion specifies the distance that must separate two spectral lines for them to be called resolved. Essentially similar criteria apply to other fields.

Previous sections have detailed phenomena that contribute to the degradation of resolution in optical spectra. Concepts useful in specifying resolution criteria have been established. Although transfer and point-spread functions of varying shape can yield identical numbers when a simple two-point criterion is applied, this many-to-one correspondence does not diminish the criterion's usefulness. More rigorous specification of the transfer function virtually requires graphical presentation for human interpretation. Its use therefore demands far more space in text and more time for study. Frequently, the functional form of the transfer function is well known anyway; systems being compared are often of similar type. In these cases, the two-point criterion is entirely adequate.

1. Optical Instruments and Rayleigh's Criterion

In optics and spectroscopy, resolution is often limited by diffraction. To a good approximation, the spread function may appear as a single-slit diffraction pattern (Section II). If equal-intensity objects (spectral lines) are placed close to one another so that the first zero of one sinc-squared diffraction pattern is superimposed on the peak of the adjacent pattern, they are said to be separated by the Rayleigh distance (Strong, 1958). This separation gives rise to a 19% dip between the peaks of the superimposed patterns.

It might seem that Lord Rayleigh's proposal is somewhat arbitrary; even if two lines are separated by a somewhat smaller distance, it will be apparent that two lines are present rather than one. Sparrow's criterion (Strong, 1958) does, in fact, say that two lines may be termed just resolved when spaced so that the dip first appears. The Rayleigh criterion, however, gives rise to particularly simple expressions for resolving power.

Resolving power at wavelength λ (wave number $\bar{\nu}$) is defined as $\lambda/\Delta\lambda = \bar{\nu}/\Delta\bar{\nu}$, where $\Delta\lambda$ is the Rayleigh distance in wavelength units ($\Delta\bar{\nu}$ is the Rayleigh distance in wave numbers). It turns out that the theoretical resolving power of a dispersive optical grating spectrometer is given simply by the product of the total number of lines on the grating, the number of times the flux is dispersed (number of passes), and the grating order in which the spectrum is observed. The Rayleigh resolution in wave numbers resulting from an optimally triangle-apodized Fourier interferogram is given by the reciprocal of the maximum optical path difference represented in the interferogram (Bell, 1972).

Let us cite an example to help us judge the equivalence between Fourier and dispersive instruments. A grating spectrometer employing a four-passed 8×10^4-line grating in the first order has a resolving power of $4 \times 8 \times 10^4 = 3.2 \times 10^5$. At 3200 cm^{-1} in the near infrared, this instrument has a Rayleigh resolution of 10^{-2} cm^{-1}. The same resolution can be achieved by a Fourier instrument having 10^2-cm maximum optical path difference. If the physical path in one leg of the interferometer is traversed twice, the total mechanical travel required between the central fringe position and the extreme is 50 cm.

Brief reflection on the sampling theorem (Chapter 1, Section IV.C) with the aid of the Fourier transform directory (Chapter 1, Fig. 2) leads to the conclusion that the Rayleigh distance is precisely two times the Nyquist interval. We may therefore easily specify the sample density required to recover all the information in a spectrum obtained from a band-limiting instrument with a sinc-squared spread function: evenly spaced samples must be selected so that four data points would cover the interval between the first zeros on either side of the spread function's central maximum. In practice, it is often advantageous to place samples somewhat closer together.

2. Other Two-Point Criteria

Though the Rayleigh criterion has particular meaning and simplicity when used with sinc or sinc-squared spread functions, the resulting 19% dip criterion may be applied to lines and objects having other shapes. Two such nonsinc equal-intensity lines, when separated so that a 19% dip appears, are said to be just resolved according to the Rayleigh criterion. Other two-point criteria may be more meaningful in these cases, however. For Gaussian and Lorentzian shapes and for unspecified peaklike objects, we often use criteria of full width at half maximum (FWHM—the breadth of the object at half of the peak height) or half-width at half maximum (HWHM—analogously defined). The variance or square root of the second moment sometimes also has value as a descriptor of breadth, and hence resolution. Fields other than spectroscopy have employed criteria such as optical line pairs, television lines, rise time, and limiting resolution.

3. Resolution "Limits" and Deconvolution Performance

What meaning do these two-point resolution criteria have in describing the deconvolution process, that is, resolution before and after deconvolution? Although width criteria may be applied to derive suitable before/after ratios, the Rayleigh criterion raises an interesting question. Because the diffraction pattern is an inherent property of the observing instrument, would it not be best to reserve this criterion to describe optical performance? The effective spread function after deconvolution is not sinc squared anyway.

An additional question arises concerning the way the Rayleigh criterion terminology has traditionally been used. There is a vague perception that the "Rayleigh limit" is a barrier to be penetrated only by supernatural means—that it is a limit of the most fundamental kind. And so it is, but only in a sense that requires some elaboration. The "Rayleigh limit" of both Fourier and dispersive instruments implies an upper limit Ω to the Fourier frequencies that may be measured. Beyond this limit, the instrument is incapable of passing any information. When the transfer function is triangular so that frequencies fall off linearly to zero at cutoff Ω (triangular apodization), the sinc-squared spread function results. The Rayleigh criterion then applies directly and meaningfully. After deconvolution, however, the effective transfer function is anything but triangular. It is thus possible to obtain the required 19% dip between lines that are spaced a good deal closer than the "Rayleigh limit" for a given instrument. Furthermore, in Chapter 4 we shall find that *even the instrumental Fourier frequency cutoff Ω does not pose a limit to the higher frequencies that may be restored*. The needed frequency extrapolation is achieved on the strength of additional information conveyed by simple constraints of physical realizability. Spectra having Fourier frequencies thus extended beyond cutoff Ω are said to be *superresolved*.

B.   CONCLUSION

We recommend that the Rayleigh criterion and Fourier frequency cutoff Ω be used as fundamental specifications of optical performance, but that other criteria such as full width at half maximum be used in describing widths before and after deconvolution and in specifying spread functions having other than sinc-squared shape.

In this chapter we have detailed the processes that give rise to distortion of spectra. Although specifics have been limited to optical spectra, a number of the principles are easily generalized to other fields, including those for which data are acquired in more than one dimension. In the next chapter we introduce traditional methods of undoing the distortion, along with some of the concepts needed to develop the most successful modern methods described in Chapter 4.

References

Armstrong, B. H. (1967). *J. Quant. Spectrosc. Radiat. Transfer* **7**, 61–88.

Bell, R. J. (1972). "Introductory Fourier Transform Spectroscopy." Academic Press, New York.

Blass, W. E., and Halsey, G. W. (1981). "Deconvolution of Absorptions Spectra." Academic Press, New York.

Brodersen, S. (1953). *J. Opt. Soc. Am.* **43**, 1216–1220.

Dicke, R. H. (1953). *Phys. Rev.* **89**, 472–473.

Galatry, L. (1961). *Phys. Rev.* **122**, 1218–1223.

Herbert, F. (1974). *J. Quant. Spectrosc. Radiat. Transfer* **14**, 943–951.

Jansson, P. A. (1968). Ph.D. Dissertation, Florida State Univ., Tallahassee.

Jansson, P. A. (1970). *J. Opt. Soc. Am.* **60**, 184–191.

Jansson, P. A., and Korb, C. L. (1968). *J. Quant. Spectrosc. Radiat. Transfer* **8**, 1399–1409.

Kauppinen, J. K., Moffatt, D. J., Mantsch, H. H., and Cameron, D. G. (1981). *Appl. Spectrosc.* **35**, 271–276.

Kielkopf, J. F. (1973). *J. Opt. Soc. Am.* **63**, 987–995.

Korb, C. L., Hunt, R. H., and Plyler, E. K. (1968). *J. Chem. Phys.* **48**, 4252–4260.

Mielenz, K. D. (1967). *J. Opt. Soc. Am.* **57**, 66–74.

Penner, S. S. (1959). "Quantitative Molecular Spectroscopy and Gas Emissivities," Addison-Wesley, Reading, Massachusetts.

Pine, A. S. (1980). *J. Mol. Spectrosc.* **82**, 435–448.

Pliva, J., Pine, A. S., and Willson, P. D. (1980). *Appl. Opt.* **19**, 1833–1837.

Puerta, J., and Martin, P. (1981). *Appl. Opt.* **20**, 3923–3928.

Ramsay, D. A. (1952). *J. Am. Chem. Soc.* **74**, 72–80.

Rodgers, C. D. (1976). *Appl. Opt.* **15**, 714–716.

Stewart, J. E. (1967). *Infrared Phys.* **7**, 77–92.

Strong, J. (1958). "Concepts of Classical Optics." Freeman, San Francisco.

Chapter 3 | Traditional Linear Deconvolution Methods

Peter A. Jansson

College of Optical Sciences, University of Arizona, Tucson

List of Symbols

a, b, g	functions $a(x), b(x), g(x)$, with no explicit dependences shown
$B(\omega)$	factor describing modification of inverse-filter transfer function by linear deconvolution method
C	relaxation factor in continuous formulation
ESCA	electron spectroscopy for chemical analysis
$\mathbf{i, o, \hat{o}}$	column vectors containing elements $i_m, o_m, \hat{o}_m$
i_C, o_C	constant values of i_m, o_m
$i_m, s_m, o_m, \hat{o}_m^{(k)}, n_m$	discretely sampled versions of $i, s, o, \hat{o}^{(k)}, n$
$i(x), i$	"image" data that incorporate smearing by $s(x)$
$\hat{\imath}^{(k)}(x)$	kth estimate of $i(x)$
$I(\omega), \tau(\omega), O(\omega), N(\omega)$	Fourier transforms of $i(x), s(x), o(x), n(x)$

j	imaginary operator such that $j^2 = -1$
L	number of nonvanishing sampled spread-function values on either side of central maximum, spread function assumed symmetrical
N_s	number of components in composite inverse filter
$n(x)$	part of the image data $i(x)$ due to noise
$o(x), o$	"object" or function sought by deconvolution, usually the "true" spectrum, but also the instrument function when this is sought by deconvolution
$\hat{o}(x), \hat{o}^{(k)}(x), \hat{O}(\omega), \hat{O}^{(k)}(\omega)$	estimate and kth estimate of $o(x)$ and $O(\omega)$, respectively
rect(x)	rectangle function having half-width $\frac{1}{2}$
s	matrix containing elements $[\mathbf{s}]_{nm}$; where s is a convolution kernel, $[\mathbf{s}]_{nm} = s_{n-m}$
$[\mathbf{s}]_{nm}$	element of matrix **s**
sinc(x)	$(\sin \pi x)/\pi x$
$s(x), s$	spread function, usually the instrument function, but also spreading due to other causes
x, x'	generalized independent variables and arguments of various functions
$y(x)$	linear restoring filter such that $\hat{o}(x) = y(x) \otimes i(x)$; takes various forms depending on criteria of derivation
$y_F(x), Y_F(\omega)$	Wiener smoothing filter and its Fourier transform, respectively
$y_s(x), Y_s(\omega)$	component of inverse filter and its Fourier transform, respectively
$y(x, x')$	generalized shift-variant inverse-filtering kernel
$Y(\omega)$	Fourier transforms of $y(x)$ and $y_F(x)$, respectively
β	parameter specifying influence of sharpness or smoothness criteria
$\otimes$	convolution operation
$\delta(x), \delta$	Dirac δ function or impulse
ζ	$1 - \tau(\omega)$;
$\theta(x)$	spurious part of solution
κ	relaxation factor in discrete formulation
$\tau(\omega), \tau$	Fourier transform of $s(x)$
$\phi_n(\omega), \phi_n$	power spectra of $n(x), n$
$\phi_o(\omega), \phi_o$	power spectra of $o(x), o$
ψ	Gaussian line-broadening function
ω	conjugate of x; Fourier frequency in radians per units of x
Ω	cutoff frequency such that $\tau(\omega) = 0$ for $|\omega| > \Omega$
Ω_p	cutoff frequency defining Frieden's optimum processing bandwidth

A linear deconvolution method is one whose output elements (the restoration) can be expressed as linear combinations of the input elements. Until recently, only linear methods received widespread attention. These methods can be developed and analyzed in detail by use of long-standing mathematical tools. Analysis of linear methods tends to be simpler than that of nonlinear methods, and computations are shorter. This point is especially important, because deconvolution is inherently computation intensive. It is not surprising that linear methods have historically dominated deconvolution research and applications.

In spite of performance advantages in the use of nonlinear methods, it is instructive to start our deconvolution study by examining the linear methods; they will give us insight into the process. The ensuing development will also define the application domain of linear methods and reveal their limitations. We shall see that in some circumstances a linear method is the method of choice.

I.   Direct Approach

Let us begin by looking at an approach that may be the most intuitive. Adhering to the notation that we introduced in Section V of Chapter 1, we write the equation for resolution distortion,

$$i = s \otimes o, \tag{1}$$

and its discrete counterpart

$$i_m = \sum_k s_{m-k} o_k. \tag{2}$$

In this equation, the image value i_m is the mth sample of the smeared version of o_k. We transform the subscripts in an obvious way that recognizes the commutative property of convolution and considers only spread functions that are nonvanishing over some finite domain having an odd number of points. We may then write

$$i_m = \sum_{l=-L}^{L} s_l o_{m-l}. \tag{3}$$

For convenience and simplicity in this section, we assume that the spread function s_l has its peak value at $l = 0$ and that it has L nonvanishing values on either side of s_0. The development based on these assumptions may easily be generalized.

Now let us assume that the true object is zero for all subscript values less than 1. When we do the convolution, then, i_m must have a value of zero for $m < 1 - L$. When we perform the numerical convolution starting with $m \ll 1 - L$, we compute zeros as we slide the spread function toward the region of finite o_k. Finally, when the end of the spread function encounters the first nonzero o_k, we obtain the first nonzero value of i_k. We illustrate this particular sum of products with an elementary example in which $L = 1$ and $m = 0$:

$$
\begin{array}{ccccc}
\vdots & & \vdots & & \vdots \\
0 & & 0 & & 0 \\
0 & \times & s_1 & \overline{\uparrow} & 0 \\
0 & \times & s_0 & \text{nonzero spread function} & i_0 \\
\overline{\uparrow} & \times & s_{-1} & \downarrow & \\
o_1 & & 0 & & \\
\text{nonzero} \quad o_2 & & 0 & & \\
\text{object} \quad o_3 & & 0 & & \\
\big| & & \vdots & & \vdots
\end{array}
$$

Here the first value of the image i_0 is given by

$$i_0 = o_1 s_{-1}, \tag{4}$$

and we may easily obtain an estimate of the first object value,

$$\hat{o}_1 = i_0/s_{-1}. \tag{5}$$

The next value of the image is given by sliding the spread function down and forming the appropriate sum of products:

$$i_1 = o_1 s_0 + o_2 s_{-1}. \tag{6}$$

Solving for o_2, we obtain the estimate

$$\hat{o}_2 = (i_1 - \hat{o}_1 s_0)/s_{-1}. \tag{7}$$

But we already know $\hat{o}_1$ and readily obtain by substitution

$$\hat{o}_2 = (i_1 - i_0 s_0/s_{-1})/s_{-1}. \tag{8}$$

Proceeding in similar fashion, we may obtain

$$\hat{o}_3 = (1/s_{-1})i_2 - (s_0/s_{-1}^2)i_1 + (s_0^2/s_{-1}^3 - s_1/s_{-1}^2)i_0, \qquad (9)$$

and so forth.

We might generalize by noting that for this method to be considered, values of o_k (and hence of i_k) need not be vanishing for a stretch at one end but must merely be constant over an interval of breadth $2L + 1$. In this case, if i_C is the constant value of i_k, the constant value of o_k is given by

$$o_C = i_C \bigg/ \sum_{l=-L}^{L} s_l. \qquad (10)$$

When the spread function is normalized, we obtain $o_C = i_C$. The estimates of the object are thus easy to determine by starting at the ends.

We may obtain a general formula for the $(k + 1)$th object estimate in terms of the preceding $2L$ estimates and the image value i_{k-L+1} by constructing a more complete and general diagram analogous to our example for $L = 1$. By so doing we may derive

$$\hat{o}_{k+1} = \left(i_{k-L+1} - \sum_{l=-L+1}^{L} s_l \hat{o}_{k-L+1-l} \right) \bigg/ s_{-L}. \qquad (11)$$

Have we solved the problem? Is this all there is to deconvolution? Let us look a little closer at the three-point example by examining Eq. (5). Suppose that the first spread-function value s_{-1} is small relative to the other values of s, as is typical. The value i_0 would also then be small. We are dealing with data acquired from the real world, and no observation can be without error. Suppose that the observation of i_m is subject to the error n_m. We may then write our object estimate

$$\hat{o}_1 = (i_0 + n_0)/s_{-1}. \qquad (12)$$

Even for good signal-to-noise ratios (small n_0), our estimate $\hat{o}_1$ could suffer considerably, being erroneous by an amount n_0/s_{-1}, where s_{-1} is small. We have calculated only the first element of the object estimate and already there is trouble.

We see that this trouble gets worse in the next step because the object estimate $\hat{o}_2$ is given by the magnified small difference between the uncertain quantity $\hat{o}_1 s_0$ and i_1, itself uncertain. The difficulties increase as we proceed, and it is clear that *uncertainty in the observations plays a fundamental role in our ability to restore the true object.* The naive approach with which we started exemplifies the way we may be misled when we do not provide explicitly for the effects of noise.

In spite of these difficulties, the method described is useful in certain special cases. An adaptation of it has been successfully applied to the removal of base-line distortion caused by inelastically scattered electrons in electron spectroscopy for chemical analysis (ESCA) (Chapter 8). As one might suspect, the method is most useful when the end value s_L is not small. In the ESCA example, we shall see that when s is written in its continuous formulation $s(x)$, it is actually terminated by a Dirac δ function.

II. Van Cittert's Method

Van Cittert (1931) recognized that the image data $i(x)$ could be considered as a first approximation $\hat{o}^{(0)}(x)$ to the object $o(x)$. After all, in the absence of deconvolution, the spectroscopist often ignores (rightly or wrongly) instrumental spreading and uses the data as if they represent the true spectrum. This being the case, could we not blur the approximation $\hat{o}^{(0)}(x)$ to yield an "approximation" to the data, $\hat{i}^{(0)}(x)$? this is the form that the data would take if $\hat{o}^{(0)}(x)$ were the true object. Certainly we could, but we already have the data $i(x)$, and so what would be the purpose?

This blurring actually does serve a useful function. The difference $i(x) - \hat{i}^{(0)}(x)$, the easily computed error in the image estimate, is in some way related to the error in the object estimate, $o(x) - \hat{o}^{(0)}(x)$. Van Cittert recognized that this image-estimate error could be applied as a correction to the object estimate, thus producing a new object estimate:

$$\hat{o}^{(1)}(x) = \hat{o}^{(0)}(x) + [i(x) - s(x) \otimes \hat{o}^{(0)}(x)]. \tag{13}$$

This is illustrated in Fig. 1. Although $\hat{o}^{(1)}(x)$ is not the solution, the correction gives us some cause for optimism. The estimate $\hat{o}^{(1)}(x)$ is certainly narrower and taller than the data $i(x)$. One may continue the process, hoping for ever better estimates:

$$\hat{o}^{(k+1)}(x) = \hat{o}^{(k)}(x) + [i(x) - s(x) \otimes \hat{o}^{(k)}(x)]. \tag{14}$$

In this expression, $\hat{o}^{(k)}(x)$ and $\hat{o}^{(k+1)}(x)$ are the kth and $(k+1)$th approximations to the object.

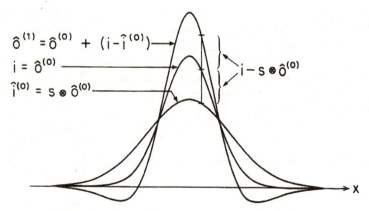

$$\hat{o}^{(1)} = \hat{o}^{(0)} + (i - \hat{i}^{(0)})$$

$$i = \hat{o}^{(0)}$$

$$\hat{i}^{(0)} = s \otimes \hat{o}^{(0)}$$

$$i - s \otimes \hat{o}^{(0)}$$

Figure 1 Corrections applied in van Cittert's method.

The fact that this method is somewhat intuitive may account for its rather wide use and apparently independent reinvention by workers following Van Cittert. The basic method has numerous variants (e.g., Herget *et al.*, 1962) and has given rise to constrained nonlinear methods that exhibit significantly better performance (Chapter 4, Section V.A). It has been firmly identified with both inverse filtering (see Section IV.D) and the solution of linear equations by relaxation (DiCola *et al.*, 1967; Jansson 1968, 1970; see Section III.C). Its convergence has been analyzed by Hill and Ioup (1976), Crilly (Chapter 5), and others (Section IV.D). Some of its variants are concerned with damping the noise that grows with each iteration when the method is applied to real data. These spurious solution components invariably have high frequency and often give rise to solutions that are not physically possible, such as transmittance less than zero or in excess of 100%.

The method has utility where only modest correction is required, such as in the determination of the spectrometer response function when a narrow line has been measured (Chapter 2, Section II.G).

III. Matrix Inversion

We have presented two deconvolution methods from an intuitive point of view. The approach that suits the reader's intuition best depends, of course, on the reader's background. For those versed in linear algebra, methods that stem from a basic matrix formulation of the problem may lend particular insight. In this section we demonstrate a matrix approach that can be related to Van Cittert's method. In Section IV.D, both approaches will be shown to be equivalent to Fourier inverse filtering. Similar connections can be made for *all* linear methods, and many limitations of a given linear method are common to all.

A. *MATRIX FORMULATION*

As we have seen earlier, we may write the convolution integral [Chapter 1, Eq. (86)] in discrete form:

$$i_n = \sum_m s_{n-m} o_m,$$ (15)

where i_n are the image values and o_m the object values; the response function is represented by the values s_{n-m}. We have omitted the limits of summation. They are understood to represent the finite range of m that is useful in describing the object over its range of interest. The values of n are usually those for which data are available. As indicated in Section I, sometimes end effects need to be considered.

Equation (15) can be expressed in a matrix formulation:

$$\mathbf{i} = \mathbf{so}.$$ (16)

In this equation, $\mathbf{i}$ and $\mathbf{o}$ are column vectors such that each element $[\mathbf{i}]_n = i_n$ and $[\mathbf{o}]_m = o_m$. The elements of $\mathbf{s}$ are given by $[\mathbf{s}]_{nm} = s_{n-m}$.

It is now possible to see that the matrix formulation has the potential to describe the more general Fredholm integral equation. This equation corresponds in spectroscopy to the situation where the functional form of $s(x)$ varies across the spectral range of interest. In these circumstances, s is expressed as a function of *two* independent variables. Although we proceed with the present treatment formulated in terms of convolutions, the reader should bear the generalization in mind.

Equation (15) reveals that, in the case of convolution, the matrix $\mathbf{s}$ has a certain symmetry. Each row is the same as the row above except that it is shifted one element to the right. Usually we choose the indices so that the largest element is on the diagonal. This shifting property may be expressed as

$$[\mathbf{s}]_{jk} = [\mathbf{s}]_{mn} \qquad \text{if} \quad j - k = m - n.$$ (17)

Under these conditions, the spread function is termed shift invariant and s is called a Toeplitz matrix. When Eq. (17) does not hold, the spread function is call shift variant, and the spreading can no longer be described by simple convolution.

If the ends of the rows wrap around so that the leftmost element of each row is the rightmost element of the row above, and so on, the matrix is called circulant. A circulant representation of $s(x)$ is one way to deal with end effects referred to earlier. Below we show a simple numerical example of Eq. (16) in which a noiseless image appears as the product of a shift-invariant circulant Toeplitz point-spread function matrix and an object column vector:

$$
\begin{pmatrix} 0 \\ 1 \\ 3 \\ 2 \\ 0 \\ 0 \\ 0 \end{pmatrix} = \begin{pmatrix} 3 & 1 & 0 & 0 & 0 & 0 & 2 \\ 2 & 3 & 1 & 0 & 0 & 0 & 0 \\ 0 & 2 & 3 & 1 & 0 & 0 & 0 \\ 0 & 0 & 2 & 3 & 1 & 0 & 0 \\ 0 & 0 & 0 & 2 & 3 & 1 & 0 \\ 0 & 0 & 0 & 0 & 2 & 3 & 1 \\ 1 & 0 & 0 & 0 & 0 & 2 & 3 \end{pmatrix} \times \begin{pmatrix} 0 \\ 0 \\ 1 \\ 0 \\ 0 \\ 0 \\ 0 \end{pmatrix}. \tag{18}
$$

In this example, a single-point object generates an image that is a reversed replica of the spread function, as it should.

Deconvolution problems for which s is Toeplitz and circulant are particularly amenable to approach via Fourier methods (Andrews and Hunt, 1977). Fortunately for computation, it is rarely necessary to numerically generate, store and operate upon s as a full matrix because it is both redundant and sparse.

Spectroscopy vs. Imaging

If a one-dimensional problem (e.g., from spectroscopy) requires a matrix expression, then what expression would a two-dimensional imaging-system problem require? Surprisingly, perhaps, the answer is still a matrix. Consider the nearly trivial example of a two-by-two discrete spread-function kernel by which a sampled object may be smeared to form a blurred image. We can visualize an example of such a kernel containing four numbers arranged as a two-dimensional spatial array:

$$
\begin{matrix} 1 & 2 \\ 2 & 3 \end{matrix} \tag{19}
$$

We can also visualize this array as being superimposed upon a two-dimensional object array. The smearing is done by sliding this kernel step-by-step across a two-dimensional object array, forming products and summing them to form each image sample in a manner completely analogous to the one-dimensional description in Chapter 1, Section II.A. Here the kernel is assumed to operate in a circulant manner upon samples from both sides, or from the top and bottom of an object whenever it straddles an edge or corner.

Now let us make the nearly absurd simplification that the object to be smeared is a spatial array having dimensions three-by-three. We may represent this object by scanning it lexicographically (left-to-right, top-to-bottom) and recording the samples in a nine-element column vector. Consider a numerical example that expresses such a two-dimensional version of Eq. (16). This example contains a nine-by-nine spread-function matrix **s** based on the four-number array shown previously as Expression (19):

$$
\begin{pmatrix} 0 \\ 1 \\ 2 \\ 0 \\ 2 \\ 3 \\ 0 \\ 0 \\ 0 \end{pmatrix} = \begin{pmatrix} 3 & 2 & 0 & 2 & 1 & 0 & 0 & 0 & 0 \\ 0 & 3 & 2 & 0 & 2 & 1 & 0 & 0 & 0 \\ 0 & 0 & 3 & 2 & 0 & 2 & 1 & 0 & 0 \\ 0 & 0 & 0 & 3 & 2 & 0 & 2 & 1 & 0 \\ 0 & 0 & 0 & 0 & 3 & 2 & 0 & 2 & 1 \\ 1 & 0 & 0 & 0 & 0 & 3 & 2 & 0 & 2 \\ 2 & 1 & 0 & 0 & 0 & 0 & 3 & 2 & 0 \\ 0 & 2 & 1 & 0 & 0 & 0 & 0 & 3 & 2 \\ 2 & 0 & 2 & 1 & 0 & 0 & 0 & 0 & 3 \end{pmatrix} \times \begin{pmatrix} 0 \\ 0 \\ 0 \\ 0 \\ 0 \\ 1 \\ 0 \\ 0 \\ 0 \end{pmatrix}. \qquad (20)
$$

The nine-element object vector **o** on the right represents a three-by-three object with only one nonvanishing element—at the right end of the middle row. The column vector **i** on the left represents the three-by-three array containing the image of the smeared object. The reader may easily verify that multiplying any lexicographically ordered column object vector by the spread-function matrix yields another lexicographically ordered column vector representing precisely the image obtained from a kernel-sliding procedure. Thus we see that two-dimensional convolutions, and larger dimensioned ones as well, collapse neatly into one-dimensional representations. Furthermore, noncirculant and shift-variant operations may be similarly treated. We therefore do not lose generality by treating one-dimensional convolutions elsewhere in the present volume.

B. DIFFICULTIES

For the present we turn our attention to a naive approach that proposes a direct solution to Eq. (16), which, after all, is only a set of linear equations in unknowns o_m. Solving this set, we should obtain the object estimate

$$\hat{o} = [s]^{-1}i. \tag{21}$$

Apparently, all we need to do is invert the matrix s.

One computational problem here is that for large one-dimensional data sets and images the matrix s is very large, and we might feel the limitation of computer time and storage. However, this problem pales in comparison with the fundamental difficulty usually encountered with s. Typically each row in the matrix is similar to the one above. Indeed, these rows do operate upon neighboring points. Each equation, therefore, brings only a small amount of added information to the system. We do not have a set of robustly independent equations. If s has no inverse at all, it is said to be *singular*. At best, s is *ill-conditioned*. In this case even the smallest noise perturbations in i cause large and possibly oscillatory fluctuations in the estimate ô. When solution is attempted by certain noniterative methods such as Gaussian elimination, small differences are computed between similar large numbers. Errors, introduced by roundoff, for example, can become a serious problem. When the inverse can be computed, the solution ô typically exhibits large spurious fluctuations that originate in the noise that must always be present in any physical measurement.

Any linear deconvolution method that is to achieve even modest success must deal with these difficulties. Numerous aspects of the matrix formulation as applied to image restoration have been reviewed by Andrews and Hunt (1977), who devote special attention to the difficulties mentioned here.

C. ITERATIVE METHODS

An iterative approach has the advantage of allowing control of spurious fluctuations by interaction with the solution as it evolves. This may be done either automatically in the algorithm or by the exercise of human judgment. Although we reserve discussion of the most-powerful methods of interaction for the next chapter, it is appropriate here to set the stage by developing a family of iterative methods that have seen much use.

1. Point Simultaneous

It is well known that large systems of linear equations can be solved iteratively provided that the diagonal element is larger than any of its neighbors in the same row. Let us rewrite Eq. (15) with the diagonal element terms pulled outside:

$$i_n = [\mathbf{s}]_{nn} o_n + \sum_{n \neq m} [\mathbf{s}]_{nm} o_m. \tag{22}$$

We have returned, for the time being, to the general case of the shift-variant spread function. Solving for o_n, we obtain

$$o_n = \frac{1}{[\mathbf{s}]_{nn}} \left\{ i_n - \sum_{n \neq m} [\mathbf{s}]_{nm} o_m \right\}, \tag{23}$$

so that we could exactly determine o_n if we knew all of its neighbors. Now suppose that by some means we had obtained estimates $\hat{o}_m^{(k)}$ of all its neighbors. We could then obtain a refined estimate of o_n,

$$\hat{o}_n^{(k+1)} = \frac{1}{[\mathbf{s}]_{nn}} \left\{ i_n - \sum_{n \neq m} [\mathbf{s}]_{nm} \hat{o}_m^{(k)} \right\}, \tag{24}$$

and cycle through all the values of n until we have a complete set of the $\hat{o}_n^{(k+1)}$. It was appropriate that we chose originally to solve for o_n and not one of its neighbors, because o_n is the single object element that is weighted most heavily by the spreading process that yields i_n; that is, the diagonal element in each row of $\mathbf{s}$ is the largest.

Because it is also presumably available, why not include the old estimate of o_n as well? This eliminates the awkwardness of a summation with a missing element and allows us to express our new estimate in terms of the old estimate and a correction term:

$$\hat{o}_n^{(k+1)} = \hat{o}_n^{(k)} + \frac{1}{[\mathbf{s}]_{nn}} \left\{ i_n - \sum_m [\mathbf{s}]_{nm} \hat{o}_m^{(k)} \right\}. \tag{25}$$

Repeated application of this equation constitutes the point-simultaneous relaxation method, also known as the point Jacobi method or the method of simultaneous displacements.

There is no guarantee that the correction term is scaled correctly to ensure the most-rapid convergence. Furthermore, if it is scaled too large, the method might not converge at all. Accordingly, we introduce a relaxation factor κ, so that

$$\hat{o}_n^{(k+1)} = \hat{o}_n^{(k)} + \frac{\kappa}{[\mathbf{s}]_{nn}} \left\{ i_n - \sum_m [\mathbf{s}]_{nm} \hat{o}_m^{(k)} \right\}. \tag{26}$$

This is called point-simultaneous overrelaxation. If we set $\kappa = [\mathbf{s}]_{nn}$, we have obtained the discrete formulation of Van Cittert's method. This connection between Van Cittert's method and the classic iterative methods of solving simultaneous equations was demonstrated in an earlier work (Jansson, 1968, 1970).

2. Point Successive

To compute a single point $\hat{o}_n^{(k+1)}$ in the estimate, we need the old estimate $\hat{o}_n^{(k)}$ at that point and a number of points to either side as well. Usually the $\hat{o}_n^{(k)}$ are computed in the order of increasing n. Why must we use old values for $\hat{o}_n$ in the summation for values of $m < n$ when more-recent estimates are available? The answer is that we need not. Accordingly, we break the summation into two parts:

$$\hat{o}_n^{(k+1)} = \hat{o}_n^{(k)} + \frac{\kappa}{[\mathbf{s}]_{nn}} \left\{ i_n - \sum_{m<n} [\mathbf{s}]_{nm} \hat{o}_m^{(k+1)} - \sum_{m \geq n} [\mathbf{s}]_{nm} \hat{o}_m^{(k)} \right\}. \tag{27}$$

As we might expect, this iteration typically converges faster because of the continual use of the most-"recent" information. It is called point-successive overrelaxation. For $\kappa = 1$, it is known as the method of successive displacements or the point-successive relaxation or Gauss-Seidel method. Note, however, that whereas Eq. (26) could be applied with either ascending or descending n, yielding identical results either way, Eq. (27) can be applied only in the order of ascending n. It is a fundamentally asymmetrical method and may produce slightly asymmetrical spectral-line artifacts when employed in the deconvolution of spectra.

Would it be possible to make use of the most-"recent" information and still preserve the symmetry of the method? This author believes that it could be done by computing the new estimates $\hat{o}_n^{(k+1)}$, not in sequence of ascending n, but in a binary interleaved fashion. Such an interleaved sampling can be obtained by a bit-reversed scheme. In this scheme, the sampling positions are generated in the sequence that results when bits are reversed in each number of a simple ascending binary count. For example, the sequences 0, 1, 2, 3, 4, 5, 6, 7 generates 0, 4, 2, 6, 1, 5, 3, 7. Allebach (1981) describes the bit-reversed scheme developed by Deutsch (1965) and other schemes that may be useful in the present application. Other alternatives might be to compute new values in order of ascending n for n odd and to compute new values in order of descending n for n even. One could also use a Monte Carlo method of selecting n. These modifications might even result in more-rapid convergence.

With all of the point-successive methods, only one array in the computer is needed to store both $\hat{o}^{(k)}$ and $\hat{o}^{(k+1)}$ because the element $\hat{o}_n^{(k)}$ is never again needed once $\hat{o}_n^{(k+1)}$ is computed. This contrasts with the point-simultaneous methods, where, for each n, the elements of $\hat{o}^{(k)}$ must be available corresponding to the full range over which $[\mathbf{s}]_{nm}$ is finite.

3. Memory-Saving Point Simultaneous

In a discrete convolution such as $a = b \otimes g$, where a and g are stored in similar-sized arrays, the ends of the a array are often not used. This occurs if a elements are computed only for those positions at which a full set of N nonvanishing g values are available, where N is the number of nonvanishing b values being used. It is thus possible to use only one array to store both a and g. One replaces g_1 with the first value of a obtained, continuing in an ascending sequence in the indices. When the operation is complete, the a result may be shifted back to the desired position. The values of g are lost by this method. The loss of g is of no consequence in some cases.

This technique is applicable in the case of Eqs. (25) and (26), where $\hat{o}^{(k)}$ values are not needed once they are outside the range of corresponding finite $[\mathbf{s}]_{nm}$. Therefore, we see that, like the point-successive method, *point-simultaneous relaxation does not require more than one principal computer array for storing the solution estimate.* Furthermore, when spectra are smoothed by polynomial convolution (see Section III.C.5), the same considerations apply, and only one array is needed to store the data before and after smoothing. If shifting the convolution product values back into position is judged undesirable, the array containing a and g, or $\hat{o}^{(k)}$ and $\hat{o}^{(k+1)}$, may be treated as a circular buffer (Knuth, 1973). Appropriate pointers are then adjusted when convolution is completed.

4. Convergence and Optimum Relaxation Factors

For all relaxation methods, theorems can be proved (Wendroff, 1966) that convergence is guaranteed provided that the diagonal of $\mathbf{s}$ is dominant:

$$[\mathbf{s}]_{nn} > \sum_{n \neq m} |\mathbf{s}_{nm}|. \tag{28}$$

In deconvolution problems, such dominance is rarely the case. It is a very modest resolution correction indeed when the response function is so narrow that its central element dominates all the others. Fortunately, the necessary conditions are far less severe, and convergence is usually not a serious problem. Equation (28) seems to tell us, however, that each new row in the matrix brings more independent information when $[s]_{nn}/\Sigma_{n \neq m}|[s]_{nm}|$ is as large as possible.

Kawata *et al.* (1978) and Kawata and Ichioka (1980a, b) discuss the convergence of these methods from both linear algebra and Fourier transform points of view. They also discuss a reblurring method that guarantees convergence. We pursue this idea in Section IV.D.3.

The optimum value of the relaxation factor κ can be obtained from the largest eigenvalue of **s**. This eigenvalue may, in turn, be obtained approximately from the relative magnitude of vectors $\hat{o}^{(k)}$ and $\hat{o}^{(k+1)}$. Improvements discussed in the next section, and amplified in the next chapter, significantly alter the method. Trial-and-error choice of κ is therefore preferable and probably necessary.

5. Band Limits and Polynomial Filters

At this point, we note that there is no mechanism presently built into the relaxation methods to prevent undesirable high-frequency noise from growing with each iteration. Any spurious solution $\theta(x)$ satisfies Eq. (1) (see also Chapter 1, Sections V.A and V.B) for ω beyond the band limit. If we know that the object $\hat{o}$ is truly band limited, with frequency cutoff $\omega = \Omega$, we can band-limit both data i and first object estimate $\hat{o}^{(1)}$. The relaxation methods cannot then propagate noise having frequencies greater than Ω into an estimate $\hat{o}^{(k)}$. (One possible exception involves computer roundoff error. Sufficient precision is usually available to avoid this problem.)

If we know that the instrument response is band limiting with frequency cutoff Ω, we may likewise process the estimate. The resulting solution will then also be band limited. *The high-frequency spectral structure beyond cutoff Ω that we would wish to restore is forever lost to these linear methods. The data contain no information about the high-frequency content.* We must wait until Chapter 4 to see how straightforward and seemingly unimportant constraints carry sufficient information to allow restoration of these "lost" frequencies.

Band limiting of the data can be accomplished either in Fourier space or, to a useful approximation, directly by convolution with least-squares polynomial filters. Although these filters had been employed in previous numerical work (Whittaker and Robinson, 1967), Savitzky and Golay (1964) are responsible for their current widespread use in spectroscopy and have developed valuable differentiating methods based on them. Numerical errors in the Savitzky-Golay tables are corrected by Steinier *et al.* (1972). Willson and Edwards (1976) and Willson and Polo (1981) give excellent spectroscopy-oriented reviews of their properties, including frequency behavior. Yamashita and Minami (1969), Porchet and Günthard (1970), Betty and Horlick (1977), Bromba and Ziegler (1981), Gans and Gill (1983), Kekre *et al.* (1989), Seah and Cumpson (1992), and others also discuss their properties. Kawata and Minami (1984) provide an interesting alternative, derived from optimal linear estimation theory, that adapts to the local statistics of the spectrum. DeNoyer and Dodd (1990) use a maximum-likelihood approach that takes advantage of a priori peak-shape knowledge.

Jansson (1972) introduces analogous two-dimensional continuous filters suitable for use with images. he bases these filters on orthogonal Zernike and Legendre polynomials and suggests extension of the concept to sampled data, particularly by use of orthogonal Gram polynomials. Edwards and Knight (1975) and Edwards (1982) derive the required sets of convolution coefficients via Cramer's rule. Kuo *et al.* (1991) derive them by use of Gram polynomials and propose a computationally efficient implementation based on sequential one-dimensional convolutions.

6. Relaxation Summary

We reemphasize that the foregoing relaxation equations containing the general shift-variant response-function element denoted by $[s]_{nm}$ are equally valid for the special case of convolution, whether discrete or continuous. Cast in the continuous notation for convolution, the relaxation methods are epitomized by the repeated application of

$$\hat{o}^{(k+1)}(x) = \hat{o}^{(k)}(x) + C[i(x) - s(x) \otimes \hat{o}^{(k)}(x)] \qquad (29)$$

to obtain successive estimates of the object, given the observed data $i(x)$ and a spread function $s(x)$. In spectroscopy, the object is usually considered to be the spectrum as it would be observed by a hypothetical, perfectly resolving spectrometer. In that case $s(x)$ is the spectrometer response function. As an alternative, $o(x)$ could be the spectrometer response function, which we seek to determine from data $i(x)$ representing observations of a spectral line of known shape. In the latter case,

$s(x)$ represents the known true spectral-line shape. In an imaging field, analogous use of a known source distribution for s could yield the two-dimensional spread function, for example, of a telescope or microscope.

For successive overrelaxation, we understand Eq. (29) to incorporate the use of $\hat{o}^{(k+1)}$ values in place of $\hat{o}^{(k)}$ values in the convolution product as soon as they are formed for preceding x values. This adaptation can be explicitly displayed by the appropriate use of the Heaviside step function in a modified version of Eq. (29). The method of Van Cittert is a special case of simultaneous relaxation in which $C = 1$.

We have also learned that the point-successive methods need not demand more computer memory than point-simultaneous methods, and that the seemingly inherent asymmetry of the point-successive methods can be overcome. Furthermore, the linear methods described in this section are incapable of restoring object Fourier frequencies beyond the cutoff of the observing spectrometer.

We defer additional analysis of the relaxation method until we have properly introduced the concept of Fourier inverse filtering.

IV. Inverse Filters

A. BASIC METHOD

We might ask whether it is possible to create a linear filter function $y(x)$ that could undo, by simple convolution, the smearing in the data caused by the spectrometer. We express the behavior of such a filter by the equation

$$o(x) = y(x) \otimes i(x). \tag{30}$$

In the previous section, we used a matrix formulation to express such a filter. It is also possible to take a Fourier transform approach.

The equation that describes instrumental spreading,

$$i(x) = s(x) \otimes o(x), \tag{31}$$

may be transformed to the Fourier domain with the aid of the convolution theorem. Doing so, we obtain

$$I(\omega) = \tau(\omega)O(\omega). \tag{32}$$

At first glance, it would seem trivial to obtain the object transform

$$O(\omega) = \frac{1}{\tau(\omega)} I(\omega). \tag{33}$$

The transfer function corresponding to the filter in Eq. (30) would then be given simply by the inverse Fourier transform of

$$Y(\omega) = \frac{1}{\tau(\omega)}. \tag{34}$$

Typically, $\tau(\omega)$ is small for ω large. A spectrometer suppresses high frequencies. If the data $i(x)$ have appreciable noise content at those frequencies, it is certain that the restored object will show the noise in a more-pronounced way. It is clearly not possible to restore frequencies beyond the band limit Ω by this method when such a limit exists. (Optical spectrometers having sinc or sinc-squared response-function components do indeed band-limit the data.) Furthermore, where the frequencies are strongly suppressed, the signal-to-noise ratio is poor, and $Y(\omega)$ will amplify mainly the noise, thus producing a noisy and unusable object estimate.

We should, however, be able to carry out the processing within some band limit $|\omega| < \Omega_p$. Frieden (1975) has shown that the processing band-width

$$\Omega_p = \Omega \left[1 - (\phi_n/\phi_o)^{1/2} \right] \tag{35}$$

is optimum in the sense of minimum mean-square error

$$\left\langle \int |o(x) - \hat{o}(x)|^2 \, dx \right\rangle \tag{36}$$

when $\tau(\omega)$ is triangular (sinc-squared response function) with cutoff Ω and the ratio ϕ_n/ϕ_o is constant. In expression (36), $\hat{o}(x)$ is the object estimate produced by such filtering, and the angle brackets denote the ensemble average. The quantities ϕ_o and ϕ_n are the power* spectra of object and noise, respectively, given by the ensemble averages

$$\phi_o(\omega) = \langle |O(\omega)|^2 \rangle, \tag{37}$$

$$\phi_n(\omega) = \langle |N(\omega)|^2 \rangle. \tag{38}$$

---

*This term may be a misnomer when it is given dimensions of $|O(\omega)|^2$. The quantity $O(\omega)$ itself has dimensions of spectral power if, for example, it is considered to be energy per unit time absorbed in a detector. On the other hand, if we let $O(\omega)$ be the transform of the detector voltage, then the power designation is appropriate.

Here we have introduced $N(\omega)$ as the Fourier transform of the noise $n(x)$. Frieden assumed a simple triangular form for $\tau(\omega)$ and constant noise-to-signal ratio for all frequencies. Goldman (1953) provides an appropriate discussion of ensemble averages and the like.

B. WIENER-TYPE FILTER

It is suitable, at this point, to repair our formulation of the problem in a realistic way that accounts explicitly for the nature of the additive noise $n(x)$:

$$i(x) = s(x) \otimes o(x) + n(x). \tag{39}$$

When we solve numerically for $o(x)$, we obtain only an estimate $\hat{o}(x)$.

Faced with the problem of noise corruption alone $[s(x) = \delta(x)]$, Norbert Wiener (Goldman, 1953) devised his well-known *smoothing* filter

$$Y_F(\omega) = \frac{\phi_o(\omega)}{\phi_o(\omega) + \phi_n(\omega)}. \tag{40}$$

In deriving the Wiener smoothing filter, one asks what filter $y(x)$ results in the smallest mean-square error between the object estimate

$$\hat{o}(x) = y_F(x) \otimes [o(x) + n(x)] \tag{41}$$

and the true object. Thus, we may obtain Eq. (40) by minimizing the ensemble average of the mean-square error,

$$\left\langle \int |y_F \otimes (o + n) - o|^2 \, dx \right\rangle. \tag{42}$$

Here the limits of integration are understood to include the entire spectrum, and we use the angle brackets to denote the average over many spectra acquired under similar conditions. For brevity, we have refrained from explicitly showing the x dependence of o, n, and y_F.

This filter is not an inverse filter of the type that we seek, being intended only for noise reduction. It does not undo any spreading introduced by $s(x)$. It is, however, an optimum filter in the sense that no better linear filter can be found for noise reduction alone, provided that we are restricted to the knowledge that the noise is additive and Gaussian distributed.

How can this approach be adapted to deconvolution? The problem is similar, but now we ask that $y(x)$ also incorporate the inverse of $s(x)$.

Both Bracewell (1958) and Helstrom (1967) have derived this variant of the Wiener filter. Accordingly, we may minimize

$$\left\langle \int |y \otimes (s \otimes o + n) - o|^2 \, dx \right\rangle \tag{43}$$

and obtain the linear *inverse* filter

$$Y(\omega) = \frac{\tau^* \phi_o}{|\tau|^2 \phi_o + \phi_n}. \tag{44}$$

Here again, and frequently henceforth, we suppress from the notation the dependence of the variables. Provided that we know only the noise and object power spectra and that the noise is additive and Gaussian distributed, no better linear filter can be found than this one.

It is true that the power spectra are not always easy to obtain and that sometimes other types of information are available. Marmolin *et al.* (1978) have optimized a restoring filter of this form for optical images, where human observers were called upon to judge the visual quality of the restorations.

C. *OTHER MINIMUM MEAN-SQUARE-ERROR VARIATIONS*

Frieden (1975) has considered a similar problem wherein he has added a sharpness criterion

$$\left\langle \int_{-\infty}^{\infty} \left| \frac{d\hat{o}}{dx} \right|^2 \, dx \right\rangle = \left\langle \int_{-\Omega}^{\Omega} \omega^2 |\hat{O}(\omega)|^2 \, d\omega \right\rangle. \tag{45}$$

In this expression Ω is the cutoff frequency above which the data contain no information about $o(x)$; that is, $I(\omega) = 0$ for $|\omega| > \Omega$. We see that the sharpness criterion is a measure of the "steepness" of the solution $o(x)$. The previous criterion, expression (43), is replaced with a sum of two terms. It includes both the mean-square-error criterion and sharpness. The filter is then sought that minimizes

$$\left\langle \int |y \otimes (s \otimes o + n) - o|^2 \, dx \right\rangle + \beta \left\langle \left| \frac{d\hat{o}}{dx} \right|^2 \, dx \right\rangle \tag{46}$$

within the allowed bandpass. This results in Frieden's sharpness-constrained filter

$$Y(\omega) = \left(\frac{\tau^* \phi_o}{|\tau|^2 \phi_o + \phi_n} \right) \left(\frac{1}{1 + \beta \omega^2} \right). \tag{47}$$

It is similar to the previous filter but contains a sharpness-controlling factor that either enhances or suppresses the high frequencies, depending on whether $\beta < 0$ or $\beta > 0$.

Other modifications are possible to the same basic approach of seeking a filter that is optimum in the sense of least mean-square error. Backus and Gilbert (1970), for example, derive a linear filter by minimizing a sum of terms in which noise and resolution criteria are separately weighted. Frieden (1975) discusses variations of this technique.

A chief advantage of all the filters described in Section IV is convenience. We have written the Fourier transfer function of each one. Certainly it is possible to perform the filtering in the Fourier space. It is also possible, however, and often even more convenient, to convolve the data with the filter function itself. This is especially true if the filter can be adequately approximated by a convolution kernel that vanishes except over a relatively small domain.

D. RELATIONSHIP TO VAN CITTERT'S METHOD

The entire discussion of relaxation methods was conducted without examining Fourier space consequences. Van Cittert's method is easy to study this way and has been treated by Burger and van Cittert (1933), Bracewell and Roberts (1954), Sakai (1962), and Frieden (1975). By applying the convolution theorem to Eq. (14), we may write

$$\hat{O}^{(k+1)}(\omega) = \hat{O}^{(k)}(\omega)[1 - \tau(\omega)] + I(\omega). \tag{48}$$

Repeated application of this formula, starting with $\hat{O}^{(0)}(\omega) = I(\omega)$, results in the series

$$\hat{O}^{(k)} = I \sum_{n=0}^{k} (1 - \tau)^n. \tag{49}$$

We may use the identity

$$(1 - \zeta) \sum_{n=0}^{k} \zeta^n = 1 - \zeta^{k+1} \tag{50}$$

with $\zeta = 1 - \tau$ to obtain

$$\hat{O}^{(k)}(\omega) = \frac{I(\omega)}{\tau(\omega)}B(\omega), \tag{51}$$

where $B(\omega) = 1 - [1 - \tau(\omega)]^{k+1}$.

For $k \to \infty$, $B(\omega)$ approaches unity, provided that $|1 - \tau(\omega)| < 1$. In this case, $\hat{O}^{(k)}$ is just the object estimated by inverse filtering. For k finite, the inverse-filter estimate is modified by a factor that suppresses frequencies for which $\tau(\omega)$ is small. The larger k is, the less is this suppression. For typical transfer functions $\tau(\omega)$ that suppress high frequencies, the factor $B(\omega)$ controls the high-frequency content of $\hat{o}^{(k)}$. In the spectrum domain, it is also possible to derive simple expressions for filters $y(x)$ that are fully equivalent to an abitrary number of relaxation iterations. Blass and Halsey (1981) have done so, but the highly useful nonlinear modifications of these methods cannot be incorporated.

1. Noise and Lost Information

With the aid of Eq. (51), we can show that $\hat{O}^{(k)}(\omega) = (k + 1)N(\omega)$ for $\tau(\omega) = 0$. The object estimate consists of noise at frequencies that τ does not pass. The noise grows with each iteration. This problem can be alleviated if we bandpass-filter the data to the known extent of τ to reject frequencies that τ is incapable of transmitting. Practical applications of relaxation methods typically employ such filtering. Least-squares polynomial filters, applied by finite discrete convolution, approximate the desired characteristics (Section III.C.5). For k finite and $\tau \neq 0$, but nevertheless small,

$$\hat{O}^{(k)}(\omega) = (k + 1)I(\omega). \tag{52}$$

We thus boost the strongly suppressed frequencies in a way that grows linearly with iteration.

Let us not forget, at this point, that well-designed bandpass filtering can only prevent the appearance of solution frequencies totally rejected by $\tau(\omega)$. Noise in the frequency range that we wish to restore cannot be thus rejected, however. In the limit of simple inverse filtering we find that

$$\hat{O}(\omega) = [1/\tau(\omega)][I(\omega) - N(\omega)] = O(\omega) - N(\omega)/\tau(\omega). \tag{53}$$

Noise is a serious problem where $\tau(\omega)$ is small. We see that the estimate $\hat{O}(\omega)$ carries with it none of the sensible noise treatment that modified filters like those of the Wiener type provide. We may, however, supply the required noise suppression between iterations by smoothing. The polynomial filters described in Section III.C.5, for example, may be used. Very little has been done, however, to determine how suppression might be accomplished optimally. Usually it is treated in an ad hoc way.

Where $\tau(\omega) = 0$—beyond the band limit Ω, for example—$I(\omega)$ contains no information about $O(\omega)$. Clearly it is impossible to restore these lost frequencies based on $\tau(\omega)$ and the information in $I(\omega)$ *alone*. We see in the next chapter how a simple modification to the various relaxation methods brings about a dramatic improvement.

2.   Relaxation Factors and Convergence

We saw earlier that Van Cittert's method converges only for frequencies at which $|1 - \tau(\omega)| < 1$. This restriction is alleviated in the case of point-simultaneous overrelaxation. Referring to Eqs. (14) and (29), we see that we may deal with the case where $C \neq 1$ by simply replacing $I(\omega)$ and $\tau(\omega)$ with $CI(\omega)$ and $C\tau(\omega)$ in the analysis of the Van Cittert method. Doing this, we obtain the new requirement that $|1 - C\tau(\omega)| < 1$ or $0 < C\tau(\omega) < 2$, for $\tau(\omega)$ real.

We may also rederive $B(\omega)$ for the case of overrelaxation, with a choice of $\hat{O}^{(0)}(\omega) = I(\omega)$, as Kawata *et al.* (1978) and Kawata and Ichioka (1980a) have done:

$$B(\omega) = 1 - [1 - \tau(\omega)][1 - C\tau(\omega)]^k. \tag{54}$$

Alternatively, we point out that the choice of $\hat{O}^{(0)}(\omega) = CI(\omega)$ is possible, and we may obtain the somewhat simpler

$$B(\omega) = 1 - [1 - C\tau(\omega)]^{(k+1)}. \tag{55}$$

When $\tau(\omega)$ is real, it is easy to adjust C so that we satisfy the required $C\tau(\omega) < 2$ for the largest $\tau(\omega)$, but what can we do if $\tau(\omega)$ has negative values? Being applied in the data space, the constant C is not a function of ω. We must find a single C that works. Yet, certain spreading phenomena *do* give rise to negative $\tau(\omega)$. The simple moving average is an example that arises in the case of image motion blur. It is equivalent to convolution with $\text{rect}(x)$, giving $\tau(\omega) = (1/\sqrt{2\pi})\text{sinc}(\omega/2\pi)$. Both coherent and incoherent optical response functions for spectrometers having *unequal* entrance- and exit-slit spectral widths present a similar situation. These cases may be analyzed by the methods of Section II of Chapter 2. In the next section we introduce a method of accommodating phenomena that give rise to $\tau(\omega)$ having negative values.

3. Reblurring and Smoothing

Kawata *et al.* (1978) described a concept of reblurring to guarantee that the transfer function does not have negative values. They recognized that the convolution of the spread function with a reversed version of itself yields, in Fourier space, a function that cannot have negative values:

$$s(x) \otimes s(-x) \leftrightarrow |\tau(\omega)|^2. \tag{56}$$

This may be easily proved by applying the definition of the Fourier transform. It is a version of the autocorrelation theorem. The authors elaborate in a later paper (Kawata and Ichioka, 1980b).

It is thus possible to convolve both spread function and data $i(x)$ with $s(-x)$. We may then use the relaxation methods as before. This time, however, we replace $i(x)$ with $s(-x) \otimes i(x)$ and $s(x)$ with $s(-x) \otimes s(x)$. Not only are we assured convergence, but we have also succeeded in band-limiting the data $i(x)$ in such a way as to guarantee that all noise is removed from $i(x)$ at frequencies where $i(x)$ contains no information about $o(x)$. Furthermore, Ichioka and Nakajima (1981) have shown that reblurring reduces noise in the sense of minimum mean-square error.

Replacement of the spread function by its autocorrelation has been used for similar purposes by other authors. For example, ratio-based corrections have been thus adapted to guarantee convergence (Chapter 4, Section IV.A). An early use of the idea appears in a rigorous mathematical treatment of the Fredholm integral equation of the first kind by Landweber (1951). The Fredholm equation poses the deconvolution problem in the continuous representation, generalized to shift-variant spread-function kernels. The Landweber method adapts van Cittert's method to these shift-variant kernels and assures convergence via the foregoing autocorrelation substitution.

The analysis in this chapter is not valid when constraints and/or smoothing between iterations are employed as described in Chapter 4. These valuable techniques complicate the situation considerably. We may, however, employ the analysis already given as a guide. In spite of attendant complications, the pragmatic truth is that these techniques, especially constraints, are well worth the effort. They produce better resolution and show reduced sensitivity to both noise in $i(x)$ and errors in $s(x)$. Results achieved are sometimes little short of astonishing when judged by the standards of the traditional linear methods described in this chapter.

E. STEPWISE IMPLEMENTATION

We have noted the noise-sensitivity problem of the simple inverse filter and introduced modifications to alleviate these difficulties. Modifications yielded different functional forms for $Y(\omega)$. The convenient single-step property of the basic method was nevertheless retained. This property contrasts with the need for possibly arbitrary stopping criteria when we use iterative methods, which are computationally more expensive. The iterative methods do, however, allow the user to control the signal-to-noise versus resolution trade-off by stopping the process when the growth of spurious components begins to negate the benefit of further resolution improvement. Unfortunately, the result (Section IV.D) is not usually equivalent to optimum tailoring of $Y(\omega)$. Would it be possible to achieve the linear inverse-filtered result in a preset fixed number of stages? If so, the users could obtain a true inverse-filtered result when justified by high data quality, but otherwise they could stop short of that objective, achieving smoother results with somewhat compromised resolution.

Redina and Larson (1975)[*] satisfied these requirements by employing the N_sth root of $Y(\omega)$ as the transfer function of a component filter $Y_s(\omega)$. Application of the component filter N_s times yields the result obtained by a single application of $Y(\omega)$. Specifically, the component filter is given by

$$Y_s(\omega) = [Y(\omega)]^{1/N_s}. \tag{57}$$

By converting to polar coordinates we obtain

$$Y_s = |Y|^{1/N_s} \exp[(j/N_s)\arctan(\operatorname{Im} Y/\operatorname{Re} Y)], \tag{58}$$

which is suitable for computation. In Eq. (58) Re and Im denote the real and imaginary parts of their arguments, respectively. The component filter may be applied either as written in Eq. (58) by multiplication in the Fourier domain or by convolution with the use of $y_s(x)$, the Fourier transform of $Y_s(\omega)$. In both cases, the x-domain result may be inspected after each pass.

[*]Abstract only; complete text obtained from authors in 1975.

The composite filter $Y(\omega)$ may either be the true inverse filter, truncated for ω large if necessary, or any of the variations described elsewhere in Section IV. In their original work, Rendina and Larson chose $Y(\omega) = \psi(\omega)/\tau(\omega)$, where $\psi(\omega)$ is a Gaussian line-broadening function that limits the ultimate resolution obtainable but yields a manageable $Y(\omega)$. For their studies Rendina and Larson used $N_s = 4$.

V.   Other Linear Methods

A large number of linear methods have been developed with particular characteristics that tend to suit them to specific deconvolution problems. None of these adaptations shows beneficial results nearly so profound as those resulting from the imposition of the physical-realizability constraints discussed in the next chapter. Furthermore, the present work is not intended as an exhaustive review, so we mention only a few additional linear methods, because of their popularity, the insight conveyed, or their adaptability to constraints.

The demand that the solution $\hat{o}$ be consistent with the data i results in the improved resolution that we expect from a deconvolution method. As we have explained, however, it also results in the amplification of high-frequency noise. The smoothing of this noise to some extent defeats the purpose of deconvolution. The trade-off between smoothness and consistency is explicit in the formulation of a method first described by Phillips (1962) and further developed by Twomey (1965). In this method, we minimize the quantity

$$\beta \sum_{n=1}^{N} \left(\hat{o}_{n+1} - 2\hat{o}_n + \hat{o}_{n-1} \right)^2 + \sum_{m=1}^{M} \left\{ i_m - \sum_{n=1}^{N} [\mathbf{s}]_{mn} \hat{o}_n \right\}^2 . \quad (59)$$

The first term governs smoothness through the second differences of $\hat{o}_n$. The second term imposes the consistency between the solution values $\hat{o}_n$ and the data values i_m. The trade-off is controlled by varying parameter β. Frieden (1975) explores the method briefly. Hunt (1973) applies it to images in a computationally efficient way that uses the fast Fourier transform.

Confronted with a problem in which two data sets were available, Breedlove *et al.* (1977) chose a solution that minimizes a sum of terms not unlike expression (59). Available were two images: one a blurred representation of the object, the other a superposition of sharp renderings. In this sum, the right-hand term accommodates the blurred image as in expression (59). The other term incorporates the multiple exposure via the Lagrange multiplier technique. Solutions obtained by this method illustrated the desirability of using all the available data.

Huang *et al.* (1975) and Huang (1975) introduced the optics community to a method based on iteratively correcting estimate $\hat{o}$ by adding a term that involves a hyperplane projection operation. Like the relaxation methods discussed in Sections II and III, it has the potential to be upgraded by the implementation of constraints. (See also Chap. 4, Sect. IX.B.)

The choice of sampling rate is sometimes governed by a trade-off between frequency aliasing (low rate) and noise sensitivity (high rate). Quick and Bolgiano (1976) have employed the Poisson transformation to avert this complication. Consult Piovoso and Bolgiano (1970) for additional background on the Poisson transform.

Matrix methods such as singular value decomposition and others based on pseudoinverses have been described in the digital image processing literature. A book chapter by Andrews (1975) and texts by Pratt (1978), Andrews and Hunt (1977), and Hall (1979) treat this subject and give further references. These works also contain abundant references and considerable detail regarding other linear deconvolution methods and their variations as applied to the computer restoration of images.

VI.   Overview of Linear Methods

All of the methods described in this chapter are linear methods because each element of $\hat{o}(x)$ can be obtained by a linear combination of the elements of $i(x)$. In the continuous regime, a linear estimate $\hat{o}(x)$ can always be expressed by the integral

$$\hat{o}(x) = \int y(x, x')i(x')\,dx', \tag{60}$$

where $y(x, x')$ is a filtering kernel, shift variant in general. Sometimes this formulation is not immediately apparent; many linear methods are well disguised by mathematical foliage (Wells, 1980). However well obscured, a linear method is a linear method, and it is severely limited in its capability relative to modern constrained nonlinear methods. Sometimes, however, the researcher does not have the *a priori* information required for application of the nonlinear methods, or this information may not be applicable. Under these conditions linear methods are useful. In addition, there are other occasions where linear methods are preferable. Included are situations where computational speed is of paramount importance, as in real-time deconvolution, and where linear filtering hardware is available, whether analog or digital. Performance can be satisfactory if the spectrometer transfer function allows passage of very high frequencies, even though they may be somewhat suppressed. The Wiener-type filter may be a good choice if one *must* use a linear restoring method because it deals explicitly with the problem of noise and with achieving a solution that is, at least in a limited sense, optimum.

In both imaging and spectroscopy, however, it is often the case that solutions having negative values are not physically possible. On occasion, an upper bound may also be imposed. These highly valuable adaptations take the discussion into the realm of nonlinearity and are deferred until the next chapter.

Nevertheless, certain types of prior knowledge can be introduced within the context of a linear method. Probabilities, signal and noise statistics, power spectra, and the like may be incorporated. Often this type of prior knowledge is difficult to obtain. In any case, it rarely exerts an influence nearly so profound as that of simple bounds on the amplitude of the solution. If the observing spread function obliterates all frequencies beyond the cutoff Ω, they are forever lost to the linear restoration methods. No linear filter's amplification factor can ever make anything but a zero out of the zero that the observing instrument has left us. All it can do is amplify noise at these frequencies.

It is a good rule to use all the prior knowledge in which one has high confidence. If, for example, it is known that an unresolved group of spectral lines has exactly four components and that one of the components is narrower than Δx, this knowledge may be employed for function fitting in the manner covered in Section V.C of Chapter 1. Once these assumptions are made, however, an unexpected fifth spectral line may be lost. As stated earlier, the function-fitting procedures are not true deconvolution methods, although they can be valuable under certain circumstances.

All deconvolution methods involve an implicit trade-off of signal-to-noise ratio for resolution. Even the optical spectrometer itself exhibits such a trade-off when slit width is varied. When slits are made narrow to boost resolution, much of the dispersed photon flux (and its information content) is lost to the detection system. A different situation is found in the use of multiplex spectrometry. In some spectral regions, photographic plates or multiple detectors may be employed. Multiple-slit spectrometers with various slit encoding schemes and single detectors can also be used (Stewart, 1970). Furthermore, it is possible to omit the disperser entirely and use a Fourier transform spectrometer. Would it be possible to defeat the conventional spectrometer trade-off by acquiring spectra with wider slits and then deconvolving? This is indeed possible; orders-of-magnitude improvement in the productivity of a conventional spectrometer can thus be achieved. Blass and Halsey (1981) address this question.

If at the outset the data are very noisy and if the noise predominates in the Fourier frequency range needed to effect a restoration, constraints provide the only hope for improvement. The reason is that many of the noise values in the data would "restore" to physically unrealizable values by linear deconvolution. The constrained methods are inherently more robust because they must find a solution that is consistent with both data *and* physical reality.

As a final note, we wish to caution the reader concerning evaluation of the literature. Many authors display the results that their methods produce when tested with noise-free, computer-simulated "data." Almost all linear methods will produce outstanding results in these circumstances. In particular, virtually perfect restorations are easy to obtain when these "data" are not band limited. The only uncertainty troubling such restorations is the roundoff noise in the computer. If the computer utilizes eight significant figures, the signal-to-noise ratio is on the order of 10^8. What integration time must the detection system employ to achieve this phenomenal data quality?

References

Allebach, J. P. (1981). *J. Opt. Soc. Am.* **71**, 99–105.

Andrews, H. C. (1975). In *Top. Appl. Phys.: Picture Processing and Digital Filtering*, Vol. 6 (T. S. Huang, ed.), pp. 21–68. Springer-Verlag, New York.

Andrews, H. C., and Hunt, B. R. (1977). "Digital Image Restoration." Prentice-Hall, Englewood Cliffs, New Jersey.

Backus, G., and Gilbert, F. (1970). *Philos. Trans. R. Soc. London Ser. A* **266**, 123–192.

Betty, K. R., and Horlick, G. (1977). *Anal. Chem.* **49**, 351–352.

Blass, W. E., and Halsey, G. W. (1981). "Deconvolution of Absorption Spectra." Academic Press, New York.

Bracewell, R. N. (1958). *Proc. Inst. Radio Eng.* **46**, 106–111.

Bracewell, R. N., and Roberts, J. A. (1954). *Aust. J. Phys.* **7**, 615–640.

Breedlove, J. R., Kruger, R. P., Trussell, H. J., and Hunt, B. R. (1977). *SPIE* **119**, 258–263.

Bromba, M. U. A., and Ziegler, H. (1981). *Anal. Chem.* **53**, 1583–1586.

Burger, H. C., and van Cittert, P. H. (1933). *Z. Phys.* **81**, 428–434.

DeNoyer, L. K., and Dodd, J. G. (1990). *American Laboratory* **22**, 21–27.

Deutsch, S. (1965). *IEEE Trans. Broadcast.* **11**, 11–21.

DiCola, G., Rota, A., and Bertolini, G. (1967). *IEEE Trans. Nucl. Sci.* **14**, 640–653.

Edwards, T. R. (1982). *Anal. Chem.* **54**, 1519–1524.

Edwards, T. R., and Knight, R. D. (1975). NASA Technical Memorandum NASA TM X-64949. Space Sciences Laboratory, Marshall Space Flight Center, Huntsville, Alabama.

Frieden, B. R. (1975). In *Top. Appl. Phys.: Picture Processing and Digital Filtering*, Vol. 6 (T. S. Huang, ed.), pp. 177–248. Springer-Verlag, New York.

Gans, P., and Gill, J. B. (1983). *Appl. Spectrosc.* **37**, 515–520.

Goldman, S. (1953). "Information Theory." Dover, New York.

Hall, E. L. (1979). "Computer Image Processing and Recognition." Academic Press, New York.

Helstrom, C. W. (1967). *J. Opt. Soc. Am.* **57**, 297–303.

Herget, W. F., Deeds, W. E., Gailar, N. M., Lovell, R. J., and Nielsen, A. H. (1962). *J. Opt. Soc. Am.* **52**, 1113–1119.

Hill, N. R., and Ioup, G. E. (1976). *J. Opt. Soc. Am.* **66**, 487–489.

Huang, T. S., ed. (1975). *Top. Appl. Phys.: Picture Processing and Digital Filtering*, Vol. 6, Springer-Verlag, New York.

Huang, T. S., Barker, D. A., and Berger, S. P. (1975). *Appl. Opt.* **14**, 1165–1168.

Hunt, B. R. (1973). *IEEE Trans. Comput.* **C-22**, 805–812.

Ichioka, Y., and Nakajima, N. (1981). *J. Opt. Soc. Am.* **71**, 983–988.

Jansson, P. A. (1968). Ph.D. Dissertation, Florida State Univ., Tallahassee.

Jansson, P. A. (1970). *J. Opt. Soc. Am.* **60**, 184–191.

Jansson, P. A. (1972). *J. Opt. Soc. Am.* **62**, 195–198.

Kawata, S., and Ichioka, Y. (1980a). *J. Opt. Soc. Am.* **70**, 762–768.

Kawata, S., and Ichioka, Y. (1980b). *J. Opt. Soc. Am.* **70**, 768–772.

Kawata, S., Ichioka, Y., and Suzuki, T. (1978). *Proceedings of the 4th International Joint Conference on Pattern Recognition*, 525–529.

Kawata, S. and Minami, S. (1984). *Appl. Spectrosc.* **38**, 49–58.

Kekre, H. B., Madan, V. K., and Bairi, B. R. (1989). *Nucl. Instrum. Methods Phys. Res.* **A279**, 596–598.

Knuth, D. E. (1973). "The Art of Computer Programming," Vol. 1, Fundamental Algorithms, 2nd Ed. Addison-Wesley, Reading, Massachusetts.

Kuo, J. E., Wang, H., and Pickup, S. (1991). *Anal. Chem.* **63**, 630–635.

Landweber, L. (1951). *Am. J. Math.* **73**, 615–624.

Marmolin, H., Nyberg, S., and Berggrund, U. (1978). *Photogr. Sci. Eng.* **22**, 142–147.

Phillips, D. L. (1962). *J. Assoc. Comput. Mach.* **9**, 84–97.

Piovoso, M. J., and Bolgiano, L. P., Jr. (1970). *In* "Proceedings of Symposium Computer Processing in Communications," (J. Fox, ed.), New York, April 8–10, 1969. Polytechnic Press, Brooklyn, New York.

Porchet, J. P., and Günthard, H. H. (1970). *J. Phys.* **E3**, 261–264.

Pratt, W. K. (1978). "Digital Image Processing." Wiley, New York.

Quick, L. T., and Bolgiano, L. P., Jr. (1976). Proceedings of the International Conference on Acoustics, Speech and Signal Processing," pp. 350–353. IEEE, Philadelphia, Pennsylvania.

Rendina, J., and Larson, P. (1975). Paper Number 66, Pittsburgh Conference on Analytical Chemistry and Applied Spectroscopy," March 3–7 (abstract only). Cleveland Convention Center, Cleveland, Ohio.

Sakai, H. (1962). "A Slit Function Correction and an Application to the Study of the Absorption Lines in the H_2O Pure Rotation Spectrum." U.S. Armed Services Technical Information Agency Report AD287897.

Savitzky, A., and Golay, M. J. E. (1964). *Anal. Chem.* **36**, 1627–1639.

Seah, M. P., and Cumpson, P. J. (1992). *Appl. Surf. Sci.* **62**, 195–198.

Steinier, J., Termonia, Y., and Deltour, J. (1972). *Anal. Chem.* **44**, 1906–1909.

Stewart, J. E. (1970). "Infrared Spectroscopy Experimental Methods and Techniques." Dekker, New York.

Twomey, S. (1965). *J. Franklin Inst.* **279**, 95–109.

Van Cittert, P. H. (1931). *Z. Phys.* **69**, 298–308.

Wells, D. C. (1980). *SPIE* **264**, 148–154.

Wendroff, B. (1966). "Theoretical Numerical Analysis." Academic Press, New York.

Whittaker, E., and Robinson, G. (1967). "The Calculus of Observations: An Introduction to Numerical Analysis," 4th Ed. Dover, New York.

Willson, P. D., and Edwards, T. H. (1976). *Appl. Spectrosc. Rev.* **12**, 1–81.

Willson, P. D., and Polo, S. R. (1981). *J. Opt. Soc. Am.* **71**, 599–603.

Yamashita, K., and Minami, S. (1969). *Jpn. J. Appl. Phys.* **8**, 1505–1512.

Chapter 4 | Modern Constrained Nonlinear Methods

Peter A. Jansson

College of Optical Sciences, University of Arizona, Tucson

List of Symbols

$a(x)$	real Fourier series such that $\hat{o}^+(x) = [a(x)]^2$
A, B	lower and upper bounds to $\hat{o}$, respectively
$A(\omega)$	Fourier transform of $a(x)$
ADALINE	adaptive linear neuron
ART	algebraic reconstruction technique
ASIC	application-specific integrated circuit
b	constraint operator
$B_W(\omega)$	$\tau(\omega)Y_W(\omega)$, factor by which simple inverse filter must be modified to give Wiener version
C	constant
CCD	charge-coupled device
CMAC	cerebellar model arithmetic computer
E	mean-square error within specified frequency limits
$\hat{\varepsilon}$	maximum-likelihood solution for noise
E_{res}	residual squared error due to noise alone
FFT	fast Fourier transform
FIR	finite impulse response
FWHM	full width at half maximum
$g(x), g$	component of $\hat{o}^+(x)$ such that $\hat{o}^+(x) = g^*(x)g(x)$
$G(\omega), G$	Fourier transform of $g(x)$
h	Planck's constant
H	Shannon-type entropy $-\Sigma n_m \ln n_m$
H_1	Burg-type entropy $\Sigma \ln n_m$
$H(\ )$	Heaviside step function
$\mathbf{i}, \mathbf{o}, \hat{\mathbf{o}}$	column vectors containing elements $i_m, o_m, \hat{o}_m$
$\otimes$	denotes convolution operation
$\vec{i}$	image vector
$i_m, o_m, \hat{o}_m, \hat{o}_m^{(k)}$	discretely sampled versions of $i, o, \hat{o}, \hat{o}^{(k)}$
$\hat{i}^{(t)}$	$s \otimes \hat{o}^{(t)}$, "estimate" of the data
$i(x), i$	"image" data that incorporate smearing by $s(x)$
$i(x)^{(k)}, \hat{i}^{(k)}$	kth "estimate" of $i(x)$ based on $\hat{o}^{(k)}$
$I(\omega), \tau(\omega), O(\omega), \hat{O}(\omega)$	Fourier transforms of $i(x), s(x), o(x), \hat{o}(x)$
k	Boltzmann constant

L	number of nonlinear filtering stages; index of limit of sought $A(\omega_l)$
LMS	least mean square
m	integer, usually an exponent or index
M	index of maximum s_m; measure of number of samples of $\hat{O}^+(\omega)$ over range $\pm\Omega_c$ such that $2M + 1$ samples cover the range; number of cells or elements constituting object o_m; number of columns in spread-function matrix
n	integer, usually an exponent or index
n_m	$o_m/h\nu_m$, number of photons counted in each object element
$\hat{n}_m$	estimate of n_m
N	$\sum_{m=1}^{M} n_m$, total number of photons in object; number of equations; number of planes
$N_\varrho(\omega)$	Fourier transform of noise
$\vec{o}^{(k)}$	object estimate vector
$o(x), o$	"object" or function sought by deconvolution, usually the true spectrum, but also the instrument function when this is sought by deconvolution
$\hat{o}(x), \hat{o}$	estimate of $o(x)$
$\hat{o}^{(k)}(x), \hat{o}^{(k)}$	kth estimate of $o(x)$
$\hat{o}^+(x), \hat{o}^+$	nonnegative estimate of $o(x)$
$\hat{o}_e(x), \hat{O}(\omega)$	bandwidth-extrapolating part of $\hat{o}_p(x)$ and its Fourier transform
$\hat{o}_p(x)$	physically acceptable estimate of $o(x)$
p	equation number; plane number
$p(q_1,\ldots,q_M)$	probability of given set of q_m
$p_c(q_m)$	one of M identical components of $p(q_1,\ldots,q_M)$
P	measure of number of samples required to specify $A(\omega)$ for $\hat{O}^+(\omega)$ over range $\pm\Omega_c$ $(P = M/2)$
$P(n_1,\ldots,n_M)$	probability of given number-count set $\{n_m\}$
PE	processing element
POCS	projections onto convex sets
q'	particular value that $q(x)$ may assume
q_0	peak height of Doppler-broadened emission line
q_1	height of Doppler emission line at distance Δx from center
q_m	relative probability that a photon occupies cell m of the object
$q(x)$	continuous form of prior spectrum
Q	constant value of all elements of flat prior spectrum
Q_m	values of prior spectrum in which user has highest possible conviction
r_0	constant in relaxation function

$r[\hat{o}], r$	relaxation function that weights correction terms—a function of $\hat{o}$
$r_x^{(k)}(x)$	$r[\hat{o}^{(k)}(x)]$, relaxation function as function of x
$R^{(k)}(\omega)$	Fourier transform of $r^{(k)}(x)$
$\mathbf{s}$	matrix containing elements $[\mathbf{s}]_{nm} = s_{nm}$
$\vec{s}_n$	nth row of spread-function matrix
s_m	sampled value of shift-invariant spread function
s_{nm}	general spread-function matrix element; for convolution, sampled version of function $s(x_n - x_m)$, where x_n and x_m are sampled values of x
$s(x), s$	spread function, usually the instrument function, but also spreading due to other causes
SVD	singular value decomposition
t	sequential number of grain
T	absolute temperature
VLSI	very large-scale integration
W_m	Bose-Einstein degeneracy factor
x	independent variable, typically wave number (cm^{-1}), wavelength, or spatial coordinate
x_n, x_m	discrete sampled value of x
x_p	location at which trial grain is added
X	limit of object extent such that $|x| \leq X$
$y(x)$	linear restoring filter such that $\hat{o}(x) = y(x) \otimes i(x)$
$y_W(x), Y_W(\omega)$	linear restoring filter of Wiener type and its Fourier transform
$Y(\omega)$	Fourier transform of $y(x)$
z_m	normal modes or degrees of freedom available for occupation by photons of frequency ν_m
δ_{mp}	Kronecker delta
$\Delta\hat{o}$	object estimate increment due to added grain
Δx	distance from center of model line, defining region over which probability will be computed
$\Delta\omega$	sample spacing required to specify object completely
κ_0	constant in discrete formulation of relaxation function
$\kappa[\hat{o}^{(k)}]$	relaxation function in discrete formulation
μ, λ_m	Lagrange multipliers
ν_m	photon frequency in hertz
ρ	factor affecting grain allocation
σ^2	variance of additive noise
Φ	objective function to be minimized
Φ_T	equilibrium value of function Φ at temperature T
$\Phi[\hat{o}(x_1), \ldots, \hat{o}(x_M)]$	objective function with object estimate elements as parameters
ω	conjugate of x; Fourier frequency in radians per units of x
Ω	cutoff frequency such that $\tau(\omega) = 0$ for $|\omega| > \Omega$

Ω_c limit of ω such that restoration $\hat{O}^+(\omega)$ is required to be consistent with $I(\omega)$ within $\pm\Omega_c$

Ω_H upper limit of frequencies $|\omega|$ used for input in Howard's method

Ω_p limit of extended Fourier bandwidth required of $\hat{o}^+$

★   correlation operation

I. Introduction

In preceding chapters we laid a foundation for the study of deconvolution. We presented several linear methods that exemplify the groundwork available before recent developments revolutionized the deconvolution field. Why, in their simplicity and elegance, did the linear methods fail to stimulate the wide adoption of deconvolution methods?* After all, available instrumental resolution is limiting in many applications, and the simplicity of the microcomputer makes numerical processing attractive.

The answer lies in the relatively poor performance of linear methods, especially with band-limited data. Frequently a linear restoration reveals little true structure that could not have have been seen in the original data. Even worse, noise-based artifacts often call the result into question. One might even say that the linear methods had helped to give deconvolution a bad reputation.

To be sure, linear methods have value where fast computation is necessary. They perform reasonably well when the experimental data are not band limited, and in trials with computer-generated "data" devoid of noise. Optical imaging and spectroscopic data are often band limited, however, and computation time is becoming less of a problem, given advances in computing hardware. Because of these reasons, and growing awareness of nonlinear methods in the 12 years since our previous volume (Jansson, 1984), more researchers are enjoying the benefits of nonlinear methods.

Knowing that the better nonlinear constrained methods are now available, why have some researchers been reluctant to accept them? Perhaps the linear approach has an attraction that is not related to performance. Early in a technical career the scientist–engineer is indoctrinated with the

---

*This is not to say that all linear methods are simple. Indeed even the linear nature of some methods is remarkably obscure.

principles of linear superposition and analysis. Indeed, a rather large body of knowledge is based on linear methods. The trap that the linear methods lay for us is the existence of a beautiful and complete formalism developed over the years. Why complicate it by requiring the solution to be physically possible?

Perhaps the benefits of physical-realizability constraints, particularly ordinate bounds such as positivity, have not been sufficiently recognized by some researchers. Surely everyone agrees in principle that such constraints are desirable. Even the early literature on this subject frequently mentions their potential advantages. For one reason or another, however, the earliest nonlinear constrained methods did not at first reveal the inherent power of constraints.

In the following pages, we trace the development of the nonlinear methods and two concurrent themes that underlie this work: first the physical-realizability theme already mentioned, and then the concept that one should be able to restore the Fourier frequencies obliterated by the finite Fourier bandpass of the observing instrument, that is, frequencies not present in the data. The extent to which these two themes are closely coupled was not fully appreciated in early work.

II. The Meaning of Constraints

Throughout this chapter, and indeed the entire volume, we often use the word "constraint" to indicate bounds placed on the solution. We frequently mean positivity; that is, the solution $\hat{o}(x)$ cannot have negative values. This constraint is the most powerful. It is usually the easiest of the amplitude bounds to implement and is sometimes inherent in the way the solution is represented. We also consider spatial bounds; the solution is known to vanish over certain values of the independent variable x, for example, outside a single region within well-defined limits. In practice, this type of constraint usually exerts a weaker influence on the solution than the amplitude bound.

Some authors consider nonlinearities in the processing of data values, that is, photometric nonlinearities. They term the resulting restoration nonlinear. Here we assume that such effects are either not present or have been corrected. We reserve the term "nonlinear" to describe the situation in which the solution $\hat{o}(x)$ cannot be expressed as a linear function of the irradiance data $i(x)$. For situations where it is proper to introduce radiometric nonlinearity, we refer the reader to Hunt (1975, 1977) and Andrews and Hunt (1977).

III. The Promise of Analytic Continuation

Sometimes an observing instrument completely obliterates the information at all Fourier frequencies ω beyond some finite cutoff Ω. This is specifically true of optical telescopes, microscopes, and dispersive optical spectrometers, where the aperture determines Ω. In the case of spectroscopy, the cutoff Ω may be extended to high Fourier frequencies by multipassing the dispersive element or employing the high orders from a diffraction grating. Whatever the method, the cutoff is usually finite and absolute. Similar considerations prevail in the Fourier interferometer, where the maximum path difference determines Ω. It would seem foolish to suggest that the information at these frequencies could be restored when it is not even present in the data.

Recall that inverse filtering (Chapter 3) produces the estimate $\hat{O}(\omega) = I(\omega)/\tau(\omega)$, which is undefined for $\tau(\omega) \geq \Omega$. Indeed, we cannot ever completely know $O(\omega)$ if we are restricted to using the information in $I(\omega)$ alone. If we add the modest bit of knowledge that the object $o(x)$ is finite in extent, however, the lost frequency components no longer appear irretrievable (Wolter, 1961). The reasoning goes as follows. It is known (Guillemin, 1949) that the Fourier transform $O(\omega)$ of a function $o(x)$ having finite extent is analytic for all ω. If $O(\omega)$ is known only over the interval $-\Omega$ to Ω, but is known with absolute accuracy, then it is known for all ω. This may be clarified if we note that an analytic function has a unique Taylor-series representation. Once its coefficients are specified for $\omega < |\Omega|$, they are valid for all ω. The reconstruction of the lost information may be done by the Taylor-series route (Frieden, 1975) or other means (Harris, 1964; Goodman, 1968).

Attempts to implement the technique have not met with much success. The difficulty lies in the knowledge available about $O(\omega)$ for $|\omega| < \Omega$. Like all quantities obtained from measurement, the estimate $\hat{O}(\omega)$ for $|\omega| < \Omega$ contains error. In most practical cases, $\tau(\omega)$ is small for ω close to Ω. Consequently, $\hat{O}(\omega)$ is most uncertain where its influence over the region to be restored is greatest. The method is essentially one of extrapolation and is subject to the difficulties that beset all extrapolation procedures when they are applied to uncertain data.

If, on the other hand, $O(\omega)$ were known to high accuracy over some neighborhood near the cutoff Ω, then its derivatives would also be known. Because an analytic function is uniquely specified for all ω by knowledge of all its derivatives at a single ω, a modest extrapolation should be possible for $|\omega|$ not too much larger than Ω. Unfortunately, many optical instruments severely degrade the signal-to-noise ratio near the cutoff Ω. Unapodized Fourier interferograms, however, do not show this fault.

The constraint of finite extent has in practice proved to be of only limited value for improving the quality of restorations, whether the process is one of analytic continuation or otherwise. This is borne out, for example, by the work of Howard (Chapter 12). We have now presented the idea of recovering frequencies beyond the cutoff Ω, however. We should not find it too surprising if other types of information bring with them the ability to restore these presumably lost frequencies.

IV. Ratio Methods

A. GOLD'S METHOD

It has been noted that deconvolution methods, most of which are linear, have a propensity to produce solutions that do not make good physical sense. Prominent examples were found when negative values were obtained for light intensity or particle flux. As noted previously, the need to eliminate these negative components was generally accepted. Accordingly, Gold (1964) developed a method of iteration similar to Van Cittert's but used multiplicative corrections instead of additive ones.

Given the equation describing spectrometer distortion,

$$i = s \otimes o, \tag{1}$$

and a spectrum estimate $\hat{o}^{(k)}$, this method proposes to estimate the observed data by use of

$$\hat{i}^{(k)} = s \otimes \hat{o}^{(k)}. \tag{2}$$

A ratio containing this quantity and the data may be used to construct a new estimate of the true spectrum,

$$\hat{o}^{(k+1)} = \hat{o}^{(k)}\left[i/\hat{i}^{(k)}\right]. \tag{3}$$

The iteration may be given in a single equation:

$$\hat{o}^{(k+1)} = \hat{o}^{(k)} \frac{i}{s \otimes \hat{o}^{(k)}}. \tag{4}$$

As with Van Cittert's method, we may take $\hat{o}^{(0)} = i$. We see that the successive estimates $\hat{o}^{(k)}$ cannot be negative, provided that s and i are everywhere positive. The reader should note that this assumption may be violated in a base-line or background region. Here, in the case of additive noise, data may contain negative values. Noise from experiments involving counting usually obeys Poisson statistics, in which case all values are positive. Indeed, Gold's method was developed for use with data from nuclear-physics counting experiments.

Gold's method has been used by a number of workers, including Siska (1973), who applied it to molecular-beam scattering data, MacNeil and Dixon (1977), who applied it to photoelectron spectra, and Jones *et al.* (1967), who restored infrared spectra of condensed-phase samples. In the work of Jones *et al.*, the resulting resolution is probably limited by the inherent breadth of spectral lines observed with condensed-phase samples.

In the previous chapter we found that when Van Cittert's method converges, it converges to the linear inverse filter. We are therefore led to inquire about the convergence properties of the present ratio method. Gold's original consideration of the problem was based on the matrix formulation given in the previous chapters. In that formulation, the spectrometer distortion is given by

$$\mathbf{i} = \mathbf{so}. \tag{5}$$

Here $\mathbf{o}$ and $\mathbf{i}$ are the object and image vectors, respectively, and $\mathbf{s}$ is the spread-function matrix as defined in Chapter 3. Gold was able to show that proper convergence is assured if the following conditions hold:

(1) all principal minors of $\mathbf{s}$ are positive,
(2) all eigenvalues of $\mathbf{s}$ are real, and
(3) all eigenvalues of $\mathbf{s}$ are positive.

These conditions prevail with all positive definite matrices, that is, all matrices having positive (only) elements and determinants greater than zero. Gold observed that even if $\mathbf{s}$ does not obey these conditions, $\mathbf{s}^T\mathbf{s}$ does, where s^T is the transpose of s. We may therefore multiply Eq. (5) by s^T and obtain

$$s^T i = s^T s o. \tag{6}$$

If i is positive, then $s^T i$ is also positive and therefore meets the requirement for a positive solution. The added smoothing of i imposed by s^T also helps to reduce or eliminate possibly troublesome negative components. Confident of convergence, we may seek the solution, using $s^T i$ in place of i and $s^T s$ in place of s. The rate of convergence depends on the conditioning of s and the choice of $\hat{o}^{(0)}$.

The matrix notation serves to stress that the technique is applicable to shift-variant spread functions, that is, where $s_{jk} \neq s_{lm}$ for $j - k = l - m$. Many of the deconvolution methods described here with the convolution notation may thus be generalized. In the convolution notation, the present method may be expressed by the equation

$$\hat{o}^{(k+1)}(x) = \hat{o}^{(k)}(x) \frac{s(-x) \otimes i(x)}{s(-x) \otimes s(x) \otimes \hat{o}^{(k)}(x)}. \tag{7}$$

We note that transposing s is equivalent to reversing the abscissa of $s(x)$. The reader will be struck by the similarity of this treatment to the reblurring method of Kawata $et\ al.$ (see Chapter 3, Section IV.D.3). Their method is, in fact, an adaptation of the present use of s^T to Van Cittert's method.

Because the solution $\hat{o}^{(k)}$, $k \to \infty$, is independent of choice $\hat{o}^{(0)}$, it is unique. But how close is it to the ideal solution $o(x)$? The fact that $\hat{o}^{(k)}$ is nonnegative means, at least, that the method does not converge to the inverse-filter estimate.

B. RICHARDSON–LUCY METHOD

Working independently in optics and astronomy, respectively, Richardson (1972) and Lucy (1974) developed a ratio method resembling that of Eq. (7). Their method arises from a maximum likelihood argument in which the image is modeled with Poisson statistics. Maximizing the image model's likelihood function yields an equation that is satisfied when the following iteration converges:

$$\hat{o}^{(k+1)}(x) = \hat{o}^{(k)}(x) \cdot s(-x) \otimes \frac{i(x)}{s \otimes \hat{o}^{(k)}}. \tag{8}$$

Like the modified Gold's method, this method uses convolution by an abscissa-reversed spread function. In this case, the correction ratio is processed as a whole. This contrasts with the modified Gold's method, in which numerator and denominator are processed separately.

As with other iterative methods, we may readily generalize Eq. (8) to the shift-variant case. The generalization would have proved useful in the method's best-known application to date—images from the Hubble Space telescope (Chapter 10)—because blur in Hubble's wide-field planetary camera can be characterized as a convolution only over relatively small isoplanatic image patches. Unfortunately, the computational load for the shift-variant signal-space summations inhibited its use. Piecewise application of Fourier-space computations proved an effective approximation.

V. Schell's Method

Focusing our attention once again on Fourier space, recall that the Wiener inverse filter $Y_W(\omega)$ is obtained by finding the function $Y_W(\omega)$ that minimizes the mean-square error

$$\int \langle |y_W(x) \otimes i(x) - o(x)|^2 \rangle \, dx = \int_{-\Omega}^{\Omega} \langle |Y_W(\omega)I(\omega) - O(\omega)|^2 \rangle \, d\omega. \quad (9)$$

Schell (1965) recognized that the major deficiency of the Wiener inverse filter is the nonphysical nature of the partially negative solutions that it is prone to generate. He sought to extrapolate the band-limited transform $O(\omega)$ by seeking a nonnegative physical solution $\hat{o}^+(x)$ through minimization of

$$\int_0^{\Omega} B_W^2(\omega)|Y(\omega)I(\omega) - \hat{O}^+(\omega)|^2 \, d\omega + \int_{\Omega}^{\Omega_p} |\hat{O}^+|^2 \, d\omega. \quad (10)$$

In this expression, $Y(\omega)$ is the simple inverse filter $1/\tau(\omega)$ and $B_W(\omega)$ the factor by which the simple filter must be modified to give the Wiener inverse filter, so that $Y_W(\omega) = B_W(\omega)/\tau(\omega)$. The quantity $\hat{O}^+(\omega)$ is the Fourier transform of $\hat{o}^+(x)$. The quantity Ω_p specifies the limit of the extended bandwidth required of the solution $\hat{o}^+(x)$.

We are permitted to specify the integrals for positive ω only, because of the even property of the integrand. This simplification, in turn, stems from the real nature of all the x-space components of the integrand. Minimizing Expression (10) is equivalent to asking that the physical solution conform to the Wiener inverse-filter estimate in the sense of minimum mean-square error *after* suitable weighting of the positive solution to ensure best conformance at frequencies of greatest certainty.

A positive, physically realizable spectrum can be represented as the inverse Fourier transform of an autocorrelation function. This may be verified by noting that such a spectrum $\hat{o}^+(x)$ can be represented as the modulus of a complex function $g(x)$:

$$\hat{o}^+(x) = |g(x)|^2 = g^*(x)g(x). \tag{11}$$

By the convolution theorem, its Fourier transform is

$$\hat{O}^+(\omega) = G^*(-\omega) \otimes G(\omega) = \int_{-\infty}^{\infty} G^*(\omega')G(\omega' + \omega)\, d\omega' = G \star G, \tag{12}$$

where G is the Fourier transform of g and $G \star G$ the autocorrelation of G as defined by Eq. (12). Schell (1965) found that an analytic function $G(\omega)$, and hence $\hat{o}^+(x)$, minimizing expression (10) may be obtained by the iterative procedure

$$G^{(k+1)}(\omega) = \text{const} \times \int_{-\Omega_p/2-\omega}^{\Omega_p/2-\omega} B_W^2(\omega')$$
$$\times [Y(\omega')I(\omega') - G^{(k)}(\omega') \star G^{(k)}(\omega')]$$
$$\times G^{(k)}(\omega + \omega')\, d\omega', \tag{13}$$

where $G^{(k)}$ and $G^{(k+1)}$ are the kth and $(k + 1)$th approximations to G, respectively.

The solutions illustrated in Schell's original publication were indeed entirely positive and showed some resolution improvement over the inverse-filter estimates. The improvement in these examples was not, however, as great as we have come to expect from the best of the newer methods and may not in fact demonstrate the method's real potential. The method does bring with it in a very explicit way, however, the idea that the Fourier spectrum may be extended, on the basis of a knowledge of positivity. Previous studies had focused on the finite extent constraint to achieve this objective.

VI. Jansson's Method

This method was the first to take advantage of both lower and upper bounds. It was also probably the first to demonstrate the real power of positivity—by showing truly substantial resolution improvement, even in the presence of noise that would cripple alternative methods.

Just as others who have used linear methods, this author was disappointed to note the appearance of spurious nonphysical components when he applied the linear relaxation methods (Chapter 3) to the deconvolution of infrared spectra. Infrared absorption spectra, and other types of spectra as well, must lie in the transmittance range of zero to one. Spurious peaks appeared to nucleate on specific noise fluctuations in the data and grow with successive iterations, even though the mean-square error

$$\Phi = \int [i(x) - s(x) \otimes \hat{o}^{(k)}(x)]^2 \, dx$$

became smaller and smaller. Indeed, negative-area contributions from nonphysical parts of the solution served to counterbalance the positive area erroneously allocated in the physical region. This insight and others were gleaned from the work of Ioup and Thomas (1967), who followed a similar line of reasoning.

Ioup and Thomas had favorable experience with setting nonphysical parts of the solution to zero. Thus, the author also considered forcing the solution to lie within the physical bounds. Besides eliminating the objectionable nonphysical result, this approach, it was hoped, would also improve the accuracy of the solution within the physical bounds. Because of the limited performance evidenced in previous deconvolution literature, the author was unprepared for the magnitude of the improvement that resulted. In order to obtain this result, however, more was required than simply setting the nonphysical parts to zero.

A. INITIAL ATTEMPT

The nonlinear iterative methods described here are based on the linear relaxation methods developed in Sections III.C.1 and III.C.2 of Chapter 3. Initially, the correction term was set equal to zero in regions where $\hat{o}^{(k)}$ was nonphysical. To illustrate this, we may rewrite the point-simultaneous equation [Chapter 3, Eq. (26)] with a relaxation parameter that depends on the estimate $\hat{o}^{(k)}$:

$$\hat{o}^{(k+1)} = \hat{o}^{(k)} + r[\hat{o}^{(k)}][i - s \otimes \hat{o}^{(k)}]. \tag{14}$$

For clipping or simple truncation of the nonphysical part, we define the relaxation function

$$r[\hat{o}^{(k)}] = \begin{cases} r_0 & \text{for } \hat{o}^{(k)} \text{ physical (in this case } 0 \leq \hat{o}^{(k)} \leq 1) \\ 0 & \text{elsewhere.} \end{cases}$$

It is generally desirable to choose r_0 large so that large corrections will be made to $\hat{o}^{(k)}(x)$, allowing it to approach $o(x)$ with the smallest possible amount of computation. Let us assume that r_0 is chosen fairly large, so that some initial corrections will be overcorrections. This procedure is normal in the linear relaxation method. The overcorrections typically "damp out" with continued iteration and a solution results. If r_0 is chosen too large, however, the method fails to converge. Either trial and error, or theorems concerning the eigenvalues of s serve to establish the optimum r_0.

With the present definition of r, however, an overcorrection that would normally disappear gradually through ensuing iterations results in a value of $\hat{o}^{(k)}(x)$ that vanishes for *all* subsequent iterations. This behavior occurs because further corrections to that value are prohibited. To use the method, the investigator is compelled to take small values for r_0. Even in this case, erroneously nonphysical values of $\hat{o}^{(k)}$ that have been forced to zero are never allowed to return to the finite range that might better represent the true spectrum $o(x)$. *This form of the method therefore demands excessive computation and yields a solution that, although physically realizable, is not the best achievable estimate.*

B. AN EFFECTIVE METHOD AND ITS FIRST APPLICATION

Instead of clipping the correction term the author introduced the concept of a relaxation function. This function modulates the correction term so that it assumes full value in the center of the physical region (transmittance = 0.5) and falls to zero at the amplitude bounds. The author conceived this approach in response to the need for a means of gently coaxing the nonphysical part back toward the bound. Previous experience had shown that noise nucleated nonphysical peaks that grew with successive iteration. Would it not be possible just to reverse the sign of the correction for these nonphysical components? To refine the idea further, we considered that larger nonphysical elements require larger "reverse" corrections. Accordingly, we adopted a functional form for r that varied linearly from a maximum value at transmittance = 0.5 to zero at the bounds, and negative beyond. Specifically,

$$r[\hat{o}^{(k)}(x)] = r_0\left[1 - 2|\hat{o}^{(k)}(x) - \tfrac{1}{2}|\right]. \tag{15}$$

This relaxation function is shown in Fig. 1. It yielded what was perhaps the first deconvolution result that demonstrated the real power of the physical-realizability constraint, and the first to make use of both upper and lower bounds.

In its first application, this relaxation function gave an excellent estimate *after only 10 iterations.* Figure 2 illustrates the result obtained from processing the Q branch of N_2O found at about 2798 cm^{-1} in the infrared. This spectrum is a natural resolution target when the sample gas pressure is low; the ever more-closely spaced lines in the band head pose the challenge. Two small "hot-band" lines are clearly resolved, as are the most closely spaced lines in the band head. Some of the spectral-line shapes are flat topped in absorptance or transmittance, as one would expect for these strongly absorbed, Doppler-broadened lines. (Remember that inherently Gaussian shape prevails only in the absorption coefficient regime; see Chapter 2, Section I.E.)

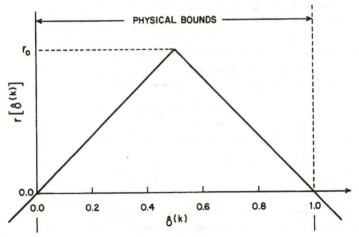

Figure 1   Relaxation function $r[\hat{o}^{(k)}(x)]$. Note the importance of extending $r[\hat{o}^{(k)}(x)]$ to negative values beyond the bounds.

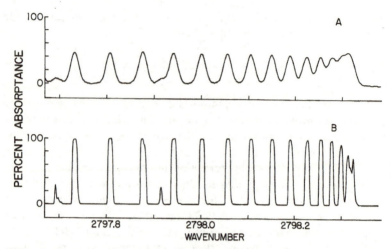

Figure 2 Deconvolution of the N_2O infrared Q branch at 2798 cm^{-1} by Jansson's method: (A) raw data, (B) restoration. Flat-topped lines in the restoration faithfully represent the saturated absorption present (Jansson, Hunt, and Plyler, 1970; reprinted with permission).

For this work, the spectrometer function $s(x)$ was determined by the method outlined in Section II.G.3 of Chapter 2. In digitizing the data, a sample density was chosen to accommodate about 70 samples taken across the full width at half maximum of $s(x)$. A 25-point cubic polynomial smoothing filter was used in the deconvolution procedure to control high-frequency noise. Instead of the convolution in Eq. (14), the point-successive modification described in Section III.C.2 of Chapter 3 was employed. In Eq. (27) of Chapter 3, we replaced κ with the expression

$$\kappa\left[\hat{o}_n^{(k)}\right] = \kappa_0\left(1 - 2|\hat{o}_n^{(k)} - \tfrac{1}{2}|\right). \tag{16}$$

This resulted in the iterative procedure described by

$$\hat{o}_n^{(k+1)} = \hat{o}_n^{(k)} + \frac{\kappa\left[\hat{o}_n^{(k)}\right]}{[s]_{nn}}\left\{i_n - \sum_{m<n}[s]_{nm}\hat{o}_m^{(k+1)} - \sum_{m\geq n}[s]_{nm}\hat{o}_m^{(k)}\right\}. \tag{17}$$

In this equation i_n is the nth element of the sampled image or observed spectrum, $\hat{o}_n^{(k)}$ and $\hat{o}_n^{(k+1)}$ are the nth elements of the kth and $(k + 1)$th estimates of the object or true spectrum, and s is the matrix characteristic of instrumental spreading. The quantity κ_0 is a constant. Like any other method that drives the error $i - s \otimes \hat{o}$ toward a minimum, this procedure tends to conserve the area under impulsive objects. That is, deconvolved spectral lines contain the same area as their counterparts in the raw data.

In the first application of this method we adopted the following procedure. The data were smoothed once and used as $i(x)$. The initial estimate $\hat{o}^{(0)}(x)$ was taken to be $i(x)$ with one additional pass of smoothing. Estimates $\hat{o}^{(1)}$ through $\hat{o}^{(5)}$ resulted from using the relaxation function of Eq. (16) with the iteration of Eq. (17), taking $\kappa_0 = 0.1$ and smoothing $\hat{o}$ between iterations. Estimates $\hat{o}^{(6)}$ through $\hat{o}^{(10)}$ (the final estimate) were obtained with $\kappa_0 = 0.2$ and no smoothing. The original publications (Jansson, 1968, 1970; Jansson, Hunt, and Plyler, 1968, 1970) may be consulted for additional experimental detail.

Shortly after the original work, Barnes *et al.* (1972) applied the method to the measurement and analysis of the ν_3 band of methane. Their analysis predicted virtually every detail of the observed spectrum and verified many details of the deconvolved spectrum that were not apparent in the observed data.

C. SAMPLING, BANDWIDTH EXTRAPOLATION, AND THE RELAXATION FUNCTION

The Fourier frequency bandpass of the spectrometer is determined by the diffraction limit. In view of this fact and the Nyquist criterion, the data in the aforementioned application were oversampled. Although the Nyquist sampling rate is sufficient to represent all information in the data, it is not sufficient to represent the estimates $\hat{o}^{(k)}$ because of the bandwidth extension that results from information implicit in the physical-realizability constraints. Although it was not shown in the original publication, it is clear from the quality of the restoration, and by analogy with other similarly bounded methods, that Fourier bandwidth extrapolation does indeed occur. This is sometimes called superresolution. The source of the extrapolation should be apparent from the Fourier transform of Eq. (14) with $r(x)$ specified by Eq. (15).

It is seen that the dependence of r on x is determined by $\hat{o}^{(k)}(x)$, which changes as k increases. Consider $r_x^{(k)}(x) = r[\hat{o}^{(k)}(x)]$ and define $R^{(k)}(\omega)$ as its Fourier transform. The Fourier transform equivalent of Eq. (14) is then

$$\hat{O}^{(k+1)}(\omega) = \hat{O}^{(k)}(\omega) + R^{(k)}(\omega) \otimes [I(\omega) - \tau(\omega)\hat{O}^{(k)}(\omega)]. \quad (18)$$

Although the bracketed term contains no frequencies beyond the band limit Ω, convolution with $R(\omega)$ ensures the presence of such components.

To carry the analysis further, we need to specify the form of r so that the explicit dependence on x can be determined through its connection with $\hat{o}^{(k)}(x)$.

In the case of a lower bound only, we may select the simple relaxation function $r[\hat{o}^{(k)}] = r_0 \hat{o}^{(k)}$ so that

$$\hat{o}^{(k+1)} = \hat{o}^{(k)} + r_0 \hat{o}^{(k)}[i - s \otimes \hat{o}^{(k)}], \qquad (19)$$

which has the Fourier transform

$$\hat{O}^{(k+1)} = \hat{O}^{(k)} + r_0 \hat{O}^{(k)} \otimes [I - \tau \hat{O}^{(k)}]. \qquad (20)$$

This expression tells us that successive estimates $\hat{O}^{(k)}$ will have nonzero values for $|\omega| \geq \Omega$.

Although we have not proved that $\hat{O}(\omega)$ approaches $O(\omega)$ for $\omega > \Omega$, experimentation tends to bear this out (see Sect. XII.C.4). This is entirely consistent with current knowledge of the general properties of constraints applied to various deconvolution methods (Biraud, 1969; Frieden, 1972; Howard, 1981a, b).

Other choices are possible for the relaxation function r. For these choices, expressions similar to Eq. (20) may be derived. Although it is awkward to deal with the absolute value in Eq. (15), a parabolic form meeting the principal requirements is more readily expressed in Fourier space. A simple quadratic (Wilson, 1973) and its generalization to other powers have been used by Blass and Halsey (1981):

$$r[\hat{o}] = 4r_0[\hat{o}(1 - \hat{o})]^n. \qquad (21)$$

Here n is an integral exponent. Frieden (1975), on the other hand, generalizes the original relaxation function [Eq. (15)] to treat the case of arbitrary upper bound B and lower bound A:

$$r[\hat{o}] = r_0 \Big[1 - 2(B - A)^{-1} |\hat{o} - (A + B)/2| \Big]. \qquad (22)$$

Maitre (1981) has experimented with trapezoidal forms and powers of expressions involving the cosine. Jansson and Davies (1974) have used both exponential and piecewise linear functions in versions adapted to emission spectra (Chapter 8, Section III.B).

Xu et al. (1994) noted that letting $r[\hat{o}^{(k)}] = \hat{o}^{(k)}/(s \otimes \hat{o}^{(k)})$ in Eq. (14) converts the Jansson method to Gold's method,

$$\hat{o}^{(k+1)} = \hat{o}^{(k)} + \frac{\hat{o}^{(k)}}{s \otimes \hat{o}^{(k)}}[i - s \otimes \hat{o}^{(k)}] = \hat{o}^{(k)}\frac{i}{s \otimes \hat{o}^{(k)}}. \qquad (23)$$

Constrast this form with that used by Jansson and Davies for emission spectra,

$$\hat{o}^{(k+1)} = \hat{o}^{(k)} + \hat{o}^{(k)}[i - s \otimes \hat{o}^{(k)}]. \tag{24}$$

Both variants enforce positivity but not an upper bound.

A linear iterative method seeks an object estimate consistent with image data. Usually, noise precludes perfect consistency. Noise helps ensure the problem's ill-posedness: there is no true solution. Nevertheless, in futility, the method attempts to satisfy noise by driving estimate $\hat{o}^{(k)}$ further and further into physically impossible regions. The Jansson method reverses, at the bound, the direction in which the correction drives. It does so by reversing the sign of the correction term. The further a value lies outside the bound, the greater is the drive to return to the physically possible region. The destructive correction mechanism of the linear method is thus converted to enforcement of the bound rather than to its violation.

Though no estimate, especially one that satisfies other constraints, can truly satisfy the data, there are many estimates that *nearly* do so. It should be no surprise that shaping relaxation function $r[\hat{o}^{(k)}]$, particularly its form near the physical bound, can bias solution $\hat{o}^{(k)}$ to give it desired characteristics. By this means additional prior knowledge is injected. Xu and Wang (1992) have tailored $r[\hat{o}^{(k)}]$ to favor a particular form of object estimate. Working in infrared spectroscopy, they noted that the relaxation factor of Fig. 1 performed well with Gaussian-shaped Doppler-broadened lines, but was error-prone in restoring the Lorentzian-shaped pressure-broadened lines. From the Lorentzian function, they derived a relaxation function that satisfactorily restored pressure-broadened lines. Clearly, there are opportunities to explore further adaptation of the $r[\hat{o}^{(k)}]$ to favor features desired in $\hat{o}^{(k)}$.

Considerable imagination can be exercised in devising the relaxation function. As noted elsewhere, however, results will be inferior unless care is taken to ensure that $r[\hat{o}^{(k)}]$ does not clip the correction term when $\hat{o}^{(k)}$ goes out of bounds (Section VI.A), and that out-of-bounds $\hat{o}^{(k)}$ values are not clipped between iterations (Section VI.E). We discuss these points further in giving detailed guidance for new applications in Section VI.G.

D. RAPID-CONVERGENCE ADAPTATION

The method of Section VI.B was developed by applying a relaxation-function concept to a traditional limited-capability linear method. Most known variations and improvements upon linear methods benefit from the relaxation-function adaptation. In this section we introduce a rapidly converging algorithm by considering the method having the fastest possible convergence—inverse filtering. First let us assume the availability of an ideal inverse filter. If available, such a filter could extract the true object directly from the image in one step, independent of any prior object estimate:

$$\hat{o}^{(k+1)} = o = y \otimes i. \tag{25}$$

Because the prior estimate is no consequence in this case, we incorporate it in a vanishing term:

$$\hat{o}^{(k+1)} = y \otimes i + (\delta - y \otimes s) \otimes \hat{o}^{(k)}. \tag{26}$$

In this equation, δ is the Dirac δ function. Rearranging, we recognize a variant of Van Cittert's method having an inverse filtered correction term:

$$\hat{o}^{(k+1)} = \hat{o}^{(k)} + y \otimes (i - s \otimes \hat{o}^{(k)}). \tag{27}$$

This variant, however, converges to the true object in only a single step. Convergence can be slowed and adjusted while retaining the linearity of the method by introducing a constant relaxation factor C:

$$\hat{o}^{(k+1)} = \hat{o}^{(k)} + Cy \otimes (i - s \otimes \hat{o}^{(k)}). \tag{28}$$

We find the conditions required for convergence by examining the series that results from repeated application of Eq. (28):

$$\hat{o}^{(n)} = \{\ldots[(\hat{o}^{(0)}(1 - C) + C\hat{o})(1 - C) + C\hat{o}]\ldots\}(1 - C) + C\hat{o}$$

$$= \hat{o}^{(0)}(1 - C)^{(n+1)} + C\hat{o} \sum_{i=0}^{n} (1 - C)^i. \tag{29}$$

In this equation, $\hat{o} = y \otimes i$, the idealized estimate that results from filtering data i with the ideal inverse filter. For present purposes, we also regard the data as ideal—that is to say, noise free. In the infinite-iteration limit we find that the first term vanishes and the summation becomes $\hat{o}$ provided $0 < C < 2$. Over this range of C, we have $\hat{o}^{(\infty)} = \hat{o}$, independent of starting estimate $\hat{o}^{(0)}$.

An ideal inverse filter is, of course, almost never possible for spread functions of practical interest. We do, however, have access to a host of useful approximations, including the Wiener inverse filter (Chapter 3, Section IV.B.), the Moore–Penrose pseudoinverse (Albert, 1981), and the SVD inverse. Maeda and Murata (1984a, 1984b), apparently the first to use this method, employed the singular value decomposition (SVD) inverse as regularized by the technique of Rushforth *et al.* (1982). They also took advantage of bounds via simple clipping. Use of bounds placed the method firmly in the nonlinear category.

A further improvement is possible based on experience with related methods. The advantages of a nonlinear method that uses a nonclipping relaxation function $r[\hat{o}^{(k)}]$ in the manner of Jansson are well documented elsewhere in Section V. Combining the relaxation function with the present accelerated method, we have

$$\hat{o}^{(k+1)} = \hat{o}^{(k)} + r[\hat{o}^{(k)}] \cdot [y \otimes (i - s \otimes \hat{o}^{(k)})]. \tag{30}$$

Agard *et al.* [see Chapter 9, Section IV.A.3, and Agard *et al.* (1989)] confirmed the convergence-rate advantage of inverse filtering the correction term, though they noted some instability. Their application, however, benefited from interleaving accelerated cycles with cycles implementing other formulas. Maeda and Murata interleaved cycles of median-window filtering to suppress impulse noise. Point-successive updates in the manner of Chapter 3, Section III.C.2, and other adaptations are also possible as improvements within the same context.

E. SMOOTHING, REBLURRING, AND OTHER ADAPTATIONS

In the author's original work, data presmoothing was found to be essential for acceptable restoration. Smoothing, however, potentially limits the resolution achievable. When applied to the data (only) before deconvolution, the smoothing function enters as additional and uncompensated spreading. This undesirable effect may be alleviated by including the additional spreading in $s(x)$. This is accomplished by smoothing $s(x)$ before it is used for deconvolution (Richards *et al.* 1979). Kawata *et al.* (1978) have carried the concept a step further in their use of a reblurring method to guarantee convergence with linear iterative methods. [See also Chapter 3, Section IV.D.3, and Kawata and Ichioka (1980b).] Future applications of the present method could possibly benefit from consistent application of prefiltering both $i(x)$ and $s(x)$ with the abscissa-reversed spread function $s(-x)$. We note, however, that in the case of spectroscopy, when an idealized optical spectrometer has equal entrance- and exit-slit spectral widths, the optical part of the transfer function is guaranteed real and positive. This condition, by analysis similar to that in Section IV.D.2 of Chapter 3, assures convergence.

Thomas (1981) and Schafer *et al.* (1981) have also discussed reblurring. In addition, Schafer *et al.* have studied a generalized class of iterative deconvolution algorithms. They examined the convergence properties of the iteration

$$\hat{o}^{(k+1)} = b\hat{o}^{(k)} + r\{i - s \otimes [b\hat{o}^{(k)}]\}, \qquad (31)$$

where b is a constraint operator that, in a simple form, may supply bounds to the solution by clipping. Here, the quantity r may be either a constant parameter or a function of x or $\hat{o}^{(k)}$. The development is based on the contraction mapping theorem of functional analysis and is actually more general than that specified in Eq. (31). The results are therefore not limited to distortions of the convolution type. In addition, constraints of finite extent in x may be applied. It is shown that, under the proper circumstances, convergence to a solution consistent with both data and constraints is guaranteed. However, as noted elsewhere in this chapter, when amplitude constraints are applied to $\hat{o}^{(k)}$ by a simple clipping operator, their full benefits are typically not realized.

Because the Jansson nonlinear method is an outgrowth of the linear ones, it is not surprising that such useful techniques as the reblurring discussed in Chapter 3 may be adapted. Other elements of this Chapter 3 may be borrowed as well, including memory-conserving methods of programming (Section III.C.3) and the use of special sampling methods to gain the rapid-convergence advantage of point-successive iteration without inducing asymmetry (Section III.C.2).

F. APPLICATIONS

In spectroscopy, the line breadth remaining after deconvolution may be influenced primarily by broadening for which compensation was not attempted. Nevertheless, the maximum breadth-reduction ratio attainable when the method is really challenged indicates its potential. The original application (Section VI.B) showed reduction on the order of 2.5 times, limited as described earlier. Subsequent applications have yielded substantially greater improvements (Chapters 7 and 8).

Later chapters detail one-dimensional applications of the Jansson method to electron spectroscopy for chemical analysis (Chapter 8), high-resolution dispersive infrared spectroscopy (Chapter 6), tunable-diode-laser spectroscopy, infrared interferometric spectroscopy, γ-ray spectroscopy, gas chromatography (Chapter 7), and astronomical optical spectroscopy (Chapter 10). Hundreds of one-dimensional applications have appeared throughout the scientific literature in other diverse fields, including mass spectroscopy (Marchetti and Mignerey, 1993), Raman spectroscopy (Wang *et al.*, 1995), chromatography (Schure, 1991; Crilly 1987), particle beam microprobe analysis (Coote and Kwan, 1995), field-flow particle fractionation (Schure *et al.*, 1989), B-dot probe signals for railgaun plasma current distributions (Evans and Smith, 1992), and Hubble Space telescope spectra (R. L. Gilliland, 1990, 1992; see also Chapter 10, Section VI.A). *In these citations authors repeatedly note the method's ease of implementation, robustness versus noise, artifact suppression, accuracy in testing with known objects, rapid convergence, economical computation, and better resolving capability compared with competing methods.*

In astronomy, besides being applied to spectra, the Jansson method has proved itself for restoring images. For example a modified version enabled the discovery of the Halley's comet nucleus (Abergel *et al.*, 1988; Dimarellis *et al.*, 1989). The raw comet images were recorded by the Soviet spacecraft VEGA-1. Figure 3 shows the restoration sequence for an image acquired 7 seconds before closest encounter. Working with such restorations, the authors were able to clearly discern nucleus contours, dust jets, and jet origin locations on the nucleus (Dimarellis *et al.*, 1989). Later, Stooke and Abergel (1991) also report that "the quality of the images was spectacularly improved...". Earth-based observations suffer from limitations of atmospheric seeing. In method comparison tests working with

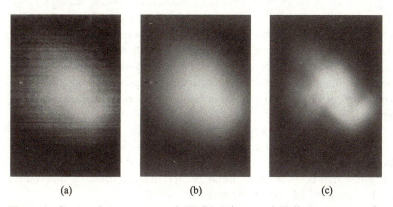

(a)   (b)   (c)

Figure 3 Restoration sequence of VEGA-1 image of Halley's comet nucleus acquired 7 seconds before closest approach: (a) raw data, (b) Fourier filtered to remove coherent noise, and (c) restoration after 17 iterations. Method detailed by Dimarellis *et al.* (1989), (Abergel *et al.*, 1988; reprinted with permission).

earth-based images of Uranus, Neptune, and the M87 jet, Heasley (1984) and Heasley *et al.* (1984) found that the Jansson method produced the sharpest restorations. Navarro *et al.* (1987) restored atmospherically blurred sunspot images with only six iterations. They cited a number of advantages, saying "its power and strength versus noise is great and advises its use."

Other applications of the method in image processing have been described by Kawata *et al.* (1978), Kawata and Ichioka (1980a), Saghri and Tescher (1980, 1984), and Gindi (1981). Mason *et al.* (1992) restored side-scan sonar images. Maitre (1981) investigated the parallelism of noncoherent optical convolution methods combined with video techniques in a totally analog application. Matsuoka *et al.* (1982) developed a hybrid optical-video-digital system based on similar principles. Pipeline processors that perform image convolution and/or Fourier transforms at high speed are ideal for digital applications.

The inherent simplicity of the method suits it to high-speed implementations in both single and multiple dimensions by other hybrid techniques as well. For example, it is possible to design a transversal filter that has all the required nonlinear properties. Even images could be processed. In Section XII, we describe such a filter in the context of neural networks.

G. GUIDELINES FOR USE

Practical experience has shown that simple bounds, where applicable, have a greater effect on the solution than a method's statistical underpinnings. The present method as yet lacks a mathematical proof of convergence. However, in practice it converges to more accurate solutions and does so with less computation than most other methods. It lacks statistical motivation; yet, because of bounds, it often outperforms statistically based methods in applications where bounds are appropriate. The method takes advantage of both upper and lower bounds and, as a bonus, is simple to apply.

Some publications, proposing to compare the present method with others, compare instead a crippled version. Beneficial modifications are both possible and desired, but misapplication based on careless reading of prior research is unfortunate. Below, in approximate order of importance, we itemize and elaborate some caveats:

- Avoid hard-clipping the correction term at the bounds (see Fig. 1).
- Avoid hard-clipping $\hat{o}^{(k)}(x)$ between iterations.
- Filter $\hat{o}^{(k)}(x)$ between iterations early in the process.
- Consider using point-successive sums instead of convolution.

1. Original Procedure

Each experiment is different; noise statistics, $s(x)$, and other factors may vary substantially. A possible starting point for first-time use is the original procedure, which was applied to spectra. To conform to this procedure, in addition to heeding the aforementioned caveats, the user should start with data $i(x)$ having root-mean-square noise of 0.75% of the maximum data range or less. Furthermore, the user should adjust sampling of $s(x)$ and $i(x)$ to a scale of about 70 samples per full width at half maximum (FWHM) of $s(x)$, prefilter $i(x)$ by convolving with a 25-point polynomial, and employ relaxation constant κ_0 values of 0.1 and 0.2 for the first and second five iterations. The same polynomial filter should be used between each of the first five iterations. The spread function $s(x)$ requires normalization to unity area. The original recipe calls for point-successive sums. A careful reading of Section VI.B will be helpful to those wishing to start work with proved conditions.

2. Filtering

The resolving capability of the original recipe, although outstanding, is limited by its overly aggressive filtering of the data, $i(x)$. The user may wish to reduce the filtering, but should be aware that some filtering of $i(x)$ prevents propagation of its noise content into the solution from outside the $s(x)$ passband [in the case of band-limiting $s(x)$]. In restoring the destroyed frequency components of $o(x)$ to achieve superresolution, the relaxation function effects a convolution in Fourier space that sends Fourier components from one frequency to another. The method then modifies them to enforce conformance to both data and constraints. The data should be filtered in a manner that eliminates all components at frequencies beyond the passband. Those components have no value and can only cause harm.

Filtering $\hat{o}^{(k)}(x)$ between iterations regularizes this method by enabling a trade-off between smoothness and data consistency. It also speeds convergence. Reducing the filter extent to pass higher frequencies may be of interest. An initial object estimate $\hat{o}^{(0)}(x)$ free of noise beyond the band limit (if one exists) provides a good start to the iteration. Absence of high-frequency components in the first estimate eliminates them as nucleation sites for artifacts in subsequent estimates $\hat{o}^{(k)}(x)$. Thus, there is an opportunity for high-frequency components' unbiased placement in $\hat{o}^{(k)}(x)$. These components will then conform solely to the requirements of the bounds and data consistency.

3. Sample Density

Note that the original recipe calls for no filtering in the final five iterations. These iterations allow the high-frequency components needed for superresolution to propagate into $\hat{o}^{(k)}(x)$ via the convolution in the correction term (Section VI.C). Although some workers do not employ filtering between early iterations, this author has found it to be more than helpful.

Lower sample density yields a computational benefit. Though this benefit would seem to decrease in importance with each year of hardware improvement, its real effect is to extend the method to ever larger data sets and into real-time applications. There are, however, better arguments for high rather than low-density sampling. First, we anticipate recovering frequency components beyond the band limit of $s(x)$, if it has one. Adequate sampling of these components necessitates sampling band-limited data $i(x)$ at a rate considerably higher than the Nyquist rate established by the band limits. Second, the Nyquist rate represents a theoretical limit; experience in practice reveals that oversampling has advantages in both achieving and displaying superior solutions.

4. Relaxation Coefficients and Successive Summation

Experimentation enables adjustment of the relaxation coefficients κ_0 to their maximum values consistent with nondivergence. Experience has shown that larger values handle noise better. The number of iterations should be adjusted to accommodate convergence-rate variations brought about by changes to the κ_0.

Use of point-successive sums [Eq. (16)] in place of convolution also profoundly affects convergence rate. The point-successive method assures use of refined $\hat{o}^{(k)}(x)$ values at the earliest possible computation stage. A potential disadvantage of point-successive sums is solution asymmetry. This effect can be subdued by adding a few iterations or, better still, altering the computation sequence of the summation terms (Chapter 3, Section III.C.2). Solution quality is usually better with the point-successive approach. Arguments in favor of this approach, where it can be used, outweigh those against it. A valid reason for the alternative, however, is availability of an equivalent Fourier-space process. In Fourier space, computational advantages accrue for large $s(x)$ kernels via the convolution theorem. The cost of abandoning the point-successive approach for this alternative is the need for more iterations and, possibly, reduced solution quality.

5. Clipping

The user should avoid hard-clipping the correction term. *An effective relaxation function* $r[\hat{o}^{(k)}(x)]$ *is the essence of this method.* The method uses it to apply bounds in a smooth and dynamic fashion. The relaxation function $r[\hat{o}^{(k)}(x)]$ should be positive in the range of $\hat{o}^{(k)}(x)$ that represents a valid solution, then continue smoothly to negative values in the undesired range, crossing the $\hat{o}^{(k)}(x)$ axis at zero. It allows gradual correction of $\hat{o}^{(k)}$ by first admitting, then suppressing, out-of-bounds components as iteration proceeds. Hard-clipping, i.e., setting $r[\hat{o}^{(k)}(x)]$ to zero in the undesired range, seriously compromises its value. Though hard-clipping forces the method to suffer many more iterations with smaller values of κ_0, its most serious effect is degradation of the solution $\hat{o}^{(k)}(x)$. Results obtained by misuse that includes clipping abound in the literature. These results are better than those obtained with strictly linear methods, and are even comparable to results possible with some other constrained nonlinear methods. When used with spectral lines, however, clipping introduces unrealistic first-derivative discontinuity at the base line. The quality of images restored through this misuse suffer as well. Not only is solution quality compromised, but many more iterations are required to produce a solution. The excess iterations result because r_0 needs to be small for convergence with clipping-based methods. Advantageous use of the method depends upon propagating refined information between neighboring regions of $\hat{o}^{(k)}(x)$. Clipping the correction term degrades the solution by eliminating the possibility of nonvanishing values over much of the $\hat{o}^{(k)}(x)$ domain early in the computation, before neighboring values are refined. Propagation of refined information should occur in a dynamic way that allows $\hat{o}^{(k)}(x)$ to change anywhere in its domain by responding to refined values of $\hat{o}^{(k)}(x)$ within the "reach" of $s(x)$. Ample flexibility for bounds enforcement exists within the context of a relaxation function that does not clip $\hat{o}^{(k)}(x)$ at the bounds.

In addition to the preceding precautions, the user should also avoid clipping $\hat{o}^{(k)}(x)$ between iterations. Filtering is sufficient. The specific relaxation function $r[\hat{o}^{(k)}(x)]$ of Fig. 1 alone does the work of bounds enforcement. Methods that (1) employ simple clipping and (2) modify $r[\hat{o}^{(k)}(x)]$ to enforce clipping yield inferior restorations.

Many more possibilities exist for improvements within the context of relaxation-function-enforced bounds. Peruse Sections VI.C–VI.E, IX, and X as a further guide.

VII.   Biraud's Method

A.   PROBLEM DEFINITION

Like Schell, Biraud (1969) forced the solution to be positive by using a representation in which negative values are not possible. He chose to represent the solution as the square of a real Fourier series. In the continuous regime, we have

$$\hat{o}^+(x) = [a(x)]^2, \tag{32}$$

where $a(x)$ is real. In Fourier space, we have

$$\hat{O}^+(\omega) = A(\omega) \otimes A(\omega), \tag{33}$$

where $A(\omega)$ is the Fourier transform of $a(x)$. Biraud formulated his method for application when the predominant noise affects the measurement before blurring. This is unlike the situation considered in Eqs. (12) and (39) of Chapter 3. Consistency of solution $\hat{o}^+(x)$ with the observed data in the presence of additive noise of this type requires minimization of

$$E = \int_{-\Omega_c}^{\Omega_c} |Y(\omega)I(\omega) - \hat{O}^+(\omega)|^2 \, d\omega \tag{34}$$

by picking the best function $A(\omega)$. In this equation $\pm\Omega_c$ defines the frequency limits within which consistency is desired. The quantity Ω_c is usually chosen to be slightly smaller than Ω to ensure that the inverse-filter estimate $Y(\omega)I(\omega)$ is not excessively affected by the noise. The discrete representation is central to Biraud's method of solution. Accordingly, we let the finite frequency band of the data be represented by $2J + 1$ samples so that we have $J\Delta\omega = \Omega$, where $\Delta\omega$ is the sampling interval required to specify the object completely. This, in turn, in established by knowledge of the object's finite extent with limits $\pm X$, so that $|x| \leq X$. We set $\Delta\omega = \pi/X$, consistent with the sampling theorem. Let us define $\omega_n = n\,\Delta\omega$ so that the $2J + 1$ samples of the solution over the finite frequency band are given by $\hat{O}^+(\omega_n)$. Now, a function that is nonvanishing only over a finite interval has a self-convolution that is contained in an interval twice as large. In this case we have only $2P + 1$ samples of $A(\omega_n)$, where $P = J/2$.

B. PROCEDURE

Biraud's method starts by considering the centermost value of $\hat{O}^+(0)$ as known, because it is at $\omega = 0$ that the inverse filter gives its most reliable estimate. Furthermore, this assumption guarantees that the total flux in the restoration is the same as that measured. Thus we find

$$\hat{O}^+(0) = Y(0)I(0) = A(\omega) \otimes A(\omega)|_{\omega=0}. \tag{35}$$

The centermost values $A(\omega_{-1})$, $A(\omega_0)$, and $A(\omega_1)$ are considered to be unknown, and all other $A(\omega_n)$ are taken as zero. Three cubic equations for determining these unknowns may be derived by using Eq. (35) as a constraint term and applying Lagrange's method of undetermined multipliers to Eq. (34) with limits $\Omega_c = \Omega$ (Frieden, 1975). The three partial derivatives $\partial/\partial A(\omega_{-1})$, $\partial/\partial A(\omega_0)$, and $\partial/\partial A(\omega_1)$ are set equal to zero, and the resulting equations are then solved. In the next step, $A(\omega_{-1})$ and $A(\omega_1)$ are taken as known, and $A(\omega_{-2})$, $A(\omega_0)$, and $A(\omega_2)$ are sought, all other A's being taken as zero. The process is continued until $A(\omega_{-P})$, $A(\omega_0)$, and $A(\omega_P)$ are obtained.

The entire procedure is now repeated, starting again with the centermost trio, this time using the previous iteration's $A(\omega_l)$ values instead of zeros for all the "outer" A's. For a time, each such cycle through the $A(\omega_l)$ produces ever smaller minima. Eventually, however, saturation is reached and little reduction is noted with continuing iteration. Even in the ideal case of a perfect solution, noise restricts the minimum achievable to

$$E_{\text{res}} = \int_{-\Omega_c}^{\Omega_c} |N_o(\omega)Y(\omega)|^2 \, d\omega, \tag{36}$$

where $N_o(\omega)$ is the Fourier transform of the noise, here considered to affect measurement of the true object or spectrum before blurring. The expected value $\langle E_{\text{res}} \rangle$ of this quantity is known if estimates of the statistics of $N_o(\omega)$ can be obtained. If this expected minimum is reached before saturation, the process is terminated and the solution is taken as the self-convolution of the prevailing $A(\omega_l)$. If saturation occurs before $\langle E_{\text{res}} \rangle$ is attained, however, additional degrees of freedom are needed to satisfy the concurrent conditions of positivity, finite extent, and consistency with the data.

The required degrees of freedom result from allowing the A's *beyond the data band limit* Ω to vary. First $A(\omega_0)$ and $A(\omega_{\pm(P+1)})$ are sought in a full repetition of everything up to this point. Because the error specified by Eq. (34) is evaluated for frequencies less than the band limits $\pm\Omega$, the added degrees of freedom are required within this band. One might well ask how degrees of freedom can be increased within the error computation band $\pm\Omega_c$ by adding coefficients to A for $|\omega| > \omega_P$, when ω_P was chosen to provide $\hat{O}(\omega)^+$ values just up to the band limit Ω, but no further. The answer lies in the mixing of the band-extending A coefficients by the self-convolution operation. Values of $\hat{O}^+(\omega)$ within the band $\pm\Omega$ are in fact formed by products containing the new coefficients of A, thus allowing E to be further reduced.

After coefficients $A(\omega_{\pm(P+1)})$ are determined, the process is continued by adding values of A until the expected residual $\langle E_{\text{res}} \rangle$ is achieved. The procedure is summarized in the accompanying structured flowchart.

Set limit of sought $A(\omega_l)$ to data band limit ($L = P$).
DO until error E is less than expected residual $\langle E_{\text{res}} \rangle$.
    DO until saturation or until $E \leq \langle E_{\text{res}} \rangle$.
        DO for $l = 1$ to sought limit L.
            Fix $A(\omega_m) = 0$ for $m > l$.
            Solve cubic equations for $A(\omega_0)$ and $A(\omega_{\pm l})$.
    Increase solution frequency band ($L = L + 1$).

C. PERFORMANCE

The method employs a gradual increase in frequency beyond the data band limit. High-frequency components are not sought until the best values of low-frequency components are found. Because frequencies are not sought above the lowest needed to satisfy the data, the method is inherently smooth. Furthermore, Biraud's method appears to be the first to have simultaneously utilized both the constraint of positivity and that of finite extent with specific limits, the latter being inherent in the sampling. These facts are probably responsible for the impressiveness of the restoration in the original publication (Biraud, 1969), which is reproduced in Fig. 4.

The original publication, in contrast to the treatment given here, which follows Frieden (1975), described the use of an assumed trial function $A(\omega)$ together with a perturbation method needed to adjust it for consis-

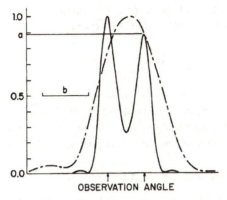

Figure 4 Restoration of Cygnus A astronomical radio source at 11 cm by Biraud's method. Dash-dotted curve, recorded measurement; solid curve, deconvolution; $a = 0.89$, b = beam width at half power. The ordinate is intensity given in arbitrary units. The peak separation is 104 arcsec (Biraud, 1969; reprinted with permission).

tency with the data. Biraud noted empirically that, for simulated test data, the method gave an $\dot{o}(x)$ estimate that was quite accurate regardless of the chosen trial $a(x)$. He also observed that the need for a high signal-to-noise ratio was greatly reduced as compared with the requirements of a linear method. These observations are consistent with those that one obtains in working with all of the better deconvolution methods constrained to physically realizable solutions.

A study and evaluation of Biraud's method was the subject of a thesis by Wong (1971). A method developed by Robaux and Roizen-Dossier (1970) is similar to those of Schell and Biraud in that it seeks to extrapolate $\hat{O}(\omega)$ for $|\omega| > \Omega$ by fitting over $|\omega| < \Omega$ a function the Fourier transform of which must be everywhere positive. More recently, Biraud's method has been applied to plasma-emitted spectra (Lesage *et al.*, 1995).

VIII. Frieden's Method of Maximum Probability

So far, we have asked only that solution $\hat{o}(x)$ satisfy the data plus bounds on amplitude and/or bounds on spatial extent. Although bounds help considerably in reducing the number of possible solutions, we are still left with the possibility that more than one solution may satisfy the data. By satisfying the data, we mean that the expression

$$\Phi = \int (i - s \otimes \hat{o})^2 \, dx \qquad (37)$$

represents a local minimum in the space having coordinates that are the elements of $\hat{o}$. In discrete form, these coordinates are the $\hat{o}(x_i)$. Criteria other than minimum mean-square error may be employed, but the difficulty remains: which of the numerous potential solutions should we pick?

A. *THE MAXIMUM PROBABLE ESTIMATE*

Frieden (1971, 1972) attacked this problem by seeking the solution that has the greatest probability of occurrence. The lower-bound condition [$\hat{o}(x)$ positive] was a natural consequence of the physical random-grain model that he used. This work was unique in that separate solutions were found for $o(x)$ and the noise. In its simplest form, the original method required the data to be all positive. Even though $o(x)$ satisfies this requirement, $i(x)$ often does not, because of noise excursions that, in spectra, may be found in the base-line region. A bias term was added to the data to ensure positivity.

Subsequent research by Herschel (1971), Kikuchi and Soffer (1977), and Frieden (Chapter 11) has refined the concept in a way that provides an explicit and sensible accounting for noise contributions. This work also provides solutions that incorporate a type of prior knowledge not used before. Spectroscopists, for example, may express their bias by proposing a prior spectrum or guess as to what the true spectrum $o(x)$ might look like. Furthermore, they may express their relative confidence in the guess by specifying a probability of occurrence for each value that may be assumed by an element of the estimate $\hat{o}(x)$. Both the prior spectrum and its associated user-conviction probability function may be obtained from past experience by statistical analysis. In Chapter 11 Frieden examines the possibilities of maximum and minimum conviction in connection with the types of prior knowledge available. He even considers the case of user bias, with high user conviction in that bias, but no empirical data. (Can we discard our spectrometers?) The survey of the various limiting cases serves to reveal important properties of the estimates obtained under practical circumstances. We now summarize Frieden's approach and consider one case of interest to the spectroscopist.

In his derivation, Frieden uses the concept of a number-count set $\{n_m\}$, each member representing the number of photons counted in a spectral interval. The total number of photons $\sum_{m=1}^{M} n_m$ is taken as known to be N. In terms of frequencies ν_m, the values of the object spectrum are given by $o_m = n_m/h\nu_m$, where h is Planck's constant. The number of normal modes or degrees of freedom available for occupation by photons of frequency ν_m is labeled z_m. The Bose-Einstein degeneracy factor

$$W_m = (n_m + z_m - 1)!/n_m!(z_m - 1)! \tag{38}$$

is introduced to express the number of ways that n_m photons may be distributed among z_m degrees of freedom. For the case of electrons, the reader may wish to consider the corresponding Fermi-Dirac case. For either case, in the limit of sparsely populated degrees of freedom, the Boltzmann factor results.

Frieden finds the probability of a given number-count set $\{n_i\}$ in the absence of empirical data to be given by an M fold integral:

$$P(n_1, \ldots, n_M)$$
$$= \left(\prod_{m=1}^{M} W_m \right) \int_{\Sigma q_m = 1} \cdots \int q_1^{n_1} \cdots q_M^{n_M} p(q_1, \ldots, q_M) \, dq_1 \cdots dq_M. \tag{39}$$

Here the values q_m represent the prior spectrum and $p(q_1, \ldots, q_M)$ is the known probability of a given spectrum (set of q_m). The number-count-set estimate $\{\hat{n}_m\}$, and hence the object estimate $\{\hat{o}_m\}$, is obtained by maximizing P by the usual methods of finding extrema.

The influence of data i_m and total photon count $N = \Sigma n_m$ is exerted by adding the appropriate constraints to $\ln P$ in proportions controlled by Lagrange multipliers λ_m and μ. The most probable solution $\{n_m\}$ is then found by maximizing

$$\ln P(n_1, \ldots, n_M) + \mu \left(\sum_{m=1}^{M} n_m - N \right) + \sum_{m=1}^{\dot{M}} \lambda_m h \sum_n (n_n \nu_n s_{mn} - i_m). \tag{40}$$

Suppose that the user has the highest possible conviction that his or her prior spectrum is flat. Specifically, let the user have the greatest certainty that each probability q_m has constant value Q. Furthermore, suppose that the degrees of freedom are sparsely populated and that the noise is additive, independent Gaussian, and strict-sense stationary of order 1 having variance σ^2 (Frieden, 1983). Then, maximizing expression (40), with added terms resulting from noise, results in solutions

$$\hat{n}_m = Q \exp\left(-1 - \mu - h\nu_m \sum_n \lambda_n s_{nm}\right), \tag{41}$$

$$\hat{\varepsilon}_m = -\sigma^2 \lambda_m \tag{42}$$

for signal and noise, respectively, expressed in terms of the Lagrange multipliers μ and λ_m. To determine these quantities the solutions are substituted into the data and total-photon-count constraint equations. Accordingly, we have

$$i_m = hQ \sum_n \nu_n s_{mn} \exp\left(-1 - \mu - h\nu_n \sum_k \lambda_k s_{kn}\right) - \sigma^2 \lambda_m, \quad m = 1, \ldots, M, \tag{43}$$

and

$$N = Q \sum_n \exp\left(-1 - \mu - h\nu_n \sum_k \lambda_k s_{kn}\right), \tag{44}$$

where s_{kn} is the spread function.

These $M + 1$ equations in $M + 1$ unknowns μ and $\{\lambda_m\}$ may be solved by the Newton-Raphson method, in which the unknowns are iteratively adjusted until the right and left sides of the equations agree. The object spectrum number-count set $\{\hat{n}_m\}$ and noise $\{\hat{\varepsilon}_m\}$ are then computed by substitution of μ and the $\{\lambda_m\}$ into Eqs. (41) and (42).

B. ENTROPY

In the derivation of the foregoing formulas by the methods used by Frieden in Chapter 11, the first term of expression (40) becomes

$$\sum_m n_m \ln(Q/n_m) = \ln Q \sum_m n_m - \sum_m n_m \ln n_m. \tag{45}$$

The quantity $-\sum n_m \ln n_m$, one term of the expression to be maximized, is the Shannon entropy H as used by Jaynes and familiar to those who have studied thermal physics. For a given set $\{n_m\}$ obeying $\sum_{m=1}^{M} n_m = N$, maximum H is attained when all the n_m have the same value $n_m = N/M$. The requirement that H be maximum, even when other constraints are attached, tends to force the n_m toward this constant value and hence inhibits large excursions. This property of maximum-entropy restorations is certainly desirable.

Otherwise, note that $i = s \otimes \hat{o}$ can be satisfied by *large spurious* fluctuations of neighboring elements n_m provided that they are of opposite polarity, equal in area, and contain only frequencies not passed by s. That is, the smoothing influence of s will average the fluctuations of the n_m, allowing them to develop without affecting the quantity $i - s \otimes \hat{o}$. Consistency with the data is, however, effective in preventing the development of spurious fluctuations having low-frequency content. It is the maximum-entropy requirement that prevents such large excursions generally, including high-frequency behavior.

Thus, although maximum entropy inhibits large excursions, it does not by itself exert a smoothing influence in the sense of suppressing high-*versus* low-frequency components of the n_m. This fact is readily noted in that neither Σn_m nor $\Sigma_m n_m \ln n_m$ depends on the order in which the n_m are placed; they may be freely interchanged without altering these quantities. This property is also desirable, because retrieval of the suppressed, even obliterated, high frequencies is our objective.

When the degree-of-freedom sites are highly occupied ($n_m \gg z_m$), the entropy term becomes $\Sigma(z_m - 1)\ln n_m$. This is a weighted form of the entropy $H_1 = +\Sigma \ln n_m$ that was maximized by Burg (1967, 1975) in his restoration scheme. Burg's entropy also discriminates against large excursions in n_m as described previously, although to a lesser degree (Lacoss, 1971). To guarantee a positive solution, Burg regarded the n_m, and hence $o(x)$, as the square of another function in a manner reminiscent of the approaches of Schell (1965) and Biraud (1969). He also assumed the first two moments of that function to be constant. As in the approaches of Schell and Biraud, inverse-filter estimates were used as data inputs.

A closed-form result involving only matrix solution and substitution operations was obtained. Under certain circumstances, particularly in the low-noise case, application of the Burg method results in spurious resolution or "spontaneous line splitting" (Fougere *et al.*, 1976). One solution to this problem (Fougere, 1977) introduces iteration.

C. CHOOSING THE PRIOR SPECTRUM

A strength of the maximum-probable approach that was not fully exploited in the foregoing analysis is the possibility of specifying the user's bias toward a nontrivial preferred prior spectrum $\{Q_m\}$ *and* the user's conviction about that choice through the probability $p(q_1, \ldots, q_M)$. Through choice of p it is possible to input the predisposition of the spectrum ordinates toward values in the wing and flat peak-top regions of the spectral lines. There is always a suitable p function, because it describes the user's state of knowledge in a given experiment. Examples are given in Chapter 11.

One way that probability p may be estimated is by statistical analysis of previous experimental data. Various spectral models could also be used to specify p. Both regular models such as that of Elsasser and random models are described by Goody (1964). They have long been used in predicting overall band contours resulting from the superposition of the large numbers of lines in atmospheric IR absorption spectra. We choose to illustrate the estimate of p by using a very simple model composed of a single emission line.

We ask what is the probability of a given object spectrum value when the functional form of the spectrum $o(x)$ is known, but its x displacement is unknown. Because of the unknown displacement, the probability distribution at each location is the same. Call this component probability p_c such that

$$p(q_1, \ldots, q_M) = \prod_{m=1}^{M} p_c(q_m). \tag{46}$$

The component p_c may be evaluated by examining a typical spectrum. The desired probability $p_c(q')$ at a given ordinate value q' is just proportional to the "amount" of x for a given "amount" of q in the typical spectrum, or the absolute value of the increment in x corresponding to the increment in q where q has the value q'. Because the value q' may generally occur more than once in the spectrum, a summation is in order. The desired probability is thus $\sum |dx/dq|$, where the summation indicates contributions from all locations at which $q(x)$ crosses q'.

For simplicity, suppose that we have an emission spectrum consisting of a single, predominantly Doppler-broadened line. This spectrum may be taken as an approximation to the case of widely separated and nonoverlapping lines of equal intensity. Again for simplicity, now consider the typical emission spectrum to be continuous, not sampled. Thus $q(x)$ is given by $q(x) = q_0 \exp(-x^2)$ for a typical line, where q_0 is the peak height. We have chosen the abscissa to be measured in either wavelength or wave number relative to a typical line center. For illustrative purposes only, the

x interval for the observation is taken to be $2 \Delta x$. If the line is centered in this interval at $x = 0$, we can never have $q(x) < \exp(-\Delta x^2)$.

The derivative dq/dx is given by

$$dq/dx = -2xq(x) = -2q\sqrt{\ln(q_0/q)}, \tag{47}$$

and the probability of q is given by

$$p_c(q) = C\,dx/dq \tag{48}$$

$$= C[\ln(q_0/q)]^{-1/2}/2q, \tag{49}$$

where C is a normalization constant chosen so that

$$\int_{q_0}^{q_1} p_c(q)\,dq = 1, \tag{50}$$

where $q_1 = q_0 \exp(-\Delta x^2)$. For simplicity we have selected half of the line as representative.

A plot of $p_c(q)$ is shown in Fig. 5. It is clear that consideration of the integral only over $x < \Delta x$ is critical to having a finite contribution to the normalization of $p(q)$ in the base-line region. At the peak(s) of the

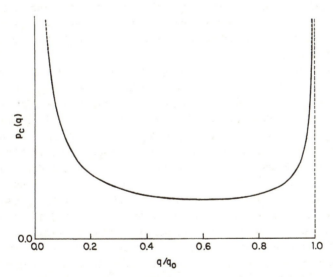

Figure 5 Component probability $p_c(q)$ for a single Doppler-broadened emission line. The ordinate scale depends on the q domain chosen for normalization.

spectral line(s) there is a large contribution due to the flatness of $q(x)$ there. This causes no problem in the evaluation of the integral in Eq. (50) provided that Eq. (48) is used. The value of C displayed implicitly in Fig. 5 is obtained as follows:

$$\frac{1}{C} \int_{q_0}^{q_1} p_c(q) \, dq = x \Big|_0^{\Delta x} = \Delta x, \qquad (51)$$

finally resulting in

$$p_c(q) = [\ln(q_0/q)]^{-1/2}/2q \, \Delta x. \qquad (52)$$

The analysis may be extended to the use of absorption lines, other line shapes, and, most important, spectra comprising many lines of varying intensity whose intensity statistics are available.

D. ADDITIONAL REMARKS

Like the other nonlinear constrained methods, the maximum-entropy method has proved its capacity to restore the frequency content of $\hat{o}$ that has not survived convolution by s and is entirely absent from the data (Frieden, 1972; Frieden and Burke, 1972). Its importance to the development of deconvolution arises from the concept of selecting a preferred solution from the multiplicity of possible solutions on the basis of sound statistical arguments.

Both varieties of maximum entropy have been discussed by Frieden (1975). Frieden's original work was stimulated by an important earlier paper by Jaynes (1968). Frieden's original report (Frieden, 1971) and journal publication (Frieden, 1972) are well worth consulting. Hershel (1971) provided an excellent early treatment of the method. Maximum entropy has been applied to restoration of images of planetary objects (Frieden and Swindell, 1976) and nebulae in the presence of Poisson noise (Frieden and Wells, 1978), as well as data from numerous other fields. Gull and Daniell (1978) have used a computationally efficient version to restore radio maps of galaxies. Kikuchi and Soffer (1977) have contributed significantly to rigorous development of the method from quantum-statistical principles. Their work is summarized, further developed, and adapted to the spectroscopic case in Chapter 11. Donoho et al. (1992) explore the special advantages of maximum entropy and other bounded methods in restoring objects having values near the bound.

IX.   Alternating Projection Methods

This chapter describes methods that facilitate superior restoration through the use of bounds. Sometimes bounds are implemented in the object domain, for example, positivity or finite extent, and sometimes in the Fourier domain, for example, band limitedness. Early work with constraints made the effectiveness of this type of prior knowledge apparent. A number of researchers therefore focused their efforts on making use of all possible combinations of constraints and partial data. What could be more suggestive than direct application of bounds to trial solutions by truncation, first in the object domain and then in the Fourier domain?

A.   EARLY APPLICATIONS IN OPTICS

The author first learned of this approach in a paper given by J. L. Harris (1971), who was concerned with the reconstruction of a three-dimensional microscopic object (the complex refractive index) from optical interference microscope image data. Subsequent to Harris' talk, Gerchberg and Saxton (1972) independently published a similar algorithm designed to solve the "phase problem" of electron microscopy. This was the very same problem that stimulated Gabor's discovery of holography, which, ironically, has never seen much use in its original field of intended application. The problem of reducing speckle in holograms generated by computer provided the motivation for a version of the method disclosed a year earlier in a patent filed by Hirsch *et al.* (1971). Gerchberg (1974) later applied the principle to the problem of obtaining resolution beyond the diffraction limit. Papoulis (1975) analyzed this application with the aid of prolate spheroidal wave functions. Later we shall see that the basic method and its variations can be placed in the context of a powerful general theory, that of alternating projections onto convex sets (POCS).

In the Gerchberg–Saxton application, random numbers constitute an initial estimate of the phase of the emerging wave front. They are combined with the observed image amplitudes and then transformed by fast Fourier transform (FFT). The amplitudes resulting from the transform are discarded. The phases of the transformed function are combined with the observed diffraction plane amplitudes and then inverse transformed. The amplitude of the transformed function is then discarded and its phases are combined with the observed image amplitudes. The cycle of transforming and combining the most recent phase estimate with observed amplitude is repeated until convergence is achieved.

Subsequently, other workers developed numerous variations and generalizations of the basic method. Most methods can be summarized by the accompanying generalized error-reduction algorithm.

> Assume a trial solution (possibly by inverse filtering).
> DO until error is acceptably small.
>
> > Make smallest possible change needed
> > to satisfy constraints and/or data in present domain.
> >
> > Fourier transform.
> >
> > Make smallest possible change needed
> > to satisfy constraints and/or data in present domain.
> >
> > Inverse Fourier transform.

This algorithm has been called an error-reduction algorithm because it was shown by Liu and Gallagher (1974) for a particular case that the error can only decrease with succeeding iterations. Note, however, that the error does not necessarily decrease to zero, or even to the desired minimum residual value that is based on noise. Subsequent research has put convergence conditions on a firm and elegant footing (see following Section IX.C).

For a basic deconvolution problem involving band-limited data, the trial solution $\hat{o}^{(0)}$ may be the inverse- or Wiener-filtered estimate $y(x) \otimes i(x)$. Application of a typical constraint may involve chopping off the nonphysical parts. Transforming then reveals frequency components beyond the cutoff, which are retained. The new values within the bandpass are discarded and replaced by the previously obtained filtered estimate. The resulting function, comprising the filtered estimate and the new superresolving frequencies, is then inverse transformed, and so forth.

Projection algorithms, though frequently guaranteed to converge, often converge with excruciating slowness. J. R. Fienup (1979, 1980) has a similar approach, which he calls the input–output method, that converges more quickly. He compared his method with the earlier algorithms and gradient search methods as well (Fienup, 1982). His conclusion favors the input–output method. Fienup has suggested to Frieden (private communication) that switching strategies between iterations is often effective in improving the rate of convergence because it avoids a saturation or plateau effect that has been noted (Fienup, 1982). Schafer *et al.* (1981) review and analyze constrained iterative algorithms and include the alternating projection methods as a special case.

B. KACZMARZ METHOD

Huang *et al.* (1975) introduced the optics community to a projection method that contemplates each row of the imaging equation [Eq. (1)] as a hyperspace plane. Originally developed as a means of solving sets of linear equations (Kaczmarz, 1937), this method begins by projecting an initial object estimate onto one such plane to establish a new estimate for projection on subsequent planes. The process repeatedly cycles through all the planes corresponding to the complete set of equations. To illustrate, consider the first step of the simplest case of two equations in two unknowns, in which each plane takes the form of a straight line:

$$i_1 = s_{11}o_1 + s_{12}o_2,$$
$$i_2 = s_{21}o_1 + s_{22}o_2. \tag{53}$$

Denoted by a vector from the origin to a point in the hyperspace, an initial object estimate $\vec{o}^{(0)}$ may be projected onto the first plane to establish a next estimate. Using the inner product denoted by the dot, we compute it as

$$\vec{o}^{(1)} = \vec{o}^{(0)} + \left(\frac{i_1 - \vec{s}_1 \cdot \vec{o}^{(0)}}{\vec{s}_1 \cdot \vec{s}_1} \right) \vec{s}_1. \tag{54}$$

This first projection is depicted in Fig. 6. Generally, each vector $\vec{s}_n$ is defined as the nth row of the spread-function matrix s so that the elements of $\vec{s}_n$ are the elements of $[s]_{nm}$, $m = 1, \ldots, M$, where M is the number of columns in the spread-function matrix, here taken as two. Typically the image vector $\vec{i}$ is taken as the initial estimate $\vec{o}^{(0)}$ or, in the notation previously used, $\hat{o}^{(0)} = i$. We readily see that successive projection from one to the other of the two planes of Eqs. (53) finds the intersection of the

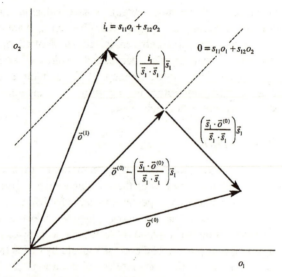

Figure 6 First projection in application of the Kaczmarz method to a two-dimensional system of equations.

planes if the planes do indeed intersect. We express the general case of $(k + 1)$th object estimate for N planes in M-dimensional space as

$$\hat{o}_n^{(k+1)} = \hat{o}_n^{(k)} + \frac{[\mathbf{s}]_{pn}}{\sum\limits_{m=1}^{M} [\mathbf{s}]_{pm}^2} \left\{ i_p - \sum_{m=1}^{M} [\mathbf{s}]_{pm} \hat{o}_m^{(k)} \right\}, \qquad (55)$$

where subscript n denotes the nth dimension of the M-dimensional space, $n = 1, \ldots, M$. The index p is the equation number in the N-equation set, or it denotes the pth plane of N planes. The index $l = 1, \ldots$ specifies the number of the pass through the full set of N equations, so that the superscript $k = N(l - 1) + p - 1$ denotes the sequential number of the object estimate $\hat{o}^{(k)}$.

Noting that the Gauss–Seidel and Jacobi methods (Chapter 3, Section II.C) do not necessarily converge, Tanabe (1971) proved that the Kaczmarz method converges for *any* system of linear equations, even when they are singular and inconsistent. When the equations have a unique solution, the method converges to that solution. Tanabe also noted that, in the case of an overdetermined set, the method converges to a generalized inverse. This result obtains regardless of choice of initial estimate $\hat{o}^{(0)}$. When an infinite number of solutions are possible, the method finds the solution that minimizes the Euclidean distance between it and the initial estimate. Herman *et al.*(1973, 1978, 1981) examined the convergence of the method, generalized and applied to tomographic reconstruction. In computed tomography, the method is considered to be a special case of the algebraic reconstruction technique (ART). It is sometimes also known as the row-action projection method (Podilchuk and Mammone, 1990). Podilchuk (1992a) has also noted that the block-action version is identical to Landweber's method.

Typical in deconvolution is the case of inconsistent equations. In this case the planes do not share at least one common point. Ill-conditioned equations typical of deconvolution problems give rise to near parallelism of numerous planes. Even small uncertainties associated with measured values $\vec{i}$ introduce undesired normal displacements that cause large coordinate shifts of the intersections. Marks (Chapter 14) describes the "greedy limit cycles" that occur when applying projection methods to such cases. In these cases, once "converged," successive projections running through the equation set serve to alter the estimate, but return it to the same "converged" coordinates with each completed pass of N projections.

Bodewig (1956) had observed that convergence is speeded by strengthening the diagonal coefficients. This observation is consistent with experience from other relaxation methods such as those of Jacobi, and Gauss and Seidel, in which diagonal dominance is also desirable. In the Kaczmarz method, diagonal dominance implies a set of planes that tend toward orthogonality. Ramakrishnam *et al.* (1979) further pursued this idea by sequentially orthogonalizing the successive equations before beginning the iterated projections. They noted a 40% computational savings. Podilchuk and Mammone (1990) saved computations by using a three-step method to accelerate the estimate toward the intersection between each set of two planes. More significantly, they noted that the Kaczmarz method is a special case of POCS, or projections onto convex sets. This observation invokes a body of powerful theory for analyzing this method.

C. CONVEX SETS

Although the Kaczmarz method is linear, we introduced it in the present chapter because it readily adapts a variety of nonlinear constraints. Furthermore, it serves as a convenient descriptive bridge to the more general theory of projections onto convex sets. Besides presenting useful technique for proving convergence of existing methods, POCS provides a means of modifying such methods to assure or speed convergence, and yields insight that can enable discoveries.

Originally developed by Bregman (1965) and Gubin *et al.* (1967), POCS was popularized and analyzed by Youla and Webb (1982). Sezan and Stark (1982) applied it to imaging problems. Its elegant theory focuses on a vivid geometrical concept: a set is convex if every line segment with endpoints in the set is totally subsumed in that set. Given a number of intersecting sets, sequential projection of a solution onto each of the sets in repeated cycles converges to a point in the intersection of the sets, provided that the intersection is not empty. An object known as belonging to two or more convex sets can be estimated by such a projection sequence. Examples of convex sets include the set of all band-limited objects, the set of all positive objects, the set of all objects having a known maximum value (upper bound), and the set of all objects having finite extent. In the Kaczmarz method, each equation defines a plane that may be viewed as a convex set. Cheung *et al.* (1988) proved that Howard's minimum-negativity algorithm (Chapter 13) is also a special case. Trussell and Civanlar (1984) demonstrate the use of both hard constraints, and statistical ones as well, in obtaining feasible solutions in a spectroscopy application. A substantial portion of an excellent volume edited by Stark (1987) is devoted to POCS. Marks provides a concise and clear tutorial development of POCS in Chapter 14 of the present work.

Alternating projections have been analyzed in the electron microscopy application by Ferwerda and colleagues [see references cited by VanToorn and Ferwerda (1977)]. Convergence properties, uniqueness of solutions, and generation of the algorithm have been investigated with prolate spheroidal wave functions, the properties of Hilbert spaces, and Von Neumann's alternating projection theorem. These developments appear in papers by Gerchberg (1974), DeSantis and Gori (1975), Papoulis (1975), Youla (1978), Sato *et al.* (1981), Lent and Tuy (1981), Fiddy and Hall (1981), Montgomery (1982), Marks (1981, 1982), Stark *et al.* (1981a, b, 1982), Cahana and Stark (1981), and Marks and Smith (1981a). Rushforth and Frost (1980) analyzed and compared these algorithms by using prolate spheroidal wave functions. They also introduced a version that extrapolates the bandwidth by a small explicit increment for each successive iteration. In this sense, their version is similar to Biraud's method.

Projection methods have been a valuable tool for researchers interested in "signal recovery and synthesis with incomplete information and partial constraints." This subject has been the focus of a triannual series of Optical Society of America conferences of similar name starting in 1983. Special issues of the Journal of the Optical Society of America contain the major conference contributions that should have value to readers interested in these and related topics; see references by Fienup (1983), Fienup and Rushforth (1987), Fienup and Gallagher (1990), and Gallagher and Sweeney (1993).

X.   Howard's Methods

In previous sections we have seen how constraints of boundedness have enabled recovery of frequencies beyond the band limit of the observing instrument. Starting with the inverse-filter estimate

$$\hat{o}(x) = y(x) \otimes i(x), \tag{56}$$

Howard (1981a, b) asked precisely what additional frequency components were required to ensure a physically acceptable result $\hat{o}_p(x)$. Specifically, he represented this result in the following way:

$$\hat{o}_p(x) = \hat{o}(x) + \hat{o}_e(x), \tag{57}$$

where $\hat{o}_e(x)$ is the bandwidth-extension function that supplies the needed but missing components. In Howard's method, only the most-certain components of the inverse-filter estimate $\hat{o}$ are used, say, for frequencies ω such that $|\omega| < \Omega_H$. Cutoff Ω_H may be Frieden's optimum processing bandwidth (Frieden, 1975) or it may be chosen by other means. It could be as large as band limit Ω, but is usually less because of relative suppression by $\tau(\omega)$ near Ω.

Solutions for objects of finite extent are obtained through the familiar minimum mean-square-error criterion by selecting the Fourier components that minimize

$$\Phi = \int [\hat{o}(x) + \hat{o}_e(x)]^2 \, dx, \tag{58}$$

where the integral covers the region outside the finite object extent.

In the practical numerical case $\hat{o}(x)$ and its transform $\hat{O}(\omega)$ are both sampled. Thus, setting equal to zero the partial derivatives of Φ with respect to the Fourier coefficients results in a linear set of normal equations that can easily be solved in closed form. The FFT is then used to obtain the sampled $\hat{o}_e(x)$ from its Fourier coefficients. Estimate $\hat{o}_p(x)$ results from combining $\hat{o}_e(x)$ and $\hat{o}(x)$ via Eq. (57). Only those coefficients are sought that are needed to produce an acceptable solution. By ignoring components of very high frequency, the ill-posed problem of deconvolution is regularized (Rushforth *et al.*, 1982) through smoothing. Note that owing to the retention of only a finite number of Fourier coefficients, the method does not produce $\hat{o}_p(x) = 0$ for x outside the known finite object extent. The mean-square value of $\hat{o}_p(x)$ is, however, the smallest that it can be with a finite number of Fourier coefficients.

Although restorations by this method are superior to those obtained by inverse filtering alone, it should be no surprise that they do not show the dramatic improvements obtainable through the positivity constraint. When, in fact, the integration limits of Eq. (58) are taken well outside the true extent of the object, very little improvement is noted. The applied finite-extent bound must literally butt against the true object for greatest effect. How, then, can the more powerful and useful constraint of positivity be incorporated?

Howard (1981b) attacked this problem by considering the minimization of

$$\Phi = \int_{-\infty}^{\infty} \left[\hat{o}_p H(-\hat{o}_p) \right]^2 dx. \tag{59}$$

Here H is the Heaviside step function. Physical realizability now implies positivity for $\hat{o}_p$. With this modification, only negative values of $\hat{o}_p$ contribute to the integral. Unfortunately, Φ is no longer differentiable. The restriction on Φ expressed in Eq. (59) inhibits derivation of the least-squares normal equations. Instead, Howard considered the Heaviside step as a limit:

$$H(-\hat{o}_p) = \lim_{K \to \infty} \left[1 - \exp(K\hat{o}_p) \right]^{-1}. \tag{60}$$

By making the substitution for H and after some algebra taking the limit, he obtained a set of simultaneous nonlinear equations for the Fourier coefficients. Howard solved these equations by an efficient iterative method. His solutions are comparable to the best that we have come to expect from the positive-constrained restoring methods.

Note that other constraints may be added. Finite extent is particularly easy. Combining the two ideas in this section is done merely by using the finite-extent limits with the integral in Eq. (59).

Howard has applied his method to high-resolution infrared spectroscopic data obtained by dispersive techniques and to both experimental and simulated Fourier interferograms. The method in the latter application explicitly renders the data as they would be observed by an interferometer having a path difference exceeding the mechanical limits of the instrument used for the observation. Details on both the method and its application constitute Chapters 12 and 13.

Zhou and Rushforth (1982) and Rushforth *et al.* (1982) have explored Howard's finite-extent-bounded method via matrix techniques. Zhou and Rushforth demonstrated the effectiveness of utilizing the maximum amount of prior knowledge. Cheung *et al.* (1988) showed that the minimum negativity variation is a special case of POCS.

XI.   Monte Carlo Methods

We first mentioned the applicability of optimization (minimization) methods in Section V.C of Chapter 1. Constraints pose no particular problem to many of these methods. It would seem that the deconvolution problem with object amplitude bounds should be a straightforward application. The most general case, however, deals with each sampled element $\hat{o}_m$ of the estimate as a parameter of the objective function and hence the solution. Excessive computation is then required. The likelihood is great that only local minima of the objective function Φ will be found. Nevertheless, the optimization idea may be teamed with a Monte Carlo technique and a decision rule to yield a method having some promise.

A.   *A DECISION RULE*

The decision-rule approach tried by Frieden (1974, 1975) will serve to introduce the concept of an object built up from grains. In this approach, *both $\hat{o}$ and x* are taken to be discrete. That is, at a particular value of independent variable x_m, we permit $\hat{o}$ to be only an integral multiple of $\Delta\hat{o}$, the grain size. We may consider the use of random numbers to select locations for grain placement. A decision rule provides the basis for acceptance or rejection of a given grain. Frieden's decision rule places the tth grain of intensity increment $\Delta\hat{o}$ at location x_p if it results in a spectrum estimate $\hat{\imath}^{(t)}$ such that

$$\hat{\imath}^{(t)}(x_m) \leq \rho i(x_m) \qquad \text{for all } m. \qquad (61)$$

The estimate is appropriately incremented by the relation

$$\hat{\imath}^{(t)}(x_m) = \hat{\imath}^{(t-1)}(x_m) + \Delta\hat{o}s(x_m - x_p). \tag{62}$$

The factor ρ is understood to have the smallest possible value permitting grain allocation to *any* solution element. The object is updated at location x_p (only) via

$$\hat{o}^{(t)}(x_p) = \hat{o}^{(t-1)}(x_p) + \Delta\hat{o}. \tag{63}$$

Because ρ is made to be minimum, the object tends to be built up in its brightest and broadest regions first. The grains that form the tails are placed last. Because the requirement on ρ assures allocation of the grain where it is *most* needed, chance really does *not* play a role in the use of the decision rule.

Other decision rules are, of course, possible. As noted earlier, local minima in the objective function surface tend to trap optimization algorithms into globally nonoptimum solutions. When an algorithm is trapped, small perturbations in the estimate elements $\hat{o}(x_n)$ only raise the value of the objective function $\Phi[\hat{o}(x_1), \ldots, \hat{o}(x_M)]$. In conventional optimization algorithms, large perturbations are sometimes used in an attempt to jump from the trap. A perturbation is accepted only if diminished Φ results. A particular Monte Carlo approach having strong parallels in statistical physics goes against this basic philosophy. The principles are detailed in the following section.

B. ANNEALING

If we increment solution $\hat{o}$ grain by grain and if the algorithm is trapped in a local minimum, it is impossible for a single grain to produce the jump needed to yield a reduced Φ. This barrier accrues from the definition of a grain as a small increment, and the concept of a local minimum having finite breadth. A means of provisionally accepting grains that *raise* Φ is needed.

When machine computation was still young, Metropolis *et al.* (1953) developed an algorithm that overcame the local-minimum trapping problem encountered in the search for the equilibrium configuration of a collection of atoms at a given temperature. This algorithm accepted steps that both decreased and *increased* the hamiltonian. Upward steps were accepted with a probability that was given by the Boltzmann factor: the higher the temperature, the more likely the acceptance. Downward steps were always accepted. After many computer cycles at a given temperature, the system evolved toward thermal equilibrium. Its macroscopic parameters had assumed the Boltzmann distribution appropriate to the specified temperature.

Annealing is a physical process by which one gradually lowers the temperature of a system until its parameters "freeze" into their final values. Kirkpatrick *et al.* (1983) have simulated this process by starting with a "melted" system and stepwise-lowering the temperature in the algorithm of Metropolis *et al.* (See also *Physics Today*, May 1982, pp. 17–19.) They applied this method to finding the minima of "hamiltonians" describing problems of optimum integrated circuit design and computer wiring, and to the classic traveling salesman problem. Smith *et al.* (1983) have used the method to attack the problem of reconstructing radio-labeled objects imaged in two projections by aperture arrays.

We propose here to apply the technique to the bounded deconvolution problem. Trial grain positions may be selected at random or by some other process, such as bit-reversed sampling (Allebach, 1981; Deutsch, 1965). The trial solution after t grains have been placed is $\hat{o}_m^{(t)}$. The corresponding objective function may be, but is not limited to, a sum of squares:

$$\Phi^{(t)} = \sum_n \left[i(x_n) - \sum_m s(x_n - x_m)\hat{o}_m^{(t)} \right]^2. \tag{64}$$

The influence of trial grain t at location x_p is assessed by evaluating the change in Φ. We may express the object change at an arbitrary location x_m by the relation

$$\hat{o}^{(t)}(x_m) = \hat{o}^{(t-1)}(x_m) + \Delta\hat{o}\delta_{mp}, \tag{65}$$

where δ_{mp} is the Kronecker delta. This relation may be employed with the definition of Φ given in Eq. (64) to evaluate the change

$$
\begin{aligned}
\Delta\Phi &= \Phi^{(t)} - \Phi^{(t-1)} \\
&= \Delta\hat{o}\left\{ \sum_n \left[s(x_n - x_p) \right]^2 \right\} - 2\,\Delta\hat{o}\left[\sum_n i(x_n)s(x_n - x_p) \right] \\
&\quad + 2\Delta\hat{o}\left[\sum_n s(x_n - x_p)\hat{i}^{(t-1)}(x_n) \right]. \tag{66}
\end{aligned}
$$

In the last term we identify

$$\hat{\imath}^{(t-1)}(x_n) = \sum_m s(x_n - x_m)\hat{o}_m^{(t-1)}. \tag{67}$$

In optimizing the computation we note that only this term has a quantity that changes with iteration. Furthermore, $\hat{\imath}$ needs to be updated only over the range of finite s via

$$\hat{\imath}^{(t)}(x_n) = \hat{\imath}^{(t-1)}(x_n) + s(x_n - x_p)\Delta\hat{o}. \tag{68}$$

Having now a value for $\Delta\Phi$, we may express the "Boltzmann factor" as $\exp(-\Delta\Phi/kT)$. The "Boltzmann constant" k is given the units of $T/\Delta\Phi$. In the manner of Kirkpatrick *et al.*, we start with T large, a "melted" system. If a trial grain produces a reduction in Φ and if it does not give rise to a constraint violation, it is unequivocally accepted. If Φ grows, the grain's probability of acceptance is $\exp(-\Delta\Phi/kT)$, contingent on the value of a random number.

In one possible implementation, the temperature may be held fixed at temperature T until little further overall reduction in Φ is observed. The objective function will fluctuate about an equilibrium value Φ_T. (Note that, for this process to work, grains must have both positive *and negative* values.) The temperature may then be repeatedly dropped in fixed increments, this process being repeated according to an annealing schedule. Alternatively, very small decrements in T may be made with each trial grain. At some point, lowering T will provide no further reduction of Φ. This occurrence indicates that $\hat{o}$ is in the "ground state"—that we have found the $\hat{o}$ best accommodating both data and constraints. How are constraints implemented? Very simply, a grain is rejected whenever it would give rise to a violation.

Like its cousin the genetic algorithm (Goldberg, 1989), annealing is computationally intensive. Both algorithms are best suited to combinatorial optimization problems, in which a continuous cost-function surface is not available for computing derivatives. Nevertheless, annealing's inherent simplicity and flexibility are appealing. It would seem that there is much room for development. Many interesting questions can be posed. Would varying the grain size be useful? Perhaps large grain sizes would be useful in the beginning, at high temperatures, to approximate the solution while economizing on computation. What annealing schedules are best? How can the rich parallel with statistical mechanics be best exploited? Some of these questions were addressed in the original work by Kirkpatrick *et al.* (1983); others remain unexplored.

XII. Neural Networks

As early as 1890, the noted psychologist William James published works that presaged the field we have come to know as neural networks. Taking their cue from burgeoning knowledge of brain anatomy and physiology, interdisciplinary researchers bridged neurobiology, electrical engineering, applied mathematics, computer science, and other fields to fashion a promising discipline. Today, although we remain ignorant of many details of brain design and function, we have learned that a highly interconnected collection of simple processing elements (PEs) can exhibit many functional features of the brain. Among these features are knowledge acquisition by learning from examples and application of that knowledge to solve new problems. This is the remarkable property of generalization that, in part, characterizes intelligence. In recent years, the neural-network field has enjoyed the contributions of many talented researchers. Some have applied this discipline to deconvolution. Although it is beyond the scope of this volume to engage the larger field in detail, we find it appropriate to describe one relevant application to deconvolution. Several others are referenced. We begin by briefly sketching the field. For a neural-network overview, consult Jansson (1991). Rumelhart and McClelland (1986), Kosko (1990), and Hertz *et al.* (1991) provide more thorough introductory material; Anderson and Rosenfeld (1988) have collected the classic papers; Simpson (1990) provides a clear and concise taxonomy of paradigms.

A. *ARCHITECTURE AND LEARNING RULES*

In the neural-network field, we speak of paradigms having various characteristics with respect to training, generalization, parallelism, feedback, and linearity, for example. A network itself consists of a collection of processing elements whose inputs and outputs are connected by weights. A paradigm comprises both the network architecture and a learning rule.

The learning rule details the means by which the weight values are to be set. Architecture, in turn, is specified by PE characteristics, interconnection topology, and associated signal flow. Biology purists might object that the characteristics of a typical PE are but an overly simplified caricature of a living neuron, and that an artificial network has interconnection topology that poorly represents a real brain network. Nonetheless, these simplified representations have proved themselves by their astonishing utility.

1. Perceptrons

One type of processing element, the linear PE, is like an operational amplifier that uses weighting resistors to form the sum of electrical signals. The sum signal fans out to drive other PEs through additional weights that correspond, roughly at least, to synapse strengths in biology. A second type, the McCullough–Pitts PE, is similar but thresholds the signal to yield a binary output before passing it on. A simple perceptron network employs a single layer of such thresholded PEs. It was proved by Minsky and Papert (1969) that simple perceptron networks are fundamentally incapable of solving linearly inseparable problems. That is, such networks are not capable of dividing clusters in an n-dimensional hyperspace unless the clusters can be separated by a hyperplane. Because this limitation excluded many important potential applications, interest in neural networks temporarily subsided.

2. Hopfield Networks

Hopfield (1982) helped to revitalize the field by publishing a landmark paper that proposed a new paradigm rich in its analogy to both biological and physical systems. This paradigm, which drew heavily on ideas from prior research, displayed interesting and useful properties such as retrieval of stored memories by content rather than by address. It could even retrieve a memory based on a partial or erroneous description that was close, in a measurable sense, to the desired memory. The network was also capable of solving optimization problems. A necessary component of the Hopfield network was a third type of PE, one having a nonlinear transfer characteristic, specifically a sigmoidal or "S" shape. We can think of these PEs as linear summing amplifiers having positive (excitatory) and negative (inhibitory) weighted inputs. A nonlinear element follows that has a linear middle range but curves smoothly to map sums outside the linear range asymptotically to saturated constant values.

3. Backpropagation

Equally significant was the development of the backpropagation algorithm, which facilitated training of layered perceptron-like networks (Rumelhart and McClelland (1986). Such so-called backpropagation networks differ from perceptrons in that, like Hopfield networks, they employ sigmoidal transfer characteristics in their PEs. Furthermore, they are trained by the process of repeated backpropagation of error. This process is nothing more than iterative application of the calculus chain rule for derivatives to assign network error credit among the network weights in proportion to their respective contributions to that error so that they can be corrected. The correction process is similar to, but not the same as, gradient-descent minimization of the sum-square-error function. Layered networks trained by this method have proved useful in mathematical modeling, function approximation, process control, pattern recognition, and many other applications. Backpropagation was discovered and rediscovered by several researchers. It is, in fact, a generalization of the elegant and effective Widrow–Hoff least mean square (LMS) algorithm widely used in adaptive signal processing (Widrow and Stearns, 1984; Widrow and Hoff, 1960).

B. A LINEAR DECONVOLVING NEURAL NETWORK

An early application of a neural network to deconvolution used Widrow's adaptive linear neuron (ADALINE) to equalize frequency response and cancel echoes over data communication lines. Because high-frequency attenuation and echo phenomena can be represented as a convolution, they are amenable to compensation by an appropriate linear inverse filter. ADALINE uses the LMS algorithm to adapt its filter coefficients to line conditions. Not only was ADALINE one of the first applications of neural networks to deconvolution, but it has enjoyed enormous commercial success as the dominant method of line equalization and filtering used in computer modems (Widrow *et al.*, 1975; McCool *et al.*, 1980). Podilchuk (1992b) has noted that the LMS algorithm is the same as the Kaczmarz method of solving linear-equation sets by successive projections (Section IX.B).

C. A SUPERRESOLVING NEURAL NETWORK

We sketched above how nonlinear PEs helped realize greater potential from neural networks via Hopfield and backpropagation paradigms. In previous sections of this chapter, we described how nonlinear algorithms improved results obtained from deconvolution. It is therefore tempting to ask what might result from applying a network having nonlinear PEs to the deconvolution problem. Though there are many possible approaches, Eq. (17) suggests an obvious one. Not only does it contain weighted sums of the type familiar in the neural network field, but it has a nonlinear component as well: the relaxation function $\kappa[\hat{o}_n^{(k)}]$. The research described in the following sections has been presented earlier (Jansson, 1990).

1. Spread Function, Weights, Sums, and Products

For simplicity of implementation, and in many practical cases, spread-function s can be taken as zero outside some limited range of x. This facilitates implementation of the sum in Eq. (17) as a finite-duration impulse-response (FIR) transversal filter in either analog or digital hardware. Input time-series data are clocked into a shift register from the right side as illustrated by the top row of triangular symbols in Fig. 7. Weights, represented by connections shown below the shift register, implement spread-function convolution. For ease of illustration, the elementary case of a three-sample spread function is shown. Also note that a discrete convolution employs the connections designated as "A" in the figure (*simultaneous* overrelaxation). The summations of Eq. (17) use contributions from *both* the (k)th iteration samples and the ($k + 1$)th iteration samples (*successive* overrelaxation). Substitute the "B" connections for the "A" connections to make this network recurrent and to illustrate accurately the preferred method. As noted in Section VI.B, this method converges faster and therefore requires fewer layers, an advantage typical of recurrent networks.

With either simultaneous or successive methods, a linear PE, shown at the joining of the connections from the shift register, forms the sum and passes it to a sigma–pi PE (Rumelhart and McClelland, 1986), which is used in higher order network architectures. This sigma–pi PE has the capability to both multiply and add signals. In the sigma–pi unit, the sum is multiplied by a signal from the shift register that has passed through a PE having the nonlinear transfer characteristic illustrated in Fig. 1. Also in this unit, the product is summed with a signal from the input data shift

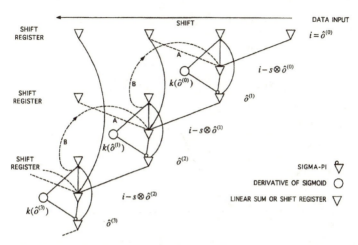

Figure 7 Superresolving neural network—simplified representation. Connections "A" for simultaneous overrelaxation; connections "B" for successive overrelaxation.

register. Resulting values enter a new shift register. The values in this register are available for the convolution required by the stage below, which represents the next iteration of Eq. (17). The design of the rest of the neural network follows this example, one shift-register layer corresponding to each iteration required.

2. Sigmoids

This network contains only one PE type that appears novel to the neural-network field—the PE having the nonlinear transfer characteristic. But, is it so unusual? Its nonlinearity (Fig. 1) is the first derivative of a sigmoid-shaped function made up of spliced parabolas (Fig. 8). Specifically, we may write

$$\kappa[\hat{o}] = d\sigma(\hat{o})/d(\hat{o}), \tag{69}$$

where

$$\sigma(\hat{o}) = \begin{cases} \kappa_0[\hat{o}^2], & \hat{o} \le \frac{1}{2}, \\ -\kappa_0[\hat{o}^2 - 2\hat{o} + \frac{1}{2}], & \hat{o} > \frac{1}{2}. \end{cases}$$

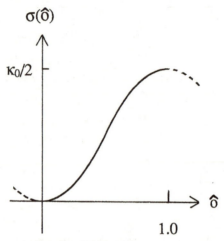

Figure 8   Spliced parabolic sigmoid. (Jansson, 1990; reprinted with permission of the International Neural Networks Society, Woodbury, New Jersey; Lawrence Erlbaum Associates, Publishers, Mahwah, New Jersey and Hove, UK.)

Over part of its range, $\sigma(\hat{o})$ is qualitatively similar to exponential-based sigmoids used in backpropagation and other network paradigms. However, it loses its monotonicity in the region representing nonphysical solutions. This characteristic is responsible for its effectiveness in enforcing the bounds. We speculate that such first-derivative sigmoids may be useful in other neural-net paradigms for similar purposes.

3.   Biological Analogies

In the implementation described, there is one sigma–pi unit and one shift register per layer. Alternatively, the top row of Fig. 7 could represent a row of linear buffers operating in parallel and receiving parallel data. Beneath them would be a layer of linear summers, a layer of nonlinear PEs, a layer of sigma–pi units, and so forth. Although we have thus far considered one-dimensional data, the analysis is readily extended to higher dimensionality. In two dimensions, for example, we draw the analogy with the retina. The first layer of linear summing PEs yielding $i - s \otimes \hat{o}^{(k)}$ is indeed not unlike the visual receptive field. Thus, even in this first stage some sharpening of $\hat{o}^{(k)}(x)$ occurs. It is interesting to speculate on the degree to which the biological analogy could be carried to higher stages of vision.

Humans *do* have the capacity to infer the underlying sharp detail in blurred images (Bottini, 1981; Bex *et al.*, 1995).

4. Simulation

We simulated the network by computer and synthesized an absorption-spectrum object containing two pairs of closely spaced lines (Fig. 9). We chose a Gaussian absorption-coefficient profile for each of the lines. The large absorption coefficients of the leftmost lines caused them to saturate and flattened the peaks. The spectrum was then blurred by rigorously band limiting its Hartley transform (see Chapter 1) with a triangular transfer function that could describe blur by diffraction in an optical spectropho-

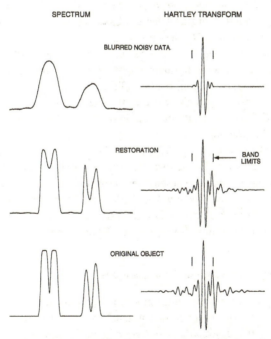

Figure 9 Neural-network deconvolution of noisy saturated Gaussian absorption line data. (Jansson, 1990; reprinted with permission of the International Neural Networks Society, Woodbury, New Jersey; Lawrence Erlbaum Associates, Publishers, Mahwah, New Jersey and Hove, UK.)

tometer. (Do not confuse this transfer function with the coincidentally similar PE characteristic and relaxation function in Fig. 1.)

Deconvolution simulations using noise-free "data" can produce unrealistically good solutions, even with inferior methods. Therefore, uncorrelated Gauss-distributed random noise having a root-mean-square value of 0.25% of full absorptance range was added to the object. We show in Fig. 9 the results of a 17-layer simulation that used polynomial smoothing convolution layers between the first 15 layers of the type illustrated in Fig. 7. These smoothing layers are easily implemented by analogy with components of the design in the figure. For clarity, we chose not to clutter the figure by illustrating them.

5. Superresolution

Results in Fig. 9 show clear resolution of both pairs of lines. The steepness of the rising Gaussian profile near the base line is well represented. The figure also shows the Hartley transforms of the simulated data, the restoration, and the original synthetic spectrum. These transforms demonstrate restoration of some frequencies lost by band limiting. Naturally, the high frequencies farther from the band limit exhibit increasing error. Even restoration of lost frequencies close to the band limit is notable, however, in view of the profound attenuation imposed by the triangular transfer function in this region.

It has not been widely appreciated that the early constrained method on which this network is based owes its success to the restoration of lost frequencies. Nevertheless, it should be no surprise that bandwidth extrapolation results (see also Section VI.C). Based on the experience of many workers, this method equals or betters other known bandwidth extrapolation methods in its ability to resolve fine detail in spectra.

6. Learning Deconvolution from Examples

Above we specified weight values in advance. We obtained these values by first determining the point-spread function $s(x)$. Other means, however, should be applicable to weight determination. For example, the network could be trained via backpropagation by presenting it with known sets of $i(x)$ vs. $o(x)$ prepared by simulation. Comparing output estimates $\hat{o}^{(k)}(x)$ with true $o(x)$ would provide the errors needed for correction of weight values by backpropagation. More compact architectures should be usable with this method of training within the context of the basic cascaded shift-register design.

7. High-Speed Filter Hardware

Either digital or analog shift registers can be used. The latter are based on either sample-and-hold circuits or charge-coupled device (CCD) "bucket brigades" for charge-carrier packets. Analog voltages derived from the tapped delay stages are easily weighted by appropriate coefficient resistors and summed into an operational amplifier. The products and sums required are thus performed rapidly and in parallel. Many hybrid analog–digital variations on this theme are possible, including those that employ DSP (digital signal processor) and application-specific integrated circuit (ASIC) chips.

The network implements a continuous process, in contrast to the batch process implied by the iterative representation. For each new $i(x)$ sample provided to the input, a lagged $\hat{o}(x)$ sample appears at the output. Because this nonlinear filter implements Eq. (17), and because the iteration of Eq. (17) finds a solution that satisfies the data $i(x)$ via the convolution $s \otimes \hat{o}^{(k)}(x)$, the filter is inherently flux preserving. That is, the area under $\hat{o}^{(k)}(x)$ well approximates the area under $i(x)$ via reasoning analogous to that in Chapter 2, Sections II.E and II.F.

Instead of an application-specific circuit, a general-purpose laboratory computer could apply a purely software version of the filter continuously as data are acquired. Or, the filter may be packaged in firmware as part of a microprocessor-based instrument. With suitable buffers and delays, the filter may be applied to scanned two-dimensional image data in real time. Any of these implementations has the added advantage of permitting real-time interactive parameter adjustment.

In data communications, bandwidth extrapolation offers the opportunity to make better use of channel capacity. Bounded methods produce their most impressive restorations when $o(x)$ is at or near its bound over much of its domain. The method of Eq. (17) was designed for use with both upper and lower bounds, which makes it ideal for the bilevel signals sometimes used in digital transmission. Substantial bandwidth extrapolation could result from implicit application of the powerful prior knowledge that the signal can assume only one of two values.

D. ADDITIONAL NEURAL-NETWORK APPLICATIONS TO DECONVOLUTION

The Hopfield paradigm and its variants have been responsible for the largest number of neural-network papers in the recent deconvolution literature. In these contributions, network weights take the role of $\hat{o}(x)$ parameters. The network is configured so that its energy, or Lyapunov function, represents a cost to be optimized. Optimization is an area in which the Hopfield network has found useful prior applications. In the deconvolution application, various means are employed to regularize the process by controlling the trade-off between a solution's smoothness and its consistency with the data. Some utilize maximum entropy (ME) for this purpose. Maximum entropy methods also implicitly use a positivity constraint because entropy is undefined for negative $\hat{o}(x)$. Such methods have potential for superior performance. Others may employ complex and sometimes nonlinear methods of computing coefficients that are used to *linearly* transform and restore the data. In this case, all the performance limitations of linear methods apply. Such nonlinearly derived but linearly applied networks may, however, offer insight, and an opportunity for modification to incorporate *a priori* information such as physical bounds, or possibly provide other advantages, such as real-time adaptation.

Abbiss and Brames (1988), Zhou *et al.* (1988), Abbiss *et al.* (1989, 1991), Paik and Katsaggelos (1990), Paik and Katsaggelos (1992), Wang and Mendel (1992), Vemuri and Jang (1992), and Bilgen and Hung (1995) have considered or experimented with Hopfield implementations. Luo and Bao (1991) use a Hopfield network to determine weights for a deconvolving finite impulse response (FIR) transversal filter. Lee *et al.* built a massively parallel custom-VLSI hybrid analog–digital circuit to deconvolve images (Lee and Sheu, 1990; Lee *et al.*, 1992). Jeffrey and Rosner (1986) and Bedini and Tonazzini (1990) described Hopfield networks using maximum entropy constraints, and Li *et al.* (1995) adapted the method of Zhou *et al.* by adding a maximum-entropy constraint. Wang and Frieden (1995) used a Hopfield network in a minimum-entropy approach. Marrian and Peckerar (1987, 1989) describe a custom artificial neural network that finds maximum-entropy solutions to deconvolution and other ill-posed problems.

A cerebellar model arithmetic computer (CMAC), originally developed for robotic applications (Albus, 1975), was applied to deconvolution by Glanz and Miller (1989). Trained by least mean square (LMS), it learned deconvolution from examples. Gallant *et al.* (1993) used a CMAC network to deconvolve overlapping chromatographic peaks. A cellular neural network (CNN) was applied to image and volume deblurring by Miller *et al.* (1994). Examples of layered designs are given by Kulkarni (1990) and Garcia-Gomez *et al.* (1989). The work reported in the latter publication takes advantage of sigma–pi units. Kulkarni effected continuous processing with shift registers. Sivakumar and Desai (1992) used a multilayer perceptron to restore images suffering indeterminate blurring (blind deconvolution) via backpropagation and multilevel sigmoids. Caudell and Soffer (1990) and Blass and Crilly (1992) have applied a standard backpropagation network to the problem of learning superresolution from examples.

Finally, Bottini (1981) invoked known anatomy and physiology to hypothesize a mechanism for the brain's known capability to remove spatial blur from moving-object images. This blur is a necessary consequence of visual persistence, which involves integration times on the order of 0.1 second.

XIII. Other Methods

The linear methods have no provisions for bounds on $\hat{o}(x)$. In principle, solutions are free to assume any value, as long as they satisfy the data $i(x)$. We have shown that a problem of uniqueness results. Many functions are thus possible candidates—they all satisfy the data, or nearly so. Noise, which is always present in measurements, unfortunately provides the basis by which a linear method prefers one solution over another.

It has been demonstrated both theoretically and computationally that a simple bound of positivity produces superior solutions, by eliminating many of the undesired solutions from the selection of possible solutions. Another way of reducing the selection is to employ a norm or solution criterion, typically a quantity to be extremized. In the foregoing we used norms of minimum mean-square error, maximum likelihood, and maximum entropy. All these norms force a unique answer to the problem. Frieden (1979, 1980, 1981) also developed the norm of maximum information. This norm selects the $\hat{o}(x)$ that conveys the most information about itself into the data $i(x)$. Frieden found empirically that solutions based on maximum information closely resemble the maximum-entropy solution for impulsive objects (such as spectral lines), but show differences for objects more typically found in images, such as those having edges or plateaus.

Since the value of bounds has come to be widely accepted, numerous other effective bounded methods have appeared. Linear programming has provided the basis for a method presented by Mammone and Eichmann (1982a, b). In a method loosely related to linear programming. MacAdam (1970) exploited the relationship between polynomial multiplication and convolution. His method is particularly suited to human interactive adjustment of constraints.

Constraints were used by Ichioka *et al.* (1981) in an application of the method of steepest descent. The rapid-convergence properties of the conjugate gradient method have been applied to deconvolution by Angel and Jain (1978). Maitre (1981) achieved superresolution by adapting constraints to the conjugate gradient method. Lawson and Hanson (1974) developed the algorithm NNLS to solve the least-squares problem in the presence of the nonnegativity constraint. Chambless and Broadway (1981) have successfully applied it to nuclear radiation spectra.

Deconvolution is an example of an ill-posed problem (Chapter 1, Section V.B). Tikhonov was one of the earliest workers to deal with ill-posed problems in a mathematically precise way. He developed the approach of regularization (Tikhonov, 1963; Tikhonov and Arsenin, 1977) that has been applied to deconvolution by a number of workers [see, for example, papers by Abbiss *et al.* (1983), Chambless and Broadway (1981), Nashed (1981), and Bertero *et al.* (1978)]. Some of the methods that we have previously described fall within the context of regularization (e.g., the method of Phillips and Twomey, discussed in Section V of Chapter 3). Amplitude bounds, such as positivity, are frequently used as key elements of regularization methods.

Probability methods based on maximum likelihood or Bayes' theorem were introduced previously in this chapter. One was the Richardson–Lucy method (Section IV.B); the other was Frieden's method of maximum probability. Additional methods based on Bayes' theorem have been described by Hunt (1975, 1977), Mendes and DePolignac (1975), Mendes *et al.* (1975), Trussell (1976), Kennett *et al.* (1978a–c), Hunt and Trussell (1976), and Trussell and Hunt (1979).

There is much diversity within the realm of Bayesian methods and Shannon's information theory. Working with related concepts, more recently Terebizh (1995) examined the feasible-solution region of the object's parameter space. He determined the shape of this region by combining Fisher information matrix eigenvectors, an image randomness test, and known physical bounds. To obtain his object estimate, Terebizh invoked the principle of Occam's razor by selecting the simplest possible object consistent with the data; his estimate retained the smallest number of principal components allowing statistically satisfactory data representation.

Earlier, Pina and Puetter (Pina and Puetter, 1993; Puetter and Pina, 1993), also invoking the "razor" terminology, had used another Bayesian idea. They represented image degrees of freedom with irregularly shaped image regions called pixons. Pixon shapes were adjusted so that they contained maximum information. An object estimate was then sought in cycles of a two-phase iteration: first the most probable image was determined given a fixed model, then the image was fixed and the model was refined. As in the Terebizh approach, Pina and Puetter sought the solution that used the fewest of degrees of freedom consistent with a statistically adequate fitting of the data.

XIV. Reviews and Reference Works

Useful reference material is available in a number of books and journal articles. They include information on both linear and nonlinear methods. Because the deconvolution literature is voluminous, these works tend to focus on specific areas such as linear algebraic methods, or POCS. Consult excellent review papers by Marks and Smith (1981), Schafer *et al.* (1981), Meinel (1986), Demoment (1989), Katsaggelos (1989), Biemond *et al.* (1990), and Sezan and Tekalp (1990b). Valuable books include works by Andrews and Hunt (1977), Huang (1984), Bates and McDonnell (1986), Stark (1987), Lagendijk and Biemond (1991), and Mammone (1992). Excellent book chapters by Frieden (1979) and Podilchuk (1992c) are especially helpful. A number of special issues of journals and proceedings important to this field have appeared. Among these are a volume of selected papers edited by Sezan (1992), and special sections of the *Journal of the Optical Society of America* edited by Fienup (1983), Fienup and Rushforth (1987), Fienup and Gallagher (1990), and Gallagher and Sweeney (1993). These journal sections document in detail much of the research presented at the ongoing series of Optical Society of America topical meetings on signal recovery and related topics. A special issue of *Optical Engineering* edited by Sezan and Tekalp (1990a) also contains a number of useful papers. For material specific to the Jansson method, including experimental comparisons of algorithms, consult the references cited in Section VI.F of this chapter.

XV. Concluding Remarks

This chapter has given some flavor of the historical development of nonlinear methods. Early investigators of these methods expended great effort in overcoming the popular notion that bandwidth extrapolation was not possible or practical. It was, for example, believed that the Rayleigh limit of resolution was a limit of the most fundamental kind—unassailable by mathematical means. To be sure, the popular notion was reinforced by a long history of misfortune with linear techniques and hypersensitivity to noise. Anyone who still needs to be convinced of the virtues of the nonlinear methods would benefit from reading the paper by Wells (1980); the nonlinear point of view is nowhere else more clearly stated.

Now that the efficacy and potential of the nonlinear methods have gained substantial acceptance, we need not fear that a few cautionary notes will spoil the sensible trend in their direction.

A. CAUTIONS

First let us deal with deconvolution in general. We have a few admonitions to readers of a literature report on a new method. They should ask, does the writer deal fairly with noise? Even the most volatile of the linear methods can produce a reasonable restoration when noise is limited to roundoff error in the seventh significant figure of the "data." A method's capability of yielding acceptable restorations in the presence of realistic noise is critical to its practicality.

There is also the question of convergence. Proof of convergence does not constitute proof that a method is practical. A method that converges quickly and consistently in trials with real data wins hands down over a proved converging method that requires hundreds or even thousands of iterations, even though the former method may lack the required proof. And, failure to find a proof does not necessarily mean that such a proof does not exist. Even the most rapid convergence is of little value unless the correct solution results. Employment of the proper norm and/or bound goes a long way toward assuring validity.

An amplitude bound has the added virtue of producing solutions having reduced noise sensitivity, fewer artifacts, superior resolution, and possible bandwidth extrapolation. In contrast, methods having an output that is linear in the irradiance data $i(x)$ either produce artifacts or trade off resolution to suppress artifacts. If a bound makes physical sense and can be computationally afforded, use it. Simple clipping of nonphysical parts typically does not work well, however. Subtle techniques usually work better.

Much attention is given to the statistical basis for deconvolution methods. It is true that probability and statistical arguments may aid in finding a solution. Remember, however, that even the *most* probable solution may not be *very* probable. There may be many solutions to an ill-posed deconvolution problem, all quite different, and all having nearly equal probability. Statistical arguments have special value when they represent the only prior information available. Experience has shown, however, that the benefit obtained from a method's statistically pure foundation is small compared with that obtained from bounds, where bounds are applicable. This is especially true with objects having values near the bounds.

Next, we ask whether anything is lost by use of the bounded methods. Although the answer is yes, the loss is almost always vastly outweighed by the benefits. In using bounded methods, we do, in fact, lose the ability to observe a certain kind of unexpected result. Consider, for example, a fluorescing sample placed between a source and an optical absorption spectrometer. A deconvolution method imposing an upper transmittance bound of 100% will be confused by the data, to say the least. The moral? Be sure of the validity of constraints before you apply them.

Far too few papers objectively compare method performance. For the potential user in search of a method, however, pitfalls await even in the published tests. This author has frequently noted reports that misuse one or more of the compared methods, thereby compromising that method's performance and invalidating results.

In spite of most authors' claims, their newest algorithm is not necessarily the best. Nevertheless, a new algorithm, even if it performs poorly, may contain the seed of an idea that can be adapted and improved. Such an algorithm, however, is not for the reader who needs a proved effective method. It is material for the deconvolution researcher.

Finally, we suggest that a backlog of data on objects having known properties be analyzed until the characteristics of the deconvolution method are completely familiar. Only then should results from unknown objects be judged. As in any experimental work, the experiment should be repeated to verify reproducibility and develop confidence in results. In this sense, the deconvolution process may be treated just like a laboratory apparatus. Indeed, it takes on that identity when packaged in a laboratory microcomputer.

B. THE FUTURE

Linear deconvolution methods served to educate us to the pitfalls of the deconvolution problem. Their occasional successful applications both tantalized and discouraged us. Gradually, there were fewer and fewer circumstances in which use of linear methods was justified. The more generally useful nonlinear methods teamed with powerful hardware to enhance prospects for wide application of deconvolution.

There is no doubt that the next few years will continue to see further improvements through incorporation of very general bounds and the fullest utilization of statistical prior knowledge. Computing time and cost will also be reduced through improvements in both algorithms and hardware.

We anticipate a proper answer to the question—how far can we go?—theoretically and experimentally. We might expect that instruments will be specifically designed to produce data for subsequent deconvolution. Parameters of interest in such a design are high signal-to-noise ratio and precise control of the independent variable x. There will be interest in optimum tailoring of the instrument response $s(x)$. Integrated hardware–software systems will increasingly define the state of the art.

Future experimentalists will come to regard deconvolution as an essential tool in the standard repertoire. Through deconvolution they will make otherwise expensive observations with inexpensive instruments. The limits of state-of-the-art experimental apparatus will be extended. This progress, in turn, will produce data that cannot fail to reveal phenomena that would otherwise lie hidden. The new experimental results will stimulate advances in physical theory, thereby providing incentives for better and better apparatus. Because of its applicability across disciplinary lines, deconvolution will play a key role in the advancement of all branches of science that yield to the tools of measurement.

References

Abbiss, J. B., and Brames, B. J. (1988). *Proc. SPIE, Int. Soc. Opt. Eng.* **880**, 100–106.

Abbiss, J. B., DeMol, C., and Dhadwal, H. S. (1983). *Opt. Acta* **30**, 107–124.

Abbiss, J. B., Brames, B. J., and Fiddy, M. A. (1989). *Proc. SPIE, Int. Soc. Opt. Eng.* **1058**, 138–146.

Abbiss, J. B., Brames, B. J., and Fiddy, M. A. (1991). *IEEE Transactions on Signal Processing* **39**, 1516–1523.

Abergel, A., Bertaux, J.-L., and Dimarellis, E. (1988). *In* "Proceedings Pixim 88 Computer Graphics in Paris, October 24–28, 1988" (B. Peroche, ed.), pp. 425–444. Hermes, Paris.

Agard, D. A., Hiraoka, Y., Shaw, P., and Sedat, J. W. (1989). *Methods Cell Biol.* **30**, 364.

Albert, A. (1981). "Regression and the Moore-Penrose Pseudo-Inverse." Academic Press, New York.

Albus, J. S. (1975). *Trans. ASME: J. Dyn. Syst. Meas. Control* **97**, 220–227.

Allebach, J. P. (1981). *J. Opt. Soc. Am.* **71**, 99–105.

Anderson, J. A., and Rosenfeld, E. (1988). "Neurocomputing: Foundations of Research." MIT Press, Cambridge, Massachusetts.

Andrews, H. C., and Hunt, B. R. (1977). "Digital Image Restoration." Prentice-Hall, Englewood Cliffs, New Jersey.

Angel, E. S., and Jain, A. K. (1978). *Appl. Opt.* **17**, 2186–2190.

Barnes, W. L., Susskind, J., Hunt, R. H., and Plyler, E. K. (1972). *J. Chem. Phys.* **56**, 5160–5172.

Bates, R. H. T., and McDonnell (1986). "Image Restoration and Reconstruction." Oxford Univ. Press (Clarendon), Oxford.

Bedini, L., and Tonazzini, A. (1990). *Image and Vision Computing* **8**, 108–113.

Bertero, M., DeMol, C., and Viano, G. A. (1978). *Opt. Lett.* **3**, 51–53.

Bex, P. J., Edgar, G. K., and Smith, A. T. (1995). *Vision Res.* **35**, 2539–2546.

Biemond, J., Lagendijk, R. L., and Mersereau, R. M. (1990). *Proc. IEEE* **78**, 856–883.

Bilgen, M., and Hung, H.-S. (1995). *Opt. Eng.* **33**, 2723–2727.

Biraud, Y. (1969). *Astron. Astrophys.* **1**, 124–127.

Blass, W. E., and Crilly, P. B. (1992). *Proceedings of the IEEE Instrumentation and Measurement Technology Conference (IMTC'92)*, pp. 631–634. New York, 12–14 May 1992.

Blass, W. E., and Halsey, G. W. (1981). "Deconvolution of Absorption Spectra." Academic Press, New York.

Bodewig, E. (1956). "Matrix Calculus." North-Holland, Amsterdam.

Bottini, S. (1981). Proceedings of the Second International Conference on Visual Psychophysics and Medical Imaging, pp. 143–149. Brussels, July 2–3, 1981, IEEE, New York.

Bregman, L. M. (1965). *Dokl. Akad. Nauk SSSR* **162**, 487–490.

Burg, J. P. (1967). Proceedings of the 37th Meeting Society Exploration Geophysicists, Oklahoma City, Oklahoma.

Burg, J. P. (1975). Ph.D. Thesis, Stanford Univ., Stanford, California.

Cahana, D., and Stark, H. (1981). *Appl. Opt.* **20**, 2780–2786.

Caudell, T. P., and Soffer, B. H. (1990). Northcon Conference Record, Seattle, pp. 257–262. Oct. 9–11.

Chambless, D. A., and Broadway, J. A. (1981). *Nucl. Instrum. Methods* **179**, 563–571.

Cheung, K. F., Marks II, R. J., and Atlas, L. E. (1988). *J. Opt. Soc. Am. A* **5**, 2008–2009.

Coote, G. E., and Kwan, B. P. (1995). *Nucl. Instrum. Methods Phys. Res. Sect. B* **104**, 228–232.

Crilly, P. B. (1987). *J. Chemom.* **1**, 175–183.

Demoment, G. (1989). *IEEE Trans. Acoust. Speech, Signal Process.* **37**, 2024–2036.

DeSantis, P., and Gori, F. (1975). *Opt. Acta* **22**, 691–695.

Deutsch, S. (1965). *IEEE Trans. Broadcast.* **11**, 11–21.

Dimarellis, E., Bertaux, J. L., and Abergel, A. (1989). *Astron. Astrophys.* **208**, 327–330.

Donoho, D. L., Johnstone, I. M., Hoch, J. C., and Stern, A. S. (1992). *Journal of the Royal Statistical Society B* **54**, 41–81.

Evans, B. J., and Smith, L. M. (1992). *IEEE Trans. Plasma Sci.* **20**, 432–438.

Fiddy, M. A., and Hall, T. J. (1981). *J. Opt. Soc. Am.* **71**, 1406–1407.

Fienup, J. R. (1979). *Opt. Eng.* **18**, 529–534.

Fienup, J. R. (1980). *Opt. Eng.* **19**, 297–305.

Fienup, J. R. (1982). *Appl. Opt.* **21**, 2758–2769.

Fienup, J. R., ed. (1983). *J. Opt. Soc. Am.* **73**, 1412–1526.

Fienup, J. R., and Rushforth, C., eds. (1987). *J. Opt. Soc. Am. A* **4**, 101–297.

Fienup, J. R., and Gallagher, N. C., eds. (1990). *J. Opt. Soc. Am. A* **7**, 379–526.

Fougere, P. F. (1977). *J. Geophys. Res.* **82**, 1051–1054.

Fougere, P. F., Zawalick, E. J., and Radoski, H. R. (1976). *Phys. Earth Planet. Inter.* **12**, 201–207.

Frieden, B. R. (1971). "Restoring with Maximum Likelihood." Technical Report Number 67, Optical Sciences Center, University of Arizona, Tucson.

Frieden, B. R. (1972). *J. Opt. Soc. Am.* **62**, 511–518.

Frieden, B. R. (1974). *J. Opt. Soc. Am.* **64**, 561 (abstract).

Frieden, B. R. (1975). In *Top. Appl. Phys.: Picture Processing and Digital Filtering*, Vol. 6 (T. S. Huang, ed.), pp. 177–248. Springer-Verlag, New York.

Frieden, B. R. (1979). *SPIE Applications of Digital Image Processing III* **207**, 14–25.

Frieden, B. R. (1980). *Opt. Eng.* **19**, 290–296.

Frieden, B. R. (1981). *J. Opt. Soc. Am.* **71**, 294–303.

Frieden, B. R. (1983). "Probability, Statistical Optics, and Data Testing: A Problem Solving Approach." Springer-Verlag, New York.

Frieden, B. R., and Burke, J. J. (1972). *J. Opt. Soc. Am.* **62**, 1202–1210.

Frieden, B. R., and Swindell, W. (1976). *Science* **191**, 1237–1241.

Frieden, B. R., and Wells, D. C. (1978). *J. Opt. Soc. Am.* **68**, 93–103.

Gallagher, N. C., and Sweeney, D., eds. (1993). *J. Opt. Soc. Am. A* **10**, 987–1120.

Gallant, S. R., Fraleigh, S. P., and Cramer, S. M. (1993). *Chemom. Intell. Lab. Syst.* **18**, 41–57.

Garcia-Gomez, R., Gomez-Mena, J., and Diez del Rio, L. (1989). *Proceedings of the International Conference on Acoustics, Speech, and Signal Processing (ICASSP-89), May 23–26, Glasgow* **4**, 2368–2371.

Gerchberg, R. W. (1974). *Opt. Acta* **21**, 709–720.

Gerchberg, R. W., and Saxton, W. O. (1972). *Optik* **35**, 237–246.

Gilliland, R. L. (1990). *In* Proceedings of the Workshop on the Restoration of HST Images and Spectra, Baltimore, Maryland, August 20–21, 1990" (R. L. White and R. J. Allen, eds.). pp. 7–12. Space Telescope Science Institute, Baltimore, Maryland.

Gilliland, R. L., Morris, S. L., Weymann, R. J., Ebbets, D. C., and Lindler, D. J. (1992). *Publ. Astron. Soc. Pac.* **104**, 367–382.

Gindi, G. (1981). Ph.D. Dissertation, Univ. Arizona, Tucson.

Glanz, F. H., and Miller, W. T. (1989). *Proceedings of the International Conference on Acoustics, Speech, and Signal Processing (ICASSP-89), May 23–26, Glasgow,* **4**, 2349–2352.

Gold, R. (1964). "An Iterative Unfolding Method for Response Matrices." AEC Research and Development Report ANL-6984, Argonne National Laboratory, Argonne, Illinois.

Goldberg, D. E. (1989). "Genetic Algorithms: In Search, Optimization, and Machine Learning." Addison-Wesley, New York.

Goodman, J. W. (1968). "Introduction to Fourier Optics." McGraw-Hill, New York.

Goody, R. M. (1964). "Atmospheric Radiation." Oxford Univ. Press, London.

Gubin, L. G., Polyak, B. T., and Raik, E. V. (1967). *USSR Comput. Math. Phys.* **7**, 1–24.

Guillemin, E. A. (1949). "The Mathematics of Circuit Analysis," p. 288. Wiley, New York.

Gull, S. F., and Daniell, G. J. (1978). *Nature (London)* **272**, 686–690.

Harris, J. L. (1964). *J. Opt. Soc. Am.* **54**, 931–936.

Harris, J. L., Sr. (1971). *J. Opt. Soc. Am.* **61**, 1563 (abstract).

Heasley, J. N. (1984). *Publ. Astron. Soc. Pac.* **96**, 767–772.

Heasley, J. N., Pilcher, C. B., Howell, R. R., and Caldwell, J. J. (1984). *Icarus* **57**, 432–442.

Herman, G. T., Lent, A., and Rowland, S. W. (1973). *J. Theor. Biol.* **42**, 1–32.

Herman, G. T., Lent, A., and Lutz, P. H. (1978). *Commun. ACM* **21**, 152–158.

Herman, G. T., Eggermont, P. B. B., and Lent, A. (1981). *Linear Alg. Appl.* **40**, 37–67.

Herschel, R. S. (1971). Ph.D. Dissertation, University of Arizona, Tucson. Available as Technical Report Number 72, Optical Sciences Center, University of Arizona, Tucson.

Hertz, J., Krogh, A., and Palmer, R. G. (1991). "Introduction to the Theory of Neural Computation." Addison-Wesley, Redwood City, California.

Hirsch, P. M., Jordan, J. A., and Lesem, L. B. (1971). U.S. Patent No. 3,619,022.

Hopfield, J. J. (1982). *Proc. Natl. Acad. Sci. U.S.A.* **79**, 2554–2558.

Howard, S. J. (1981a). *J. Opt. Soc. Am.* **71**, 95–98.

Howard, S. J. (1981b). *J. Opt. Soc. Am.* **71**, 819–824.

Huang, T. S. (1984). "Advances in Computer Vision and Image Processing," Vol. 1, Image Reconstruction from Incomplete Observation. JAI Press, London.

Huang, T. S., Barker, D. A., and Berger, S. P. (1975). *Appl. Opt.* **14**, 1165–1168.

Hunt, B. R. (1975). *International Optical Computing Conference*, pp. 11–13.

Hunt, B. R. (1977). *IEEE Trans. Comput.* **C-26**, 219–229.

Hunt, B. R., and Trussell, H. J. (1976). *Proceedings of the International Conference on Acoustics, Speech, and Signal Processing, Philadelphia, April, IEEE*, 354–356.

Ichioka, Y., Takubo, Y., Matsuoka, K., and Suzuki, T. (1981). *J. Opt.* **12**, 35–41.

Ioup, G. E., and Thomas, B. S. (1967). *J. Chem. Phys.* **46**, 3959–3961.

Jansson, P. A. (1968). Ph.D. Dissertation, Florida State Univ., Tallahassee.

Jansson, P. A. (1970). *J. Opt. Soc. Am.* **60**, 184–191.

Jansson, P. A., ed. (1984). "Deconvolution: With Applications in Spectroscopy." Academic Press, New York.

Jansson, P. A. (1990). "Proceedings of the International Joint Conference on Neural Networks (IJCNN-90), Washington, D.C., Jan. 15–19," pp. 375–378. International Neural Networks Society, Woodbury, New Jersey, Erlbaum, Mahwah, New Jersey and Hove, England.

Jansson, P. A. (1991). *Anal. Chem.* **63**, 357A–362A.

Jansson, P. A., and Davies, R. D. (1974). *J. Opt. Soc. Am.* **64**, 1372.

Jansson, P. A., Hunt, R. H., and Plyler, E. K. (1968). *J. Opt. Soc. Am.* **58**, 1665–1666.

Jansson, P. A., Hunt, R. H., and Plyler, E. K. (1970). *J. Opt. Soc. Am.* **60**, 596–599.

Jaynes, E. T. (1968). *IEEE Trans. Syst. Sci. Cybernetics* **SSC-4**, 227–241.

Jeffrey, W., and Rosner, R. (1986). *Astrophys. J.* **310**, 473–481.

Jones, R. N., Venkataraghavan, R., and Hopkins, J. W. (1967). *Spectrochim. Acta* **23A**, 925–939.

Kaczmarz, S. (1937). *Bull. Acad. Pol. Sci. Lett. A*, 355–357.

Katsaggelos, A. K. (1989). *Opt. Eng.* **28**, 735–748.

Kawata, S., and Ichioka, Y. (1980a). *J. Opt. Soc. Am.* **70**, 762–768.

Kawata, S., and Ichioka, Y. (1980b). *J. Opt. Soc. Am.* **70**, 768–772.

Kawata, S., Ichioka, Y., and Suzuki, T. (1978). *Proceedings of the 4th International Joint Conference on Pattern Recognition*, 525–529.

Kennett, T. J., Prestwich, W. V., and Robertson, A. (1978a). *Nucl. Instrum. Methods* **151**, 285–292.

Kennett, T. J., Prestwich, W. V., and Robertson, A. (1978b). *Nucl. Instrum. Methods* **151**, 293–301.

Kennett, T. J., Brewster, P. M., Prestwich, W. V., and Robertson, A. (1978c). *Nucl. Instrum. Methods* **153**, 125–135.

Kikuchi, R., and Soffer, B. H. (1977). *J. Opt. Soc. Am.* **67**, 1656–1665.

Kirkpatrick, S., Gelatt, C. D., and Vecchi, M. P. (1983). *Science* **220**, 671–680.

Kosko, B. (1990). "Neural Networks and Fuzzy Systems: A Dynamic Systems Approach to Machine Intelligence." Prentice-Hall, Englewood Cliffs, New Jersey.

Kulkarni, A. D. (1990). *Proceedings of the ACM Eighteenth Annual Computer Science Conference, Washington, D.C., February 20–22*, 373–378.

Lacoss, R. T. (1971). *Geophysics* **36**, 661–675.

Lagendijk, R. L., and Biemond, J. (1991). "Iterative Identification and Restoration of Images." Kluwer, Dordrecht, The Netherlands.

Lawson, C. L., and Hanson, R. J. (1974). "Solving Least Squares Problems." Prentice-Hall, Englewood Cliffs, New Jersey.

Lee, J.-C., and Sheu, B. J. (1990). "Proceedings of the IEEE International Conference on Computer Design," pp. 126–129. IEEE Comput. Soc. Press, Los Alamitos, California.

Lee, J.-C., Sheu, B. J., Choi, J., and Chellappa, R. (1992). *IEEE Transactions on Circuits and Systems for Video Technology* **2**, 319–324.

Lent, A., and Tuy, H. (1981). *J. Math. Anal. Appl.* **83**, 554–565.

Lesage, A., Depiesse, M., and Richou, J. (1995). Proceedings of the 12th International Conference on Spectral Line Shapes, pp. 93–94. 13–17 June 1994, Toronto, Ontario, American Institute of Physics, New York.

Li, H. D., Kallergi, M., Qian, W., Jain, V. K., and Clarke, L. P. (1995). *Opt. Eng.* **34**, 1431–1440.

Liu, B., and Gallagher, N. C. (1974). *Appl. Opt.* **13**, 2470–2471.

Lucy, L. B. (1974). *Astron. J.* **79**, 745–754.

Luo, F.-L., and Bao, Z. (1991). *Proc. Intl. Conf. Industrial Electronics, Control and Instrumentation (IECON '91), Xian, China, Oct. 28–Nov. 1*, **2**, 1449–1453.

MacAdam, D. P. (1970). *J. Opt. Soc. Am.* **12**, 1617–1627.

McCool, J. M., Widrow, B., Zeidler, J. R., Hearn, R. H., and Chabries, D. M. (1980). "Adaptive Line Enhancer." U.S. Patent 4,238,746.

MacNeil, K. A. G., and Dixon, R. N. (1977). *J. Electron Spectrosc. Relat. Phenom.* **11**, 315–331.

Maeda, J., and Murata, K. (1984a). *J. Opt. Soc. Am. A* **1**, 28–34.

Maeda, J., and Murata, K. (1984b). *Appl. Opt.* **23**, 857–861.

Maitre, H. (1981). *Computer Graphics Image Proces.* **16**, 95–115.

Mammone, R. J., ed. (1992). "Computational Methods of Signal Recovery and Recognition." Wiley, New York.

Mammone, R., and Eichmann, G. (1982a). *Appl. Opt.* **21**, 496–501.

Mammone, R., and Eichmann, G. (1982b). *J. Opt. Soc. Am.* **72**, 987–991.

Marchetti, A. A., and Mignerey, A. C. (1993). *Nucl. Instrum. Methods Phys. Res. A* **324**, 288–296.

Marks II, R. J. (1981). *Appl. Opt.* **20**, 1815–1820.

Marks II, R. J. (1982). *Opt. Lett.* **7**, 376–377.

Marks II, R. J., and Smith, M. J. (1981a). *Opt. Lett.* **6**, 522–524.

Marks II, R. J., and Smith, D. K. (1981b). *Proc. SPIE Int. Soc. Opt. Eng.* **373**, 161–178.

Marrian, C. R. K., and Peckerar, M. C. (1987). *Proceedings of the IEEE First International Conference on Neural Networks, June 21–24, San Diego*, **3**, 749–757.

Marrian, C. R. K., and Peckerar, M. C. (1989). *IEEE Trans. Circuits Syst.* **36**, 288–294.

Martirosyan, V. R. (1984). *Avtometriya* **5**, 87–92.

Mason, D. C., LeBas, T. P., Sewell, I., and Angelikaki, C. (1992). *Marine Geophysical Research* **14**, 125–136.

Matsuoka, K., Shigematsu, Y., Ichioka, Y., and Suzuki, T. (1982). *Appl. Opt.* **21**, 4493–4499.

Meinel, E. S. (1986). *J. Opt. Soc. Am. A* **3**, 787–799.

Mendes, M., and DePolignac, C. (1975). *Nucl. Instrum. Methods* **127**, 413–419.

Mendes, M., DePolignac, C., and Delestre, C. (1975). *Nucl. Instrum. Methods* **127**, 405–411.

Metropolis, N., Rosenbluth, A. W., Rosenbluth, M. N., Teller, A. H., and Teller, E. (1953). *J. Chem. Phys.* **21**, 1087–1092.

Miller, J. P., Roska, T., Sziranyi, T., Crounse, K. R., Chua, L. O., and Nemes, L. (1994). *Proceedings of the Third IEEE International Workshop on Cellular Neural Networks and Their Applications (CNNA-94), Rome, December 18–21*, 237–242.

Minsky, M., and Papert, P. (1969). "Perceptrons." MIT Press, Cambridge, Massachusetts.

Montgomery, W. D. (1982). *Opt. Lett.* **7**, 1–3.

Morawski, R. Z., Szczecinski, L., and Barwicz, A. (1995). *J. Chemom.* **9**, 3–20.

Nashed, M. Z. (1981). *IEEE Trans. Antennas Propag.* **AP-29**, 220–231.

Navarro, R., Santamaria, J., and Gomez, R. (1987). *Astron. Astrophys.* **174**, 344–351.

Paik, J. K., and Katsaggelos, A. K. (1990). *Proc. SPIE Int. Soc. Opt. Eng.* **1246**, 298–307.

Paik, J. K., and Katsaggelos, A. K. (1992). *IEEE Transactions on Image Processing* **1**, 49–63.

Papoulis, A. (1975). *IEEE Trans. Circuits Syst.* **9**, 735–742.

Pina, R. K., and Puetter, R. C. (1993). Publications of the Astronomical Society of the Pacific **105**, 630–637.

Podilchuk, C. I., and Mammone, R. J. (1990). *J. Opt. Soc. Am. A* **7**, 517–521.

Podilchuk, C. I. (1992a). *In* "Computational Methods of Signal Recovery and Recognition" (R. J. Mammone, ed.), p. 79. Wiley, New York.

Podilchuk, C. I. (1992b). *In* "Computational Methods of Signal Recovery and Recognition" (R. J. Mammone, ed.), p. 73. Wiley, New York.

Podilchuk, C. I. (1992c). *In* "Computational Methods of Signal Recovery and Recognition" (R. J. Mammone, ed.), pp. 55–107. Wiley, New York.

Puetter, R. C., and Pina, R. K. (1993). Proceedings of the SPIE—The International Society for Optical Engineering **1946**, 405–416.

Ramakrishnam, R. S., Mullick, S. K., Rathore, R. K. S., and Subramanian, R. (1979). *Appl. Opt.* **18**, 464–468.

Richards, M. A., Schafer, R. W., and Mersereau, R. M. (1979). *IEEE International Conference on Acoustics, Speech, and Signal Processing*, 401–404.

Richardson, W. H. (1972). *J. Opt. Soc. Am.* **62**, 55–59.

Robaux, O., and Roizen-Dossier, B. (1970). *Opt. Acta* **17**, 733–746.

Rumelhart, D. E., and McClelland, J. L. (1986). "Parallel Distributed Processing: Explorations in the Microstructure of Cognition." Vol. 1, Foundations, 4-Vol. Ed. MIT Press, Cambridge, Massachusetts.

Rushforth, C. K., and Frost, R. L. (1980). *J. Opt. Soc. Am.* **70**, 1539–1544.

Rushforth, C. K., Crawford, A. E., and Zhou, Y. (1982). *J. Opt. Soc. Am.* **72**, 204–211.

Saghri, J. A., and Tescher, A. G. (1980). *Proceedings of the SPIE Conference on Advances in Image Transmission II* **249**, 71–77.

Saghri, J. A., and Tescher, A. G. (1984). *Opt. Eng.* **23**, 675–677.

Sato, T., Norton, S. J., Linzer, M., Ikeda, O., and Hirama, M. (1981). *Appl. Opt.* **20**, 395–399.

Schafer, R. W., Mersereau, R. M., and Richards, M. A. (1981). *Proc. IEEE* **69**, 432–450.

Schell, A. C. (1965). *Radio Electron. Eng.* **29**, 21–26.

Schure, M. R. (1991). *J. Chromatogr.* **550**, 51–69.

Schure, M. R., Barman, B. N., and Giddings, J. C. (1989). *Anal. Chem.* **61**, 2735–2743.

Sezan, M. I., ed. (1992). "Selected Papers on Digital Image Restoration," Vol. 47, SPIE Milestone Series. SPIE, Bellingham, Washington.

Sezan, M. I., and Stark, H. (1982). *IEEE Transactions on Medical Imaging MI-1*, 95–101.

Sezan, M. I., and Tekalp, A. M., eds. (1990a). *Opt. Eng.* **29**, 391–574.

Sezan, M. I., and Tekalp, A. M. (1990b). *Opt. Eng.* **29**, 393–404.

Simpson, P. K. (1990). "Artificial Neural Systems: Foundations, Paradigms, Applications, and Implementations." Pergamon, San Diego.

Siska, P. E. (1973). *J. Chem. Phys.* **59**, 6052–6060.

Sivakumar, K., and Desai, U. B. (1992). *IEEE Transactions on Signal Processing*, **41**, 2018–2022.

Smith, W. E., Barrett, H. H., and Paxman, R. G. (1983). *Opt. Lett.* **8**, 199–201.

Stark, H., ed. (1987). "Image Recovery: Theory and Application." Academic Press, Orlando, Florida.

Stark, H., Cahana, D., and Webb, H. (1981a). *J. Opt. Soc. Am.* **71**, 635–642.

Stark, H., Cahana, D., and Habetler, G. J. (1981b). *Opt. Lett.* **6**, 259–260.

Stark, H., Cruze, S., and Habetler, G. (1982). *J. Opt. Soc. Am.* **72**, 993–1000.

Stooke, P. J., and Abergel, A. (1991). *Astron. Astrophys.* **248**, 656–668.

Strickland, R. N., and Chandler, D. W. (1991). *Appl. Opt.* **30**, 1811–1819.

Tanabe, K. (1971). *Numer. Math.* **17**, 203–214.

Terebizh, V. Y. (1995). *Phys.-Usp.* **38**, 137–167.

Thomas, G. (1981). *Proceedings of the IEEE Conference on Acoustics, Speech and Signal Processing*, **1**, 47–49.

Tikhonov, A. N. (1963). *Sov. Math. (Providence)* **4**, 1624–1627.

Tikhonov, A. N., and Arsenin, V. Y. (1977). "Solutions of Ill-Posed Problems." Winston, Washington, D.C.

Trussell, H. J. (1976). Ph.D. Thesis, University of New Mexico, Albuquerque.

Trussell, H. J., and Hunt, B. R. (1979). *IEEE Trans. Comput.* **C-27**, 57–62.

Trussell, H. J., and Civanlar, M. R. (1984). *IEEE Transactions on Acoustics, Speech, and Signal Processing* **ASSP-32**, 201–212.

Van Toorn, P., and Ferwerda, H. A. (1977). *Optik* **47**, 123–134.

Vemuri, V., and Jang, G. S. (1992). *J. Franklin Inst.* **329**, 241–257.

Wang, L.-X., and Mendel, J. M. (1992). *Geophysics* **57**, 670–679.

Wang, Q., Allred, D. D., and Knight, L. V. (1995). *J. Raman Spectrosc.* **26**, 1039–1043.

Wang, Y., and Frieden, B. R. (1995). *Appl. Opt.* **34**, 5938–5944.

Wells, D. C. (1980). *SPIE* **264**, 148–154.

Widrow, B., and Hoff, M., Jr. (1960). *IRE WESCON Conv. Rec.*, PT4, 96–104.

Widrow, B., and Stearns, S. D. (1984). "Adaptive Signal Processing." Prentice-Hall, Englewood Cliffs, New Jersey.

Widrow, B., Glover, J. R., Jr., McCool, J. M., Kaunitz, J., Williams, C. S., Hearn, R. H., Zeidler, J. R., Dong, E., Jr., and Goodlin, R. C. (1975). *Proc. IEEE* **63**, 1692–1716.

Willson, P. D. (1973). Ph.D. Thesis, Michigan State University, East Lansing.

Wolter, H. (1961). *In* "Progress in Optics," (E. Wolf, ed.). Vol. 1.

Wong, H. C. F. (1971). M.Sc. Thesis, Queen's Univ., Kingston, Ontario, Canada.

Xu, J., and Wang, J. (1992). *J. Quant. Spectrosc. Radiat. Transfer* **48**, 419–426.

Xu, C., Aissaoui, I., and Jacquey, S. (1994). *J. Opt. Soc. Am. A* **11**, 2804–2808.

Youla, D. C. (1978). *IEEE Trans. Circuits Syst.* **CAS-25**, 694–702.

Youla, D. C., and Webb, H. (1982). *IEEE Transactions on Medical Imaging* **MI-1**, 81–94.

Zhou, Y., and Rushforth, C. K. (1982). *Appl. Opt.* **21**, 1249–1252.

Zhou, Y.-T., Chellappa, R., Vaid, A., and Jenkins, B. K. (1988). *IEEE Trans. Acoust. Speech Signal Process.* **36**, 1141–1151.

Chapter 5 | Convergence of Relaxation Algorithms

Paul Benjamin Crilly

Department of Electrical Engineering, The University of Tennessee
Knoxville, Tennessee

List of Symbols

c	upper bound on x
e	error constant
$\mathbf{E}$	error vector
$f(x)$	filter function to remove noise
$\mathbf{i}, \mathbf{o}, \hat{\mathbf{o}}, \mathbf{n}$	vectors of $i(x)$, $o(x)$, $\hat{o}(x)$, $n(x)$, where x is a discrete variable
$I(\omega), \tau(\omega)$	Fourier transforms of $i(x)$, $s(x)$
$i(x)$	observed spectrum, or data
j	imaginary operator such that $j^2 = -1$
M	constant, an upper bound
$n(x)$	additive noise
$\mathbf{o}^*$	version of $\mathbf{o}$ that is in error due to errors in T
$\hat{\mathbf{o}}^{*(k)}$	version of $\hat{\mathbf{o}}^{(k)}$ that is in error due to errors in T
$o(x)$	true spectrum, not "distorted" by observation
$\hat{o}^{(k)}(x)$	kth estimate of $o(x)$
$\hat{o}(x)$	estimate of $o(x)$
$\mathbf{R}$	residual vector
RMS	root mean square

$\mathbf{s}$	matrix form of $s(x)$, where x is a discrete variable
$\mathscr{S}$	a linear vector space
$s(x)$	response or spread function
$s_e(x), \tau_e(\omega)$	even part of $s(x)$ and $\tau(\omega)$
$s^{(n)}(x)$	nth derivative of $s(x)$
SNR	signal-to-noise ratio
T	an operator
T^k	operator T applied k times
T^*	version of T that is in error
x	independent variable, such as time, wave number, or index
$y(x)$	linear restoring filter
$\otimes$	denotes convolution operation
$\tau^*(\omega)$	complex conjugate of $\tau(\omega)$
ω	conjugate of x; Fourier frequency in radians per units of x

I. Introduction

In choosing to use a given iterative deconvolution method, a concern arises as to what degree does the kth estimate accurately represent the true solution. Furthermore, we are interested in what factors affect this, the robustness of the algorithm, the merits of this method versus another, and if convergence can be enhanced. The ultimate question is "what confidence should be placed in the recovered solution?"

With linear methods such as Van Cittert's, it is relatively easy to answer some of these questions mathematically. Unfortunately, convergence has not yet been proved mathematically for some of the nonlinear constrained methods, even though they perform considerably better in practice. Most methods require prior knowledge of the impulse response function, $s(x)$, of the system. What then are the adverse consequences if this is not precisely known? In the case of gas chromatographic data, this issue is especially relevant because the system is only approximately time invariant. In order to understand fully the convergence properties of these iterative, nonlinear methods, we have to rely to some degree on experimental evaluations.

Many of the factors that affect convergence of iterative methods are discussed in this chapter. The topics include the mathematical basis of convergence, factors in the data and impulse response function that affect convergence, and, finally, ways convergence can be enhanced. In Chapter 7, Crilly, Blass, and Halsey evaluate various methods and finds that the constrained Jansson method appears to have the best convergence properties in practice. Therefore, we focus primarily on that method.

The easiest and most informative way to examine convergence is to use simulated data. This way, we have complete control over the parameters that affect convergence and accurate knowledge of the true solution.

II. Error Analysis

When using iterative methods, the decision to continue iterating is usually based on achieving a specified minimum error value or convergence rate. However, there are some pitfalls with respect to error analysis that should be understood (Chapra and Canale, 1988; Crilly, 1993).

Using the matrix notation of Chapter 3, Section III.A, we can model the convolution process as follows:

$$\mathbf{i} = \mathbf{so} + \mathbf{n}. \tag{1}$$

If the exact value of **o** were known, the decision to continue iterating could be based on the value of the following error function or the rate at which it changes:

$$\mathbf{E} = \mathbf{o} - \hat{\mathbf{o}}. \tag{2}$$

In most practical cases, **o** is not known, and an alternate residual can be calculated:

$$\mathbf{R} = \mathbf{i} - \mathbf{s\hat{o}}, \tag{3}$$

where **R** contains the effects of additive noise in **i** and errors in the estimate **ô**.

Cursory inspection suggests that using **R** would seem reasonable, but this can be misleading. Rewriting Eq. (1),

$$\mathbf{n} = \mathbf{i} - \mathbf{so}. \tag{4}$$

If Eq. (4) were subtracted from Eq. (3) we would find that

$$\mathbf{R} - \mathbf{n} = \mathbf{s(o - \hat{o})} \tag{5}$$

and

$$\mathbf{o} - \hat{\mathbf{o}} = \mathbf{s}^{-1}\mathbf{R} - \mathbf{s}^{-1}\mathbf{n}. \tag{6}$$

Equation (6) implies that even if we compute an estimate of **o** that leads to small residuals in Eq. (3), and if $\mathbf{s}^{-1}$ is ill-conditioned, then the exact error as calculated by Eq. (2) or (6) may be very large. Therefore, a small residual may not imply an accurate solution.

Equation (6) also shows the two types of errors that arise with any deconvolution method. The first term on the right-hand side of Eq. (6) is the regularization error, which Chapra and Canale (1988) define as the error due to the algorithm. The second term is the error due to noise.

III.   Mathematics of Convergence

Because Van Cittert's method is one of the earliest, simplest, and easiest to analyze and is a basis for other iterative algorithms, we will first examine its convergence properties and determine what is required for convergence. It may be, and often is the case, that other iterative methods will converge to the correct solution, one that could not be found with Van Cittert's algorithm. If we understand under what circumstances Van Cittert's method does not converge, we might be able to minimize the hindrances and make it easier for another method to converge. We will then consider the general class of iterative algorithms and develop some insight into their convergence properties.

A.   CONVERGENCE OF VAN CITTERT'S METHOD

The analysis of Van Cittert's method in Chapter 3, Section IV.D, shows that as $k \to \infty$ the algorithm converges to the inverse filter if

$$|1 - \tau(\omega)| < 1. \tag{7}$$

Equation (7) can be rewritten as

$$0 < \mathrm{Re}[\tau(\omega)] < 2. \tag{8}$$

As was discussed in Chapter 3, in order for the inverse filter to provide a unique solution, the observed signal has to be noiseless at frequencies where the system function approaches zero, or

$$I(\omega) = 0 \quad \text{whenever} \quad \tau(\omega) \to 0. \tag{9}$$

Equation (7) puts additional restrictions on the spread function (Hill and Ioup, 1976). Satisfying Eq. (7) may require a shift in the origin. If so, then a corresponding change will have to be made to $i(x)$ and of course the solution will be similarly shifted.

In order to preserve the total area under the signal function, when performing the deconvolution, $s(x)$ is scaled such that it has a unit area, and thus the upper bound of Eq. (8) is met.

Hill and Ioup (1976) have done the following analysis to determine additional explicit requirements for $s(x)$ so that the lower bound of Eq. (8) is satisfied.

If $s(x)$ is a real function, then $\tau(\omega) = \tau^*(-\omega)$ and the even part of $\tau(\omega)$ has to satisfy:

$$0 < \tau_e(\omega) < 2. \tag{10}$$

If we take the magnitude of the inverse Fourier transform of $\tau_e(\omega)$ we have

$$|s_e(x)| = \left| \int_{-\infty}^{\infty} e^{j2\pi x\omega} \tau_e(\omega)\, d\omega \right| \tag{11}$$

$$= \left| \int_{-\infty}^{\infty} \cos(2\pi x\omega) \tau_e(\omega)\, d\omega \right|. \tag{12}$$

Thus we find

$$|s_e(x)| < \int_{-\infty}^{\infty} \tau_e(\omega)\, d\omega, \tag{13}$$

giving

$$|s_e(x)| < s_e(0) \qquad \text{for all} \quad x \neq 0. \tag{14}$$

The condition of Eq. (14) requires $s(x)$ to have a unique maximum at $x = 0$. This precludes systems that have spread functions of shapes such as those shown in Fig. 1.

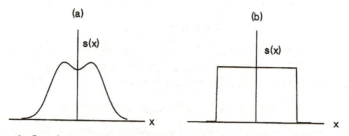

Figure 1 Impulse response functions for which Van Cittert's method will not converge because $x = 0$ is not located at a unique maximum of the even part of the function. (Hill, N. R. and Ioup, G. E. (1976). *J. Opt. Soc. Am.* **66**, p. 488. Reprinted with permission.)

Hill and Ioup (1976) have also considered the derivative property of the Fourier transform in Eq. (11) to show additional necessary conditions for convergence. These are

$$|s_e^{(n)}(x)| < |s_e^{(n)}(0)| \qquad \text{for all} \quad x \neq 0 \tag{15}$$

and

$$(j)^n s_e^{(n)}(0) > 0 \tag{16}$$

with n even. These are the conditions on the values of even derivatives of $s_e(x)$. For example, the spread functions shown in Fig. 2 would not converge because their second derivatives have singularities at $x = \pm c$ that are larger than singularities that might occur at the origin.

The analysis so far has presented a way to inspect the spread function visually and determine beforehand if convergence will be possible. Lajendijk and Biemond (1991) have shown that in order for Van Cittert's algorithm to converge it is necessary for all eigenvalues of matrix s to be positive.

B. CONVERGENCE OF A GENERAL CLASS OF ITERATIVE ALGORITHMS

Though the previous approaches toward showing convergence may be satisfactory for Van Cittert's method, they are not readily adaptable to the nonlinear constrained methods and thus far it has not been possible to show convergence in closed mathematical form. However, several authors (Biemond *et al.*, 1990; Crilly, 1992; Kang and Katsaggelos, 1994; Katsagge-

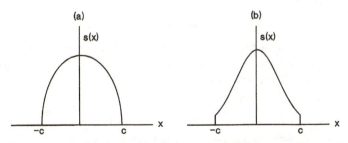

Figure 2 Impulse response function for which Van Cittert's method will not converge because of singularities in the second derivative. (Hill, N. R. and Ioup, G. E. (1976). *J. Opt. Soc. Am.* **66**, p. 488. Reprinted with permission.)

los, 1989; Schafer *et al.*, 1981; Singh and Tandon, 1986; Thomas, 1981; Wendroff, 1966; Xu *et al.*, 1994) have presented a signal-space approach as a method to analyze convergence for a limited class of iterative algorithms. Their work looks at things such as noise, errors in the spread function, and data class, which affect convergence. In particular, Singh and Tandon (1986) have made an excellent presentation on convergence, which will be summarized herein. (Their paper has also provided an excellent list of references.) Rushforth (1987) provides the mathematical definitions crucial to understanding this material.

1. Notation and Definitions

Rewriting Eq. (1),

$$\mathbf{i} = \mathbf{so} \qquad \text{(noiseless)} \tag{17}$$

and

$$\mathbf{i}^+ = \mathbf{i} + \mathbf{n} \tag{18}$$

or

$$\mathbf{i}^+ = \mathbf{so} + \mathbf{n}. \tag{19}$$

Let $\mathbf{i}$, $\mathbf{i}^+$, and $\mathbf{o}$ belong to a linear vector space $\mathscr{S}$, which is a Banach space (Rushforth, 1987). For the purposes of analyzing our iterative methods with respect to convergence, we have the following (Collatz, 1966; Singh and Tandon, 1986):

1. In space $\mathscr{S}$, let $d(\mathbf{o}, \mathbf{i})$ be the distance between any two elements of vectors $\mathbf{o}, \mathbf{i}$.

2. Let the norm of vector $\mathbf{o}$ be $\|\mathbf{o}\|$.

3. Let $\{\hat{\mathbf{o}}^{(k)}\}$ be a Cauchy convergence sequence (Rushforth, 1987) such that

$$d(\hat{\mathbf{o}}^{(k)}, \mathbf{o}) \to \mathbf{o} \qquad \text{as} \quad k \to \infty.$$

4. Let T be some operator on the elements of $\mathscr{S}$, where T may include the spread function $\mathbf{s}$, and constraint operators. The domain and range of T are subsets of $\mathscr{S}$.

5. Operator T is bounded if

$$d(T\mathbf{o}, T\mathbf{i}) \leq Md(\mathbf{o}, \mathbf{i}) \tag{20}$$

with M being a constant that can limit operator T.

6. If $0 \leq M < 1$, then T is a contraction operator (Schafer *et al.*, 1981).

7. For iterative cases, operator T^k denotes that T is applied k times.

2. Convergence Analysis

Equation (14) of Chapter 3 and Eqs. (14) and (23) of Chapter 4 can be generalized as follows:

$$\hat{o}^{(k+1)} = T\hat{o}^{(k)} \tag{21}$$

and

$$T o = o, \tag{22}$$

where T is a contraction and is a function of the deconvolution operations of Eq. (14) of Chapter 3 and Eq. (14), including Eq. (22) or (23), of Chapter 4.

If the above assumptions hold and

$$d(\hat{o}^{(k+1)}, o) = d(T\hat{o}^{(k)}, o), \tag{23}$$

then

$$d(\hat{o}^{(k+1)}, o) \leq M d(\hat{o}^{(k)}, o) \tag{24a}$$

and

$$d(\hat{o}^{(k+1)}, o) \leq M^k d(\hat{o}^{(0)}, o). \tag{24b}$$

Equations (23), (24a), and (24b) imply that if operator T is properly chosen, then there will be an upper bound on the error between $\hat{o}^{(k)}$ and o.

3. Errors in Operator T

Because operator T is a function of the impulse response s and whatever constraint function is used, errors in it are caused by corresponding errors in s and/or an improperly chosen constraint function. In practical instrumental situations, most errors are in s. For example, with gas chromatographic data, the impulse function may be in error for one or all of the following reasons: (a) instrument conditions vary (i.e., drift), (b) isothermal methods cause peak spreading with increasing retention times, and (c) some compounds have widely different tailing properties.

The question is how do we model the effects of these discrepancies on our estimate, and what are their practical consequences?

1. Assume Eq. (21) converges to the correct solution o given that T is correct.
2. Let T^* be a version of T that is in error so that $T^*\hat{o}^{*(k)}$ converges to o^*.

3. With T^* bounded such that

$$d(T^*\mathbf{o}, T\mathbf{o}) < e \qquad \text{for all } \mathbf{o} \text{ in } \mathscr{S}, \qquad (25)$$

with e being some error value.

4. Then we obtain

$$d(\hat{\mathbf{o}}^{*(k+1)}, \hat{\mathbf{o}}^{(k+1)}) = d(T^*\hat{\mathbf{o}}^{*(k)}, T\hat{\mathbf{o}}^{(k)}), \qquad (26)$$

$$\leq M d(\hat{\mathbf{o}}^{(k)}, \hat{\mathbf{o}}^{*(k)}) + e, \qquad (27)$$

with $\hat{\mathbf{o}}^{*(k)}$ obtained from $\hat{\mathbf{o}}^{(0)}$ using Eq. (21) and operator T^*.

5. From Eqs. (26) and (27) and M applied k times, we have

$$d(\hat{\mathbf{o}}^{*(k+1)}, \hat{\mathbf{o}}^{(k+1)}) \leq (1 - M^{k+1})/(1 - M)e. \qquad (28)$$

6. If $k \to \infty$, $\hat{\mathbf{o}}^{*(0)} = \hat{\mathbf{o}}^{(0)}$, and $M < 1$, then in the limit we have

$$d(\mathbf{o}^*, \mathbf{o}) \leq e/(1 - M). \qquad (29)$$

From Eqs. (27)–(29) we see even if errors occur in T but T is still a contraction, then there is an upper limit to the errors we can expect from our estimate.

4. Noise

We model the effects of noise in a way similar to the one we used to model the errors in operator T. If we let $\hat{\mathbf{o}}^{*(k)}$ be the kth estimate resulting from operator T onto $\mathbf{i}^+$, and $\hat{\mathbf{o}}^{(k)}$ be the kth estimate if the data were noiseless, we would obtain

$$\hat{\mathbf{o}}^{*(k+1)} = \hat{\mathbf{o}}^{(k+1)} + \mathbf{n}^{(k+1)}. \qquad (30)$$

Because T is a contraction operator, Singh and Tandon (1986) report that the resulting estimate's signal-to-noise ratio (SNR) at the kth step is

$$\text{SNR}^k \leq \|\mathbf{i}\|/\|\mathbf{n}\| = \text{original SNR}. \qquad (31)$$

Singh and Tandon (1986) also report that if constraints are incorporated into the operator T, then the resultant estimate's signal-to-noise ratio could actually be improved.

If the noise is white, a large portion of its content does not occupy the same spectrum as the information to be restored. Therefore, its adverse effects can be reduced if the data are prefiltered. Crilly (1990a, b) discusses the use of n-point polynomial and cross-correlation prefiltering to improve the ability of constrained methods to restore noisy data.

However, even more troublesome, if a system has $1/f$ (or $1/\omega$) noise characteristics, successful deconvolution may be extremely difficult because the noise may occupy the same spectral interval as the information.

C. DATA CLASS AND CONVERGENCE

As previously reported, the successful use of Van Cittert's method precluded use of some types of impulse response functions. However, there is another perspective to explain why certain functions cannot be used or why it is difficult to recover certain functions, $o(x)$.

Deconvolution restores the frequency components that were attenuated, or in some cases even lost due to the convolution operation. This is further described by the bandwidth extrapolation, Eq. (18) of Chapter 4. Therefore, the performance of an algorithm depends on the spectral extent of the signal, $o(x)$, to be recovered. For example, it is virtually impossible to recover the square wave signal of Fig. 3a, because it has an almost infinite bandwidth. Conversely, it is relatively easy to recover the Gaussian signal of Fig. 3b, because it has a limited frequency spectrum. Experimental testing bears this out.

There are similar difficulties in performing deconvolution using Fourier methods when the impulse response function is a pulse or other function that has a near-infinite frequency extent.

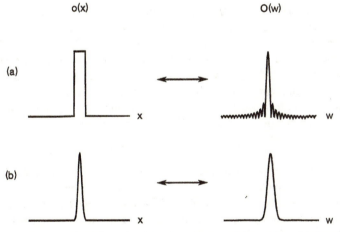

Figure 3 Fourier transform pairs of various classes of input data. (a) Rectangle pulse, (b) Gaussian pulse.

D. ENHANCEMENTS TO IMPROVE CONVERGENCE

As already mentioned, and as observed in Chapter 4, Eq. (18), the key to success in deconvolving is to restore those frequency components that were lost in the convolution without significantly restoring the noise components contained in the same passband. This can be done by pre-filtering the observed function $i(x)$ as follows:

$$i(x) \otimes f(x) \rightarrow i(x). \tag{32}$$

Function $f(x)$ should be such that it passes the essential information contained in $i(x)$, but significantly rejects the noise components. Because we have modified $i(x)$, the spread function has to be changed in the same manner:

$$s(x) \otimes f(x) \rightarrow s(x). \tag{33}$$

What is critical is that the scheme must remove the noise without significantly affecting the information content of the signal. Spectroscopic and chromatographic applications can use an n-point polynomial filter (Crilly, 1990a, b; Enke and Nieman, 1976; Madden, 1978; Savitzky and Golay, 1964; Steinier *et al.*, 1972; Willson and Edwards, 1976; Willson and Polo, 1981).

Alternatively, the reblurring method of Kawata and Ichioka (1980a, b) can be used to prefilter the data where filter function $f(x)$ is

$$f(x) = s(-x). \tag{34}$$

Operations (32) and (34) constitute auto- and cross-correlation filtering, which has the effect of eliminating those frequency components that are not in common to both functions (Crilly, 1990b; Hieftje *et al.*, 1973; Horlick, 1973; Lam *et al.*, 1982). If much of the noise is outside the passband of the information, and given that $s(x)$ is noiseless, the practical result is to eliminate most of the noise contained in $i(x)$.

In addition to noise removal, cross-correlation prefiltering can guarantee that the positive condition of Eq. (8) is met by

$$s(x) \otimes s(-x) \rightarrow \tau(\omega)\tau^*(\omega) = |\tau(\omega)|^2. \tag{35}$$

This therefore ensures convergence over a wider range of data conditions (Kawata and Ichioka, 1980a, b; Schafer *et al.*, 1981; Thomas, 1981).

The eigenalaysis of Lajendijk and Biemond (1991) shows that with reblurring, the new **s** matrix only has to have nonzero eigenvalues in order for Van Cittert's method to converge to the inverse filter.

Crilly (1990b) shows how cross-correlation prefiltering can remove a significant amount of noise from the signal and thus enable deconvolution of data whose signal-to-noise ratio is as low as 1:1.

However, reduced convergence rate is the steep price we pay to enjoy the advantage of cross-correlation prefiltering. For example, data whose signal-to-noise ratio is 100:1 can be easily and accurately deconvolved with 100 iterations when prefiltered with a 9-point polynomial filter of Savitzky and Golay (1964) (Crilly, 1990a). However, when using cross-correlation prefiltering on the same data, it took 1000 iterations to achieve the same result with respect to root-mean-square (RMS) errors.

In order to speed the rate of convergence, Morris *et al.* (1987) propose a technique where the spread function is also updated after each iteration; the new one is then used in a subsequent iteration. The authors claim they can get factors of five or more increase in convergence rates.

Maeda and Murata (1984) and Agard (1989) modify the correction term in the basic iteration function by convolving it with variations of an inverse filter giving:

$$[i(x) - s(x) \otimes \hat{o}^{(k)}(x)] \to y(x) \otimes [i(x) - s(x) \otimes \hat{o}^{(k)}(x)], \quad (36)$$

where $y(x)$ represents the impulse response function of an inverse filter such as the Wiener filter. Their work shows that with this modification, the number of iterations to achieve some RMS error goal can be decreased by as much as 10-fold.

IV.   Experimental Testing

Data used for evaluation purposes are presented in Fig. 4. Figure 4a is the ideal signal and Fig. 4b was obtained by convolving the ideal signal with a Gaussian-shaped impulse response function. Noise was added to the signal of Fig. 4b to obtain the signal of Fig. 4c such that its signal-to-noise ratio was 4:1. For our purposes, signal-to-noise ratio is defined as the ratio of the peak signal amplitude to the peak-to-peak noise amplitude.

A.   ERRORS IN THE IMPULSE RESPONSE FUNCTION

To determine how errors in $s(x)$ affect deconvolution, the signal of Fig. 4b was deconvolved using Jansson's method with $k = 1000$ iterations and an $i(x)$ of various widths (in this case, standard deviations). Standard devia-

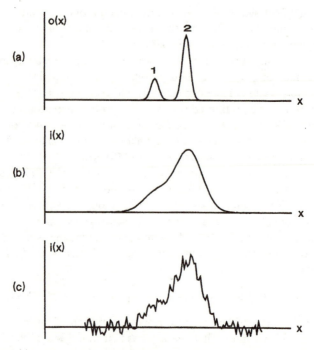

Figure 4   (a) Ideal signal, (b) distorted version, and (c) distorted and noisy version. (Crilly, 1991 © 1991 IEEE; reprinted with permission).

tions chosen were the correct value [i.e., the one used to generate $s(x)$ and ones that deviated by $\pm 15\%$]. The resultant estimates are plotted in Fig. 5. It should be noted that with this data class, it would take about 100 iterations to obtain similar results with respect to RMS errors.

The estimate shown in Fig. 5c that came about by the narrower $s(x)$ shows that despite the 10-fold increase in number of iterations, the algorithm is unable to converge fully and achieve significant resolution improvement. If we chose an extreme case, e.g., chose $s(x)$ to be a pure impulse, then there would be no change between the first and kth estimates.

Conversely, as the plot of Fig. 5d shows, if we chose $s(x)$ wider than its true width, the algorithm "overrestores" the signal and we do not have an accurate representation of its true shape.

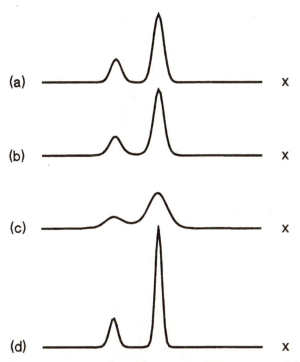

Figure 5   Results of deconvolving signal of Fig. 4b. (a) Ideal signal $o(x)$, (b) $o(x)$ using true $s(x)$, (c) $o(x)$ using $s(x)$ whose width deviated by -15%, (d) $o(x)$ using $s(x)$ whose width deviated by $+15\%$.

B. NOISE EFFECTS

The data of Fig. 4b were modified such that the SNR was 200:1 and then deconvolved with and without prefiltering by the 9-point polynomial filter of Savitzky and Golay (1964). The results are shown in Fig. 6. The data of Fig. 4c were also deconvolved after prefiltering via this 9-point polynomial filter; these results are shown in Fig. 7a. Finally, the data of Fig. 4c were deconvolved after prefiltering by the reblurring scheme of Eqs. (32)–(34), with the results shown in Fig. 7b.

As already mentioned, the prefiltering scheme chosen should be suffi- cient to remove the noise so that the deconvolved estimate has the minimum amount of spurious results, but without eliminating the impor-

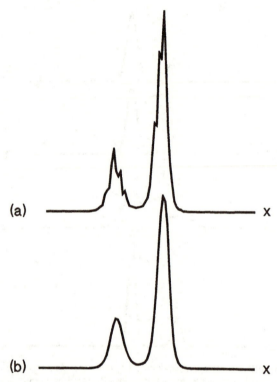

Figure 6 Deconvolution performance with and without prefiltering of image signal $i(x)$. Image signal was that of Fig. 4b, but with noise added so SNR = 200:1. (a) Signal $o(x)$ with no prefiltering; (b) $o(x)$ with 9-point polynomial prefiltering.

tant information the deconvolution process then has to restore. Too much prefiltering will require an additional number of iterations required to achieve a given RMS error. Which particular scheme chosen will depend on the signal-to-noise ratio, the types of data, and their resolution.

Many practical signals encountered have signal-to-noise ratios of at least 200:1 and thus the 9-point polynomial prefiltering scheme of Savitzky and Golay (1964) is sufficient to achieve good results, as illustrated in Fig. 6, without an increase in number of iterations. Testing using 5- and 7-point polynomial smoothing did not reduce the number of iterations, and actually increased the RMS error. However, with these same data, pre-

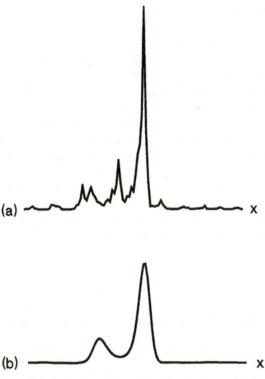

Figure 7   Effects of 9-point polynomial versus auto/cross-correlation (reblurring) prefiltering of image signal $i(x)$ on deconvolution performance. Image signal was that of Fig. 4c. (a) Signal $o(x)$ with 9-point polynomial prefiltering; (b) $o(x)$ with cross-correlation prefiltering.

filtering using cross-correlation, which is a much more severe form of prefiltering, required a 10-fold increase in the number of iterations to achieve the same RMS error results (Crilly, 1990a).

In the cases of extremely noisy data, as shown in Fig. 7, only the cross-correlation prefiltering was adequate to achieve acceptable estimates. Other tests (Crilly, 1990b) indicate that with cross-correlation prefiltering, Jansson's method will work with data whose signal-to-noise ratio is as low as 1:1.

V. Conclusion

We have examined the issues that affect convergence from both a mathematical and an experimental perspective. According to the foregoing mathematical treatment, the effects of data class, the type of algorithm, and the impulse function will determine if convergence is possible. These factors all affect our deconvolution operator T. If T is a contraction operator, then there is some upper limit to the RMS error and we can have more confidence in our solution. Furthermore, we can determine *a priori* which types of spread function or data class limit our ability to deconvolve a signal accurately. In Chapter 7, we will see that of all the conventional iterative algorithms, Jansson's method using two-sided constraints provides a solution with the least RMS error.

The other factor that hinders success is noise. But then, there are filtering schemes that can minimize even this problem.

References

Agard, D. A., Hiraoka, Y., Shaw, P., and Sedat, J. W. (1989). *Methods Cell Biol.* **30**, 363–365.

Biemond, J., Lagendijk, R. L., and Mersereau, R. M. (1990). *Proc. IEEE* **78**, 856–883.

Chapra, S. C., and Canale, R. P. (1988). "Numerical Methods for Engineers." McGraw-Hill, New York.

Collatz, L. (1966). *Functional Analysis and Numerical Mathematics* (H. Oser, trans.). Academic Press, New York.

Crilly, P. B. (1990a). *J. Chemom.* **4**, 51–59.

Crilly, P. B. (1990b). *J. Chemom.* **4**, 291–298.

Crilly, P. B. (1991). *IEEE Trans. Instrum. Meas.* **40**, 558–562.

Crilly, P. B. (1992). *Cheomom. Intell. Lab. Syst.* **12**, 291–298.

Crilly, P. B. (1993). *IEEE Trans. Instrum. Meas.* **40**, 78.

Enke, C., and Nieman, T. (1976). *Anal. Chem.* **48**, 705A–712A.

Hieftje, G., Bystroff, R., and Lim, R. (1973). *Anal. Chem.* **45**, 253–258.

Hill, N. R., and Ioup, G. E. (1976). *J. Opt. Soc. Am.* **66**, 487–489.

Horlick, G. (1973). *Anal. Chem.* **45**, 319–324.

Kang, M. G., and Katsaggelos, A. K. (1994). *Opt. Eng.* **33**, 3222–3232.

Katsaggelos, A. K. (1989). *Opt. Eng.* **28**, 735–748.

Kawata, S., and Ichioka, Y. (1980a). *J. Opt. Soc. Am.* **70**, 762–768.

Kawata, S., and Ichioka, Y. (1980b). *J. Opt. Soc. Am.* **70**, 768–772.

Lajendijk, R. L., and Biemond, J. (1991). "Iterative Identification and Restoration of Images." Kluwer, Boston.

Lam, R., Sparks, D., and Isenhour, T. (1982). *Anal. Chem.* **54**, 1927–1931.

Madden, H. (1978). *Anal. Chem.* **50**, 1383–1385.

Maeda, J. and Murata, K. (1984). *J. Opt. Soc. Am.* **A1**, 28–34.

Morris, C. E., Richards, M. A., and Hayes, M. H. (1987). *J. Opt. Soc. Am.* **4**, 200–207.

Rushforth, C. K. (1987). *In* "Image Recovery: Theory and Application" (H. Stark, ed.) pp. 2–7. Academic Press, New York.

Savitzky, A., and Golay, M. J. E. (1964). *Anal. Chem.* **36**, 1627–1639.

Schafer, R. W., Mersereau, R. M., and Richards, M. A. (1981). *Proc. IEEE* **69**, 432–450.

Singh, S., and Tandon, S. N. (1986). *Signal Processing* **11**, 1–11.

Steinier, Y., Termonia, Y., and Deltour, J. (1972). *Anal. Chem.* **44**, 1906–1909.

Thomas, G. (1981). *IEEE Trans. Acoust. Speech Signal Process.* **ASSP-29**, 938–939.

Wendroff, B. (1966). "Theoretical Numerical Analysis." Academic Press, New York.

Willson, P. D., and Edwards, T. H. (1976). *Appl. Spectrosc. Rev.* **12**, 1–81.

Willson, P. D., and Polo, S. R. (1981). *J. Opt. Soc. Am.* **71**, 599–603.

Xu, C., Aissaoui, I., and Jacquey, S. (1994). *J. Opt. Soc. Am.* **11**, 2804–2808.

Chapter 6 | Instrumental Considerations

William E. Blass
George W. Halsey

Department of Physics and Astronomy, The University of Tennessee
Knoxville, Tennessee

List of Symbols

$A(f)$	Fourier transform of signal processing function $A(t)$
$A(t)$	signal processing function
$A_M(x)$	measured absorptance
$\hat{A}_T(x)$	estimate of true absorptance $A_T(x)$
BW	full bandwidth of tuned amplifier
$B_T(x)$	flux at zero sample pressure
$\text{chop}(x)$	chopping function
d	grating spacing
$d\bar{\nu}/d\theta_d$	variation in monochromator wave number as the diffraction angle is varied

$d\bar{\nu}/d\theta_{\mathrm{g}}$	variation in monochromator "center" wave number as the grating angle is varied
$D_T(x)$	stray background flux
$e^{(k)}$	error function, also called root-mean-square error associated with kth iteration
f_0	center frequency of tuned amplifier
f_{c}	focal length of principal collimating mirror
f_{ch}	chopping frequency
FWHM	full width at half maximum
gate(t/T)	modulation function related to rect(t); gate$(t/T) = 1$ when t lies between 0 and T, and 0 otherwise
Hg:Ge	mercury-doped germanium detector
$H(t)$	1 when $x > 0$, 0 when $x \leq 0$
HWHM	half-width at half maximum
InSb	indium antimonide detector
$i(x)$	observed signal
j	integer-sequence-ordering variable
k	grating constant
m	number of passes
M	molecular weight
$M(t)$	signal modulation function (e.g., the change in center passband frequency of spectrometer as a function of time)
n	diffraction-order number
$n(t)$	noise voltage
N_j	number of counts in photon channel j
NEP	noise-equivalent power
$o(x)$	true signal as a function of x, here equivalent to time as a sequence-ordering variable
$\hat{o}^{(k)}(x)$	kth estimate of $o(x)$
P	power falling on detector
P_1, P_2	arbitrary power levels
$P(x)$	absorption coefficient
$P_{\mathrm{V}}(x)$	absorption coefficient for a Voigt profile line
Q	Q of a tuned amplifier ($\approx f_0/\mathrm{BW}$)
$r_{\max}^{(k)}$	constant coefficient of relaxation function $r^{(k)}[\hat{o}^{(k-1)}(x)]$
$r^{(k)}[\hat{o}^{(k-1)}(x)], r^{(k)}[y]$	relaxation function used at kth iteration
$r(x)$	response function
R	responsivity of a detector in volts per watt
R_{D}	detector resistance
$(R_{\mathrm{D}})_0$	R_{D} at zero signal
R_{L}	load resistance
R_{new}	data-acquisition rate with resolution degraded by multiplicative factor γ
R_{old}	data-acquisition rate prior to degradation
RAM	random access memory
RC	resistance-capacitance

$s(x)$	response function or instrumental spread function
$S(f)$	spectral density function
S/N	signal-to-noise ratio
t	time
T	absolute temperature
T_c	period of chop(x)
TDL	tunable diode laser
$U_T(x)$	transmittance; transmitted spectral flux
$v(t), V_{in}(t)$	input voltage of RC filter
V	voltage
V_0	voltage across detector at zero signal: $R_D = (R_D)_0$
$V_C(t), V_{out}(t)$	output voltage of RC filter
$V_{sig}(\alpha)$	output signal of bias circuit
w_g	ruled width of grating
W	slit width
x	independent variable
X	pressure times path length of sample
X_1, X_2	arbitrary quantities
α	parameter embodying all of the dependence of R_D on incident photon flux
γ	resolution degradation factor
δ_d	$\theta_g - \theta_d$
δ_i	$\theta_i - \theta_g$
Δt_{csam}	continuous scan sampling time interval
$\Delta \theta_d$	angle subtended by exit-slit width at principal collimating mirror
$\Delta \theta_{ssam}$	grating angle interval between data samples; if Ω_{step} is the change in grating angle for one step, $\Delta \theta_{ssam}$ is an integer multiple of Ω_{step}
$\Delta \bar{\nu}_D$	full width at half maximum of Doppler profile
$\Delta \bar{\nu}_{min}$	limit of resolution
$\Delta \bar{\nu}_{obs}$	full width at half maximum of observed line: this usage is different than in other chapters, where $\Delta \bar{\nu}$ is given a somewhat different meaning
$\Delta \bar{\nu}_{res}$	full width at half maximum of response function
$\Delta \bar{\nu}_{ssw}$	spectral slit width in cm^{-1}; note that $\Delta \bar{\nu}_{ssw}$ is a full width at half maximum equivalent (see text)
$\epsilon(\lambda)$	intermediate symbolic parameter
$\zeta(\omega)$	phase of $\Xi(\omega)$
η	signal-to-noise improvement factor
θ	angle between grating normal and optical axis of collimating mirror
θ_d	diffraction angle
θ_i	angle of incidence
θ_g	grating angle
λ	wavelength

$\bar{\nu}$ wave number $(= 1/\lambda)$
$\xi(t)$ transfer function of filter
$\Xi(\omega)$ transform of $\xi(t)$
τ amplifier time constant

I. Introduction

Application of deconvolution is at the same time seductively attractive and potentially dangerous. Deconvolution is attractive because the researcher can obtain resolution beyond that achievable without deconvolution. In addition, when resolution enhancement is not a goal, deconvolution can lead to richly enhanced acquisition rates. Deconvolution is dangerous because numerical processing may radically alter the character of experimental data and hence the conclusions drawn from them.

To grasp the flavor of this caveat, consider this: the adjustment range of an instrument's physical parameters is limited. Furthermore, many modifications, because they are time consuming and expensive, are carefully considered before implementation and fully tested afterwards. On the other hand, because deconvolution is numerical, radical changes are readily made and often not as thoroughly tested as are physical changes in the instrument. Very subtle effects are possible.

Lest the reader become disenchanted, consider the following: although the path to utilization of deconvolution contains pitfalls for the unwary, the process is testable at each step. Careful users come to view the process (and requisite computer system) as an extension of the observing device.

II. Resolution–Acquisition-Time Trade-Offs

Deconvolution of spectra, such as infrared absorption spectra, provides researchers with a tool that they can use to carry out a particular experiment. It provides an extra measure of flexibility in the design of experiments and in the observation process. In dispersive infrared spectroscopic systems, Blass and Halsey (1981) have shown that effective resolution–acquisition-time trade-offs may be made, owing to the fact that dispersive infrared spectroscopy is usually detector noise limited. Acquisition rates are therefore optical throughput dependent, which is equivalent to saying that acquisition rates are resolution dependent. Blass and Halsey (1981) show that, for a constant signal-to-noise ratio,

$$R_{new} = \gamma^5 R_{old}, \tag{1}$$

where γ is the factor by which the resolution has been degraded, R_{old} the old acquisition rate, and R_{new} the enhanced rate. For example, degrading the resolution by a factor of 2 increases the allowable acquisition rate by a factor of 32. Deconvolution can readily recover the factor of 2. Expensive scientific equipment may therefore be used more effectively. It also permits experiments that would otherwise be impractical owing to excessive observation time. Degrading resolution by a factor of 3, for example, allows an increase in the acquisition rate by a factor of 243.

When an improved signal-to-noise ratio is required, Eq. (1) becomes

$$R_{new} = (\gamma^5/\eta^2) R_{old}, \tag{2}$$

where η is the required improvement factor for the signal-to-noise ratio.

Not all forms of spectroscopy allow such productive resolution–acquisition-rate trade-offs. Where they are possible, deconvolution becomes an even more powerful experimental tool.

III.   Acquiring the Data: Dispersive Spectrophotometers

A large proportion of spectral data is acquired by dispersive spectrophotometry. The discussion that follows is restricted to instruments that use a diffraction grating as the principal dispersive element. The sense of the following also applies to systems that use a prism. In general, we treat systems using photosensitive detectors and fixed-position slits. Scanning is achieved by rotation of the diffraction grating.

To place the discussions of this and the following chapter in a clear context, we describe a specific system model. Application discussions may be seen in the light of a specific instrument. To use deconvolution techniques effectively, the instrument as well as the data acquired must be analytically well characterized.

A.   SYSTEM MODEL

For clarity and precision in the following discussion, the 5-m Littrow spectrometer at the University of Tennessee at Knoxville will serve as the prototype (Jennings, 1974). The optical diagram is presented in Fig. 1. The

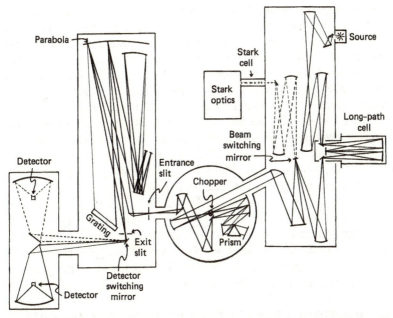

Figure 1 Optical diagram of the prototype system model. The drawing is not to scale and is not to be considered an optical ray diagram. The principal dispersing element is a coarse echelle ruled grating 20 × 40 cm wide. Theoretical double-pass resolution at four normal slits is approximately 0.0095 cm^{-1}; actual achievable resolution is approximately 0.009 cm^{-1}.

system is used as an absorption spectrometer; most of the discussion also applies to acquisition of emission spectra.

The source is a resistively heated carbon rod operating at 2600 K (Boyd *et al.*, 1974). From the source, the radiation follows the mirror path in sequence. As the optical beam enters the round vacuum tank the beam begins traversing a prism spectrometer. The need for the prism predisperser arises from the low density of rulings of the echelle gratings used in the grating monochromator (31.6 and 79/mm). The incident beam must pass through this optical bandpass filter prior to detection because of the limited free spectral range of the grating monochromator. The particular prism predisperser currently used is a Wadsworth–Littrow prism monochromator (Wadsworth, 1894; James and Sternberg, 1969; Stewart, 1970).

As the beam leaves the prism predisperser, it is focused on the entrance slit of the grating monochromator. The slit is curved, has variable width, and opens symmetrically about the chief ray (optical center line of system). The monochromator itself is of the off-axis Littrow variety (James and Sternberg, 1969; Stewart, 1970; Jennings, 1974) and uses a double-pass system described by McCubbin (1961). The double-pass aspect of the system doubles the optical retardation of the incident wave front and theoretically doubles the resolution of the instrument. The principal collimating mirror is a 5-m-focal-length, 102-cm-diam parabola.

From the exit slit, the beam is directed to either of two detectors by the movable mirror behind the exit slit. Typical detectors are InSb, a photovoltaic indium antimonide detector operated at 77 K, and He:Ge, a photoconductive Mercury-doped germanium detector maintained at $\approx$ 15 K by a closed-cycle helium refrigerator

The rotation of the grating is achieved by using a tangent arm drive. The tangent arm system is driven by a 200-step/revolution stepping motor. One step corresponds to approximately 1.45×10^{-7} rad (8.3×10^{-6} deg). Note particularly that the tangent drive results in an approximately linear relation between grating angle and drive-shaft rotation angle (Blass, 1976b). A block diagram of the entire system is shown in Fig. 2.

B. ANALYTICAL CHARACTERIZATION OF THE SYSTEM MODEL

The grating equation for both single- and double-pass Littrow monochromators is given by

$$\bar{\nu} = nk \csc \theta_g, \tag{3}$$

where $\bar{\nu} = 1/\lambda$ is the wave number, $[\bar{\nu}] = \text{cm}^{-1}$ (λ is the wavelength, $[\lambda] = \text{cm}$); $n = 1, 2, 3, \dots$ is the grating order; θ_g is the grating angle, that is, the angle between the collimating mirror axis and the grating normal; and $k = 1/2d$ is a grating constant, $[k] = \text{cm}^{-1}$ (d is the grating spacing, $[d] = \text{cm}$). Equation (3) is correct for a true Littrow in single pass and for a double-pass Littrow when the inversion of the single-pass exit image to form the double-pass entrance image is carried out symmetrically about the collimation axis (Jennings, 1974).

The following definitions are required: θ_g is the angle between a ray parallel to the collimator axis and the grating normal, the angle of incidence $\theta_i \equiv \theta_g + \delta_i$, and the diffraction angle $\theta_d \equiv \theta_g - \delta_d$. In the symmetrical Littrow case $\delta_i = -\delta_d$, and Eq. (3) follows from the general grating equation

$$\lambda = (2/nk)(\sin \theta_i + \sin \theta_d). \tag{4}$$

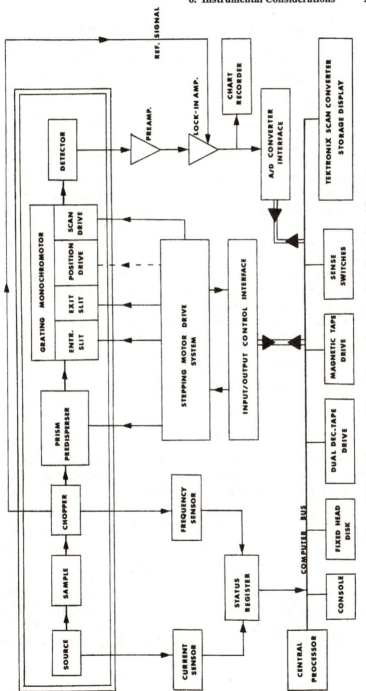

Figure 2 Block diagram of the spectrometric system at the University of Tennessee.

For the dispersion one finds for single pass (Jennings, 1974)

$$d\bar{\nu}/d\theta_{\mathrm{d}} = -(\bar{\nu}/2)\cot\theta_{\mathrm{g}}, \qquad \theta_{\mathrm{g}}, \theta_{\mathrm{i}} = \text{const}, \tag{5}$$

$$d\bar{\nu}/d\theta_{\mathrm{g}} = -\bar{\nu}\cot\theta_{\mathrm{g}}, \qquad \theta_{\mathrm{d}}, \theta_{\mathrm{i}} = \text{const}, \tag{6}$$

and for double pass

$$d\bar{\nu}/d\theta_{\mathrm{d}} = -(\bar{\nu}/4)\cot\theta_{\mathrm{g}}, \qquad \theta_{\mathrm{g}}, \theta_{\mathrm{i}} = \text{const}, \tag{7}$$

$$d\bar{\nu}/d\theta_{\mathrm{g}} = -\bar{\nu}\cot\theta_{\mathrm{g}}, \qquad \theta_{\mathrm{d}}, \theta_{\mathrm{i}} = \text{const}. \tag{8}$$

From Eqs. (6) and (8) one verifies that the angular separation of two specific wave numbers is the same in single and double pass. Equations (6) and (8) are used to determine scan rates or scanning observation intervals as discussed later.

To determine the spectral slit width, it is necessary to use Eqs. (5) and (7). Conceptually this can be viewed as the variation in the wave number passed through the monochromator as one moves from one edge of the exit slit to the other. The monochromator is assumed to be set at some center wave number given by Eq. (3).

Using Eqs. (5) and (7) and assuming $\delta_{\mathrm{i}} = \delta_{\mathrm{d}} \approx 0$, we obtain for the spectral slit width ($\Delta\bar{\nu}$, edge to edge of the slits)

$$\Delta\bar{\nu}_{\mathrm{ssw}} = |\partial\bar{\nu}/\partial\theta_{\mathrm{d}}|\Delta\theta_{\mathrm{d}}, \qquad \theta_{\mathrm{g}}, \theta_{\mathrm{i}} = \text{const}, \tag{9}$$

$$\Delta\bar{\nu}_{\mathrm{ssw}} = (\bar{\nu}W/2mf_{\mathrm{c}})\cot\theta_{\mathrm{g}}, \tag{10}$$

where W is the slit width, given by

$$W = f_{\mathrm{c}}\,\Delta\theta_{\mathrm{d}}, \tag{11}$$

m is the number of passes, and f_{c} is the focal length of the principal collimating mirror. The quantity $\Delta\theta_{\mathrm{d}}$ is the angle subtended by the slit width at the principal collimating mirror. Equations (5) and (7) have been used to simplify Eq. (9).

The theoretical Rayleigh limit of resolution is given by

$$\Delta\bar{\nu}_{\mathrm{min}} = (2mw_{\mathrm{g}}\cos\theta_{\mathrm{g}})^{-1}, \tag{12}$$

where w_{g} is the ruled width of the grating. The calculated spectral slit width of a well-designed, properly aligned system should be close to the observed resolution of the instrument. Resolution may be approximated by the full width at half maximum (FWHM) of spectral features expected to be considerably narrower in wave number than $\Delta\bar{\nu}_{\mathrm{ssw}}$. Alternatively, for high-dispersion systems, the FWHM of a known Doppler-limited single transition can be used as a measure of $\Delta\bar{\nu}_{res}$, the instrumental resolution (see Chapter 2, Sections II.G.1–II.G.3). Note that, throughout Chapters 6 and 7, $\Delta\bar{\nu}$ with any subscript is the *full width* at *half maximum* of the function specified.

The Doppler width (FWHM) is given by (Townes and Schalow, 1955)

$$\Delta\bar{\nu}_D = 7.162 \times 10^{-7}\,\bar{\nu}\sqrt{T/M}, \tag{13}$$

where T is the absolute temperature and M the molecular weight of the gas phase sample.

The properties of a gaussian result in the relationship

$$(\Delta\bar{\nu}_{obs})^2 = (\Delta\bar{\nu}_{res})^2 + (\Delta\bar{\nu}_D)^2, \tag{14}$$

where the instrument function is taken to be well approximated by a gaussian of full width (FWHM) equal to $\Delta\bar{\nu}_{res}$. Strictly speaking, Eq. (14) holds only for emission spectra. In absorption spectra, the transmitted flux is given by Eq. (9) of Chapter 2. The measured absorptance is given by Eq. (50) using Eq. (46) of Chapter 2. The measured absorptance profile for a Gaussian absorption coefficient of half-width at half maximum (HWHM) Δx_{Dop} and Gaussian response function of half-width (HWHM) Δx_{res} was calculated. The resulting HWHM was measured, as was the apparent peak absorptance. In Fig. 3 the percentage error, as $100[1 - \Delta x_{obs}(\text{meas})/\Delta x_{obs}(\text{calc})]$, is plotted against $\Delta x_{Dop}/\Delta x_{res}$. Figure 3 also presents the apparent peak absorptance versus $\Delta x_{Dop}/\Delta x_{res}$ for line strengths corresponding to true peak absorptances of 0.1 to 0.9. Note that all HWHM quantities (in the notation of Chapter 2) appear as ratios and are thus equivalent to ratios of the corresponding FWHM quantities.

For a more accurate representation of the instrumental distortion effects see Section II.G of Chapter 2. The approximation given by Eq. (14) is convenient and for high-dispersion instruments observing weakly absorbing spectral lines; it is often accurate enough for a number of experimental uses.

C. GENERAL DATA-ACQUISITION CONSIDERATIONS

There are two different methods of changing the grating angle θ_g. One method continuously changes θ_g and samples the data at sequential, equal time intervals Δt_{csam}. The alternative approach samples the data at specific angular intervals $\theta_g + j\,\Delta\theta_{ssam}$, where j is an integer-sequence-ordering

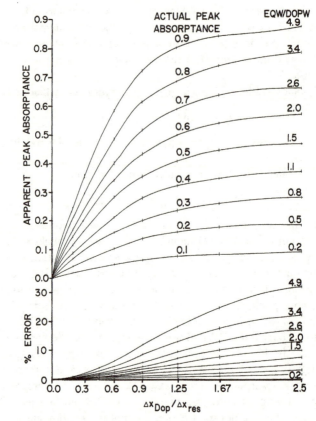

Figure 3   Upper traces, apparent peak absorptance vs. $\Delta x_{\mathrm{Dop}}/\Delta x_{\mathrm{res}}$, the Doppler width per unit resolution. Each trace is identified by the actual peak absorptance. Lower traces, percentage error incurred when $(\Delta x_{\mathrm{Dop}}^2 + \Delta x_{\mathrm{res}}^2)^{1/2}$ is used to approximate Δx_{obs} for an absorption line vs. $\Delta x_{\mathrm{Dop}}/\Delta x_{\mathrm{res}}$. The curves are labeled with the appropriate equivalent width per unit Doppler width as EQW/DOPW.

variable and $\Delta \theta_{\mathrm{ssam}}$ is the grating angle interval between samples. In both cases, the sampled data are converted from analog to machine-readable form and stored on some transportable medium such as magnetic tape or a floppy disk. [There is no reason why simple fast random access memory (RAM) storage is precluded except that all processing of the data—such as deconvolution—must be performed before the RAM can be used again to acquire data from the instrument. Such an approach is often not cost effective.]

The processes involved in data acquisition and the considerations relevant to obtaining appropriate data for deconvolution are nearly identical for both continuous scanning and step scanning instruments. Where no specific distinction is noted, no differences are significant at the current state of the deconvolution art. Sampling, aliasing, and other details of the data-acquisition process can be analyzed using the methods discussed in Chapter 1.

1. Signal-to-Noise Ratios

To increase resolution of recorded data effectively by using deconvolution, appropriately high signal-to-noise ratios are necessary. Depending on the spectral region involved, a factor that usually dominates the selection of detectors used, very different situations regarding signal-to-noise ratios may prevail.

a. Infrared Detection In the infrared region of the spectrum a variety of detectors are available (Stewart, 1970; Blass and Nielsen, 1974). The limiting source of noise is generally the detector, regardless of type, in infrared spectrophotometry using well-designed instrumentation. Most detectors in routine use in high-resolution spectroscopy are either photoconductive or photovolatic, and detection is achieved by measuring either a voltage or a current as a function of incident photon energy density. Noise in infrared spectrophotometry is generally assumed to be band-limited Gaussian white noise. (Other types of detection, including thermal detectors, are adequately modeled in this fashion.)

Because infrared spectrophotometry is detector noise limited, a profitable trade-off is possible between signal-to-noise ratio and resolution. Such a trade-off may be useful when deconvolution is used to recover resolution "lost" by opening monochromator slits to increase acquisition rates. In any system where degrading the resolution improves the signal-to-noise ratio nonlinearly, a potentially useful trade-off is possible.

b. Visible Detection Visible spectrophotometry generally makes use of photomultiplier detectors and photon-counting techniques. In this region, source noise is often the dominant noise source. Because the noise is assumed to be Poisson distributed, the signal-to-noise ratio may be represented as

$$\text{S/N} \propto N_j/\sqrt{N_j}, \tag{15}$$

where N_j represents the photon count in channel j. Data taken with equal counts in each channel are most desirable for deconvolution. Unfortunately, this places severe time restrictions on acquisition both in emission and in absorption spectrophotometry and is not often used. Equal-time-per-channel counting results in relatively poor signal-to-noise ratios in the small-signal regions in emission spectroscopy and in the vicinity of absorption lines in absorption spectroscopy.

   c. *Signal-to-Noise Calculations*   In a concise treatment of signal-to-noise ratios in Fourier spectroscopy compared with dispersive spectroscopy, Treffers (1977) shows that

$$\text{S/N} = (P/\text{NEP})\sqrt{2} \int_{-\infty}^{\infty} A(t)M(t)\, dt \bigg/ \left[\int_{-\infty}^{\infty} A(t)^2\, dt\right]^{1/2}, \quad (16)$$

where the noise is assumed to be white noise, P is the signal power falling on the input aperture of the modulator represented by $M(t)$, NEP is the noise-equivalent power of the detector, and $A(t)$ is the signal processing function described later.

   Treffer's model system consists of signal power P (in watts) falling on the input aperture of the modulator (i.e., the spectrophotometer) that modulates the input power as a function of time by a factor $M(t)$ such that $0 \leq M(t) \leq 1$. The modulation function $M(t)$ is *not* the modulation of the signal due to chopping but modulation of the signal due to scanning. The modulated signal falls on a detector with responsivity R (in volts per watt) (Kruse *et al.*, 1962; Stewart, 1970) and flat frequency response. The idealized instantaneous output voltage $PRM(t)$ (in volts) has noise $n(t)$ (in volts) added to it, and the resulting model signal is linearly transformed, producing the processed and integrated signal

$$\text{S} = PR \int_{-\infty}^{\infty} A(t)M(t)\, dt. \quad (17)$$

It is assumed that the noise voltage $n(t)$ is the result of a real stationary process (Davenport and Root, 1958) with zero mean. Because it can be shown that the spectral density function $S(f)$ is the Fourier transform of the autocorrelation function of the noise, it follows that the rms noise is given by

$$\text{N} = \left[\int_{-\infty}^{\infty} S(f)|A(f)|^2\, df\right]^{1/2}, \quad (18)$$

where $A(f)$ is the Fourier transform of the signal processing function $A(t)$.

Assuming that the noise is white noise, that is, $S(f) = S_0$, Eq. (18) is simplified and the signal-to-noise ratio is determined from Eqs. (17) and (18). Equation (18) is simplified by the definition of the noise-equivalent power (in watts per $Hz^{1/2}$)

$$\text{NEP} = (2S)^{1/2}/R, \tag{19}$$

yielding Eq. (16).

We define a function gate (t/T), which is equal to 1 when t is in the range from 0 to T, and 0 otherwise. The quantity T is thus the length of time the function is nonzero. Using the result given by Eq. (16), Treffers shows the following.

(1) For a photometer with no chopping such that

$$M(t) = A(t) = \text{gate}(t/T) = \begin{cases} 1 & \text{if } 0 \le t \le T \\ 0 & \text{otherwise,} \end{cases}$$

$$\text{S/N} = (P/\text{NEP})(2T)^{1/2}. \tag{20}$$

(2) For a square-wave-chopped photometer with up-down integration such that

$$M(t) = A(t) = \text{gate}(t/T)$$

half the time and

$$M(t) = 0 \text{ and } A(t) = -\text{gate}(t/T)$$

half the time,

$$\text{S/N} = (P/\text{NEP})(T/2)^{1/2}. \tag{21}$$

(3) For a square-wave-chopped photometer with phase-sensitive detection following a bandpass filter centered on the chopping frequency such that

$$M(t) = \begin{cases} \text{gate}(t/T) & \text{when } A(t) > 0 \\ 0 & \text{when } A(t) < 0 \end{cases}$$

and

$$A(t) = \cos 2\pi f t \, \text{gate}(t/T),$$

$$\text{S/N} = (P/\text{NEP})(4\pi/\pi^2)^{1/2}. \tag{22}$$

These results are valid for consideration of the signal-to-noise ratios of the several signal processing modes. The processor may be expanded to include other elements in the systems and the signal-to-noise ratio calculated on the basis of Eq. (16).

2. Analog Signal Processing

The detector output signal is generated by a current preamplifier for photovoltaic detectors, such as InSb, and by a simple detector bias circuit shown in Fig. 4 for photoconductive detectors, such as PbS and Hg:Ge. The voltage signal derived from the bias circuit is normally preamplified and forwarded to a phase-sensitive synchronous detector usually embodied in a lock-in amplifier (Stewart, 1970; Blass, 1976b).

The behavior of the bias circuit of Fig. 4 is of some interest. Let $(R_D)_0$ be the equivalent detector resistance at zero signal. The quantity R_L is the load resistance (preferably wire-wound or metal film; see Stewart, 1970) and V_0 is the background detector voltage that corresponds to a detector resistance $R_D = (R_D)_0$. Then the output signal of the bias circuit, $V_{sig}(\alpha)$, is given by

$$V_{sig}(\alpha) = VR_D(\alpha)/[R_L + R_D(\alpha)], \tag{23}$$

where α represents the net parametric dependence of R_D on incident photon flux. Because the radiation is modulated by the spectrophotometer, generally by a mechanical chopper, the quantity of interest, for the signal sensed across the detector, is

$$\frac{\partial^2 V_{sig}}{\partial R_L \, \partial \alpha} = \frac{V(R_D - R_L)}{(R_D + R_L)^3} \frac{\partial R_D}{\partial \alpha}, \tag{24}$$

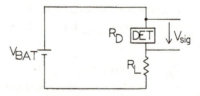

Figure 4 Simple photoconductive detector bias circuit. R_D is the detector resistance, $(R_D)_0$ the equivalent detector resistance at zero signal (i.e., background flux only on detector). R_L is the load resistance, V_0 the signal at $(R_D)_0$, V_{sig} the signal voltage at R_D. $V_0 = V_{bat}(R_D)_0/[R_L + (R_D)_0]$, $V_{sig}(t) = V_{bat}R_D/(R_L + R_D)$, and V_{bat} is the battery voltage.

which indicates that $(\partial V_{sig}/\partial\alpha)\,\delta\alpha$, the ac component of V_{sig}, is a maximum for $R_L = R_D$. It is readily verified that if the signal is sensed across the load resistance, the condition $R_L = R_D$ produces a maximum $\partial V/\partial\alpha$ in that case also. In either case, the ac component of the signal voltage is given by

$$V_{sig} = (V_{sig})_0 - (V/4R_D)(\partial R_D/\partial\alpha)\,\delta\alpha. \qquad (25)$$

If it is assumed that (i) the signal is sensed at a fixed wavelength and (ii) the chopping function is given by

$$\text{chop}(x) = \begin{cases} 1 & \text{for } T_c/4 \le x \le 3T_c/4 \\ 0 & \text{for } 0 < x < T_c/4,\, 3T_c/4 < x < T_c, \end{cases} \qquad (26)$$

where T_c is the period of the chopping function, so that

$$\delta\alpha = \gamma\,\text{chop}(x), \quad \gamma = \text{const}, \qquad (27)$$

(iii) the detector response time is negligibly short, and (iv) $\varepsilon(\lambda)$ is given by

$$\varepsilon(\lambda) = -(\partial R_D/\partial\alpha)\gamma, \qquad (28)$$

then we obtain

$$V_{sig} = (V_{sig})_0 + [\varepsilon(\lambda)V/4R_D]\text{chop}(x). \qquad (29)$$

This is the appropriate equation to Fourier transform if the frequency content of the detector output is desired.

The typical synchronous detection or demodulation system follows a tuned amplifier (tuned to optical chopping frequency f_{ch}) and is realized in the form of a lock-in amplifier. The lock-in amplifier effectively filters out the dc component of V_{sig} and produces an output proportional to the fundamental component (at f_{ch}) of the time-varying V_{sig}. As a part of the lock-in signal processing chain, the user must select an amplifier time constant τ, which is related to the bandwidth of the amplifier centered on f_{ch}. Although the bandwidth depends on the specific lock-in amplifier, a reasonable assumption is that the signal energy ($\propto V^2$) is down by 3 dB (50%) at

$$f = \pm 1/2\pi\tau, \qquad (30)$$

which is just the case for a simple signal-section RC filter (see Fig. 3 of Chapter 2) with 6 dB rolloff per octave (20 dB per decade). For the single-section RC filter we have from Eqs. (41) and (42) of Chapter 2

$$V_C(t) = V_{out}(t) = \xi(t) \otimes v(t) = \xi(t) \otimes V_{in}(t), \qquad (31)$$

where

$$\xi(t) = (1/\tau)H(t)e^{-t/\tau}, \qquad (32)$$

$$\tau = RC, \qquad (33)$$

and $H(t)$ is the Heaviside step function (see Chapter 1).

The transform of $\xi(t)$ is given by

$$\Xi(\omega) = |\Xi(\omega)|e^{j\zeta}, \qquad (34)$$

where the gain function normalized to unity at $\omega = 0$ is given by

$$|\Xi(\omega)| = (1/\tau)\left[\omega^2 + (1/\tau)^2 \right]^{-1/2} \qquad (35)$$

and the phase is

$$\zeta(\omega) = \tan^{-1}(\omega\tau). \qquad (36)$$

The 3-dB points are found by setting

$$(V_{out}/V_{in})^2 = \tfrac{1}{2} \qquad (37)$$

and solving Eq. (35) for f as a function of τ; the result is given by Eq. (30), the absolute value of which is the half-bandwidth. The decibel is a logarithmic ratio measure. The quantity X_1 is said to be 3 dB over X_2 when X_1 is twice X_2. That is, X_1 is $10 \log_{10}(X_1/X_2)$ dB greater than X_2. Because power is a function of the square of voltage, power ratios in terms of voltages are given by

$$(P_1/P_2) = 20 \log_{10}(V_1/V_2) \text{ dB}. \qquad (38)$$

Some lock-in amplifiers have switch-selectable (6 dB)–(12 dB) rolloff filters, and the specific bandwidth properties should be obtained from the manufacturer's specifications.

It is also true that some lock-in amplifiers as well as tuned preamplifiers allow the user to select the Q of the tuned amplifiers. The Q of an amplifier is given approximately by $Q \approx f_0/\text{BW}$, where f_0 is the center frequency and BW the full bandwidth. The system bandwidth must be carefully and properly selected when recording data to be deconvolved, and *all* factors that influence the system bandwidth must be carefully taken into account.

3. Digitization of Signals

The output of the lock-in amplifier is input to a sample-and-hold amplifier or directly to the analog-to-digital converter. This signal is converted to a digital and thus machine-readable form.

In considering the signal digitizing process, one must exercise a great deal of care to fully understand just how a specific system works. This understanding should include detailed characterization of both hardware *and* software functions. For example, in a continuous scanning system, "integraton" over a period of several time constants will yield different results depending on whether a number of successive "sample-and-hold" samples are averaged or a number of successive samples from the signal line are fed directly into an analog-to-digital converter (ADC). In many cases, such differences are of little consequence. To be sure of this, however, one must consider *all* aspects of the acquisition process in detail and consider carefully the impact of the approximations that are made.

a. Sample-and-Hold Amplifiers   Sample-and-hold amplifiers are operational amplifiers; the output tracks (follows) the input until a hold command is received. While the hold command is asserted, the output signal is equal to the input signal averaged over a window in time. The time at which the window begins is determined by the hold command. A typical window aperture might be 100 nsec.

Most sample-and-hold amplifiers are biased, unity gain, inverting amplifiers of a specific type—a fact that calls for some care in the subsequent processing of the digitized data. The sample-and-hold amplifier biased at 10 V converts, for example, 10 V to 0 V, 0 V to 10 V, and 3 V to 7 V. This example assumes that the input range of the amplifier is 0–10 V and that it is biased at 10 V, so the output signal is $10 \text{ V} - 1 \times V_{\text{in}}$. The need for careful and periodic adjustment of sample-and-hold amplifiers might best be communicated by reminding the reader that a sample-and-hold amplifier is essentially a dc amplifier with a very high slew rate.

b. Analog-to-Digital Conversion   Analog-to-digital converters are generally of the *n*-bit variety if they are interfaced to computers, the term "*n* bit" meaning that the maximum number that can be generated as digitized output is $2^n - 1$. (The least significant bit has a value of 2^0 and the most significant a value of 2^{n-1}.) The range of input voltage that can be converted is specified for each ADC. Some are bipolar, some are unipolar, and all have a specific, ideally linear, relationship between input analog voltage and digitized representation. For example, a 12-bit ADC, unipolar with an input range of 0 to 10 V, may represent 0 V as $000\,000\,000\,000_2$ or 0000_8, whereas 10 V becomes $111\,111\,111\,111_2$ or 7777_8, and 2 V is represented by $001\,100\,110\,011_2$ or 1463_8. For those who find such things mystical, to convert V_{10} into V_2 or V_8, calculate (for an n-bit converter)

$$\left[\frac{V_{10}}{(V_{AD\ max})_{10} - (V_{AD\ min})_{10}} \times 2^n - 1 \right]_{10} = (\text{converted value})_{10} \quad (39)$$

and convert to base 2. V_{10}, V_2, and V_8 are voltages in base 10, 2, and 8, respectively; $(V_{AD\ max})_{10}$ and $(V_{AD\ min})_{10}$ are the maximum and minimum ADC input voltages. We may then readily convert to base 8. The binary representation goes over to octal by writing down the octal value of three binary digits at a time starting from the least significant bit on the right. The specific binary representation of a converted voltage depends on the specific device. Manufacturer specifications should be consulted.

Just as sample-and-hold amplifiers must be properly adjusted, so must ADCs; both range and linearity are generally adjustable. As with sample-and-hold amplifiers, periodic validation of adjustment should be carried out.

c. Digital Sampling   The analog signal must be sampled at a frequency sufficiently high to avoid aliasing in the deconvolution process. Although the Jansson algorithm is carried out in signal space, there is nevertheless potential for aliasing. Note that the term "aliasing" describes a consequence of the sampling process, which may be considered in either signal or Fourier space (see Chapter 1). The sampling of a spectrum of Gaussian lines has been previously discussed in detail (Blass, 1976a; Blass and Halsey, 1981). These results are summarized in Section III.D.

d. Digital Filtering   In addition to analog filtering in the lock-in amplifier, we use some modest digital filtering prior to recording the raw data. The value in using digital filtering is twofold: high-frequency noise suppression is improved over the analog result, and the bandwidth may be more precisely set than with a typical discrete switch-selected time constant on a lock-in amplifier.

Jansson further discusses electrical and digital filtering in Section II.D of Chapter 2 and Section III.C.5 of Chapter 3.

D. ACQUISITION WITH CONTINUOUS SCANNING

Blass (1976a) and Blass and Halsey (1981) discuss data acquisition for a continuous scanning spectrometer in detail. The principal concept is that as a system scans a spectral line at some rate, the resulting time-varying signal will have a distribution of frequency components in the Fourier domain. The sampling theorem tells us that we must sample the signal at a rate equal to or greater than twice the highest-frequency component in the Fourier transform of the signal. For an actual spectrum of absorption lines, this maximum frequency depends on the selected scanning rate in a direct way. In addition, the bandpass of the electronics must be established in such a way that the maximum signal frequency will be minimally attenuated.

If there were no noise in the world, these factors would control the required sampling rate. However, because noise exists and noise can be aliased as well as data, the maximum noise frequency passed by the electronics system actually determines the required sampling rate.

The truth of this is readily seen in the following hypothetical example. Assume that a spectrum of Gaussian lines is to be scanned at a rate such that the maximum Fourier component is 100 Hz. We might then establish the electronic bandpass such that a 100-Hz component is attenuated less than 3 dB. Without noise we would sample at 200 Hz. However, significant noise signals exist out to at least six times the passband frequency of 100 Hz, which means that we must sample at 1200 Hz to avoid aliasing the noise as we deconvolve. This is an extremely conservative approach, and one might well sample less frequently without difficulty.

The net result of these considerations is that, for a spectrum of Gaussian lines, one should sample 10 times per resolution element ($\approx$ FWHM of isolated lines) and that the scan rate should be adjusted to yield a scan rate of 10 times constants to scan one resolution element (Blass, 1976a).

E. ACQUISITION WITH STEP SCANNING

Although one can treat data acquisition with discrete or step scanning quite apart from the continuous scanning case, we generally opt for treating our data acquisition as if we were scanning continuously. This is reasonable because the stepping time consumed by step scanning is negligible. We therefore acquire data as described in the preceding section. The only significant difference is that we stop and digitally integrate at one grating position for each sample acquired. The integration time is approximately one time constant (determined by the system bandwidth) and, as in the continuous case, we acquire 10 samples per resolution element.

F.   CALIBRATION OF SPECTRA

Wavelength or wave-number calibration of infrared absorption spectra is normally accomplished by simultaneously recording the spectrum along with a spectrum of a simple molecule having precisely measured transition frequencies.* There are a number of "variations on a theme" in this process, but generally it is necessary to deconvolve both the desired absorption spectrum *and* the calibration spectrum. The stability of observed transition frequencies under deconvolution has been investigated (Willson, 1973; Blass and Halsey, 1981). Using "point-simultaneous" constrained nonlinear deconvolution (see Chapter 3, Sections III.C.1 and III.C.2), Blass and Halsey (1981) found line-center frequency stability to be excellent. [Willson (1973) found a stability problem in "point-successive" deconvolution.]

Recent work on a production basis involving 5-μm spectra of $^{12}CD_3F$ has verified line-center frequency stability under deconvolution. Large numbers of experimental records of $^{12}CD_3F$, simultaneously recorded with CO in the sample, were calibrated before and after deconvolution (point-simultaneous methods, Jansson algorithm). No systematic differences were detected in comparisons of the before and after frequencies of nonblended absorption lines. That is, the variance was consistent with the optomechanical precision of the spectrometer and the mean deviation summed to approximately zero, validating the frequency calibration of the deconvolved data.

There are some potential calibration problems with deconvolution. The principal potential problems stem from the use of calibration lines in a grating order different from that in which the target spectrum is recorded. The dispersion $\Delta\bar{\nu}/\Delta\theta_g$ is a function of the wave number $\bar{\nu}$ passed by the

---

*For simplicity of expression, the wave number associated with a given transition will often be referred to as a frequency. This is not without some physical basis, because $\bar{\nu} = cf$, where $\bar{\nu}$ is in reciprocal centimeters, c in centimeters per second, and f in hertz; that is, we often think of wave number as scaled frequency, with the proportionality constant being equal to the speed of light.

monochromator [see Eqs. (5)–(8)] and thus, for calibration lines of significantly different $\bar{\nu}$, the instrument response function has a significantly
different width (in either time or steps as we scan the spectrum). Especially in such cases, the recorded data file should be split into two files:
one for calibration lines and one for target spectrum lines. The gaps in
each file should be zero filled. The two files should then be deconvolved in
precisely the same way (i.e., identical parameterization of the process in
terms of relaxation parameter and interation number) using an instrument
response function appropriate to each spectrum file. To date, we have not
accumulated a sufficiently extensive data base in such cases to make any
definitive comments. However, we see no serious problem with the approach; the process may be tested in the same way that the deconvolution
of the 5-μm $^{12}CD_3F$ data was tested.

In the $^{12}CD_3F$ tests, there were several examples of a 10% variation in
$\bar{\nu}$ over the range of the target spectrum. In these cases we used a response
function appropriate to the target spectrum average $\bar{\nu}$ for all lines and
experienced no difficulties. However, if one were using 5-μm (2000-cm^{-1})
CO lines to calibrate 12-μm C_2H_6, for example, it is likely that not only
would the frequency dependence of the transfer function be involved but
also the slit widths used for the 5- and 12-μm observations would be
different. The splitting of the data to be deconvolved into two files, and so
forth, would be mandatory in this case.

IV. Details of Spectroscopic Experiments

In this section we offer some comments based to a large extent on eight
years of the study, application, and production use of deconvolution with
high-resolution infrared absorption spectra. In Chapter 7, a number of
specific examples and test cases are presented that illustrate many of the
following comments.

A. IMPACT OF NOISE ON DECONVOLUTION

Consider the following scenario. You have an experiment that you wish to
perform—perhaps you intend to obtain the high-resolution absorption
spectrum of ethane at 12 μm. With the average dispersive spectrometer,
this is a demanding task and also a time-consuming experimental run. As
you set up the run, you adjust experimental and instrumental parameters

to maximize the signal and minimize the run time. You are faced with a number of competing and contradictory goals.

The decisions to be made if the resulting data are to be deconvolved differ considerably from those if the data are to be measured as recorded. For example, the most significant single cause of the failure of the deconvolution process to produce physically acceptable results is *excessive noise* in the recorded data. Typically the spectroscopist seeks to maximize resolution and, to some extent, minimize the run time. For direct measurement of recorded data, this is often an acceptable decision. In many cases, the person measuring the data is capable of making informed judgments of high quality at signal-to-noise ratios as low as 20:1. Such is not the case when data are to be deconvolved. For resolution enhancements of a factor of 2.5 to 3, a minimum signal-to-noise ratio for band-limited Gaussian noise is 75:1 to 100:1 (Blass and Halsey, 1981).

As a rule of thumb, we have found that, in the range of resolution enhancements from 2 to 5, the minimum required signal-to-noise ratio can be approximately represented by an empirical relationship

$$(S/N)_{min} \approx 6 \exp(\Delta \bar{\nu}_{meas} / \Delta \bar{\nu}_{decon}). \tag{40}$$

Equation (40) is presented as an approximate guideline, and no physical significance is implied by the form or factors used. A generated data test showing the impact of various signal-to-noise ratios is available (see Blass and Halsey, 1981, Fig. 24); further tests and comments will be found in the next chapter.

The effects of inadequate signal-to-noise ratio on deconvolution range from divergence of the iterative Jansson method to the generation of spurious weak features that appear to be absorption lines. We have never experienced divergence of the Jansson method used as described (Blass and Halsey, 1981) but that is due in large part to our not attempting to deconvolve really noisy data and to the approach that we take in the deconvolution iteration process. Although this is discussed later, we remark that our approach is to use moderate underrelaxation in the initial iterations. This seems to aid in the stabilization of the process (see Blass and Halsey, 1981, pp. 43–47).

One further comment regarding noise: in absorption spectral data, it is the signal-to-noise ratio that affects the quality of the results, not the "peak-height"-to-noise or information-to-noise ratio. This statement assumes that the data to be deconvolved are principally 10–30% absorbing and that the signal-to-noise ratio satisfies the requirements of Eq. (40).

That this is a reasonable observation follows from the physically meaningful constraints that are imposed and the deconvolution process as discussed in Chapters 4 and 7.

B. IMPACT OF BOUGUER–LAMBERT LAW

As shown in Sections I.E.1 and I.E.2 of Chapter 2, the transmitted spectral flux U is given by Eq. (9), and in the case of a pressure-broadened line

$$U = U_0 \exp(-\text{natural} \otimes \text{collision} \otimes \text{Doppler}). \qquad (41)$$

From Section II.E.2 of Chapter 2 we find that the measured absorptance (where x is identified with $\bar{\nu}$, the frequency in reciprocal centimeters) is given by

$$A_M(x) = r(x) \otimes \left[1 - \frac{U_T(x) - D_T(x)}{B_T(x) - D_T(x)} \right] \qquad (42)$$

using Eqs. (49) and (52) from Chapter 2. The quantity U_T is the transmitted spectral flux and is given by Eq. (41); the quantity B_T is the maximum transmitted flux for zero sample pressure and may be identified with U_0 in Eq. (41). If we assume that $D_T(x)$, the stray background flux, is negligible, as well it might in a well-designed and maintained system, then Eq. (42) becomes

$$A_M(x) = r(x) \otimes [1 - \exp(-\text{natural} \otimes \text{collision} \otimes \text{Doppler})]. \quad (43)$$

Figure 5 illustrates the measured data values as well as the data converted to absorptance and transmittance. There are a number of implications of Eq. (43) for deconvolution, and we shall discuss each of these in turn.

1. Apparent versus Actual Absorption

Rewriting Eq. (43) in the sense of Eq. (9) of Chapter 2, we find

$$A_M(x) = r(x) \otimes [1 - \exp(-PX)], \qquad (44)$$

where P is the spectral absorption coefficient and X the pressure times the path length. For the present we ignore the fact that PX may be represented as a convolution as indicated in Eq. (43), and focus on the fact that $P(x)X$ is a typical "bell"-shaped function that gives rise to an $A_M(x)$ as illustrated in Fig. 5. Remember that as the pressure times path length increases, the peak absorption $[A_M(x)]_{\text{max}}$ does not increase linearly as a

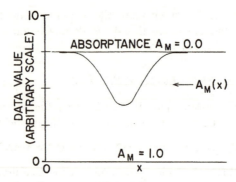

Figure 5 Absorption line as recorded with measured data values related to absorptance and transmittance.

function of X, and that for a fixed X the area under the $A_M(x)$ curve is a constant independent of the half-width of the instrument function $r(x)$ (or, loosely, the instrumental resolution).

Consider an example. Assume that we are looking at a spectral line and trying to decide how to observe the line for deconvolution giving width reduction by a factor of 3. Further assume that pressure broadening is negligible and that the Doppler full width is 0.005 cm^{-1}. We are going to adjust the pressure path for good deconvolved results. It should be easy, but it is not always obvious. There is a large difference between the choice of X for spectrometers with resolutions of 0.008 and 0.050 cm^{-1}. In the former case, deconvolving to Doppler width increases resolution only by a factor of 1.9, whereas in the latter case a full factor of 3 achieves only a FWHM of 0.017. Thus the 0.050-cm^{-1} instrument will "see" an increase in peak height $[A_M(x)]_{max}$ that exceeds the increase for the 0.008-cm^{-1} instrument simply because a greater narrowing of the observed line is achievable in the latter case.

There is, however, a further and more subtle effect. A 0.050-cm^{-1} instrument will make a 99%-absorbing 0.005-cm^{-1} Doppler line look like a well-behaved 5%-absorbing line, and even mild deconvolution reveals obviously saturated spectral lines. Examples of this effect may be found in the literature on deconvolution (see, e.g., Blass and Halsey, 1981, Fig. 22b, third and fourth traces, and Fig. 2 of Chapter 4).

A rough rule of thumb can be set forth to help avoid having your deconvolved spectral lines exhibit saturation effects. Reduce the pressure path (and thus apparent absorption) until you are sure that it is actually too low; then cut it in half again. As an alternative, one could easily generate a dictionary of simulations based on the ratio of the FWHM of the true line to the equivalent FWHM characterizing specific instrumental resolution. In our laboratory we have simply developed experience in evaluating the requirements. Our instrumental resolution of 0.009 cm^{-1} makes that fairly easy.

2. Deconvolving beyond the Spread Function

There have been a number of discussions about "deconvolving beyond the Doppler limit" (see Pliva *et al.*, 1980) but we prefer to go back a step further and simply discuss deconvolving beyond removal of the instrumental spread function. Jansson discusses deconvolving inherent broadening in Section I.E.2 of Chapter 2.

We rewrite Eq. (44) as

$$A_M(x) = r(x) \otimes \{1 - \exp[-P(x)X]\}, \tag{45}$$

where generally $P(x)$ is given by $P_V(x)$ [Chapter 2, Eq. (9)],

$$P_V(x) = P' \frac{\mu}{\pi} \int_{-\infty}^{\infty} \frac{\exp(-\zeta^2)}{\mu^2 - (x - x_0 - \zeta)^2} d\zeta, \tag{46}$$

the Voigt function discussed in Chapter 1. Though Eq. (45) looks complicated in the light of Eq. (46), remember that Eq. (45) can be represented as

$$A_M(x) = r(x) \otimes \{1 - \exp[P_N(x) \otimes P_P(x) \otimes P_D(x)X]\}. \tag{47}$$

Examining Eq. (47), we see two possible ways to go beyond a modified (i.e., fully deconvolved) $A_M(x)$, that is, beyond our best estimate $\hat{A}_T(x)$ of the true absorptance A_T:

$$\hat{A}_T(x) \approx \{1 - \exp[P_N(x) \otimes P_P(x) \otimes P_D(x)X]\}. \tag{48}$$

The most obvious approach has not been attempted to our knowledge—namely, deconvolve $A_M(x)$ until the effect of $r(x)$ has been removed, subtract both sides of Eq. (48) from 1, take the natural logarithm of the resulting equation [and data set: $1 - \hat{A}_T(x)$], and proceed. (As Jansson mentions in a slightly different context in Chapter 2, the impact of noise has not been studied in this case.) The modified data set is now of the form

$$\ln\left[1 - \hat{A}_\mathrm{T}(x)\right] = P_\mathrm{N}(x) \otimes P_\mathrm{P}(x) \otimes P_\mathrm{D}(x)X. \qquad (49)$$

As mentioned before, we know of no one who has attempted this approach. It is obvious that extremely good signal-to-noise ratios would be required.

The second approach is to expand Eq. (47) for small values of the experimental argument. This is discussed briefly by Jansson in Section I.E.2 of Chapter 2. For very-high-resolution data where very short paths and low pressures are used (i.e., where X is small) this approach is potentially feasible though signal-to-noise ratio requirements would be very demanding for weak apparent absorptions.

For tunable-diode-laser (TDL) studies where $r(x)$ is much narrower than even a modestly pressure-broadened line, one might either attempt the former method characterized by Eq. (49) or perhaps simply ignore (cautiously) the impact of $r(x)$. From the foregoing discussion, however, it should be obvious that the considerations and potential problems involved in going beyond removal of the effect of $r(x)$ are in no way trivial. Nevertheless, we have successfully deconvolved some data beyond the limit of the removal of $r(x)$. In many instances we have only taken 0.012-cm^{-1} data for a 0.005-cm^{-1} Doppler spectrum to 0.004 cm^{-1} and have encountered no serious problems. (Quantities refer to FWHM of the peaks.) In other instances, we have deconvolved 0.002-cm^{-1} TDL spectra of C_2H_6 at 12 μm to 0.0009 cm^{-1}, which is roughly one-half the Doppler limit. An example is shown in Chapter 7.

C. IMPACT OF INSTRUMENTAL RESPONSE FUNCTION

Jansson discusses the determination of the instrumental response function as well as its analytical characterization in Chapter 2.

We have found, for dispersive infrared spectrophotometry at extremely high resolution, that a Gaussian function works extremely well for critical line-position work. Most of our deconvolution experience of a production nature (as opposed to tests and trials) involves spectra showing predominantly Doppler broadening with a Doppler width FWHM on the order of 0.005 cm^{-1}. Even though we have verified that the double-pass 5-m Littrow system achieves double-pass theoretical resolution at four normal (4N) slits (Todd and McCubbin, 1977), we generally run the system at 2.5 to 3.5 times the Doppler width or 0.012 to 0.018 cm^{-1} resolution. This results in observed FWHM values of 0.0130 to 0.0187 cm^{-1} for single transitions. Our general procedure is to make several measurements of the FWHM of isolated lines and use a gaussian of approximately the same (or slightly smaller) width as an instrument function. For our system, this appears to be completely satisfactory for line-position work. M. A. Dakhil of our laboratory, in the process of taking data on $^{12}CD_3F$ in the 5-μm region, has run numerous spectra under the typical conditions just described. Using ^{12}CO, ^{13}CO, $^{12}CO_2$, and $^{13}CO_2$ absorption lines as standards he has calibrated the CD_3F spectra. Prediction of well-known frequencies observed in the spectra led to a standard deviation of errors in the range of 3×10^{-4} to 10^{-3} cm^{-1} prior to deconvolution. Following deconvolution (including the calibration lines) the standard deviations are as good or better than those of the calibration fits. Extensive comparisons of measured wave numbers in nondeconvolved and deconvolved spectra show no systematic variations. Standard deviations of $^{12}CD_3F$ lines based on comparisons are consistent with the standard deviations of the calibration fits. We believe that the residual errors are due to mechanical limitations of the grating drive mechanism. The magnitudes of the variances are consistent with this assumption.

It is also necessary to pay attention to the variation of the instrumental dispersion over the range of the data to be deconvolved. We have not experienced any practical problems with dispersion variations of up to 10% from one end to another of a 150-cm^{-1} spectrum using a fixed-width (in data points) Gaussian response function—although the final deconvolved resolution was not constant over the range of the data. We have cnsidered linearizing infrared dispersive spectrometer data in wave number prior to deconvolution, but have not implemented such procedures at this time. For situations where the dispersion varies significantly over the range of the data, either linearization should be considered or the spectrum broken into smaller segments and deconvolved piecewise using appropriate response functions.

For TDL spectra we have in fact linearized the data by using simultaneously recorded Ge étalon fringes. For such spectra this is often an absolute necessity because of the "dispersion variation" in the recorded data.

If one were to attempt any critical-intensity or line-shape-related deconvolution, we would seriously suggest that a great deal of effort be expended to determine the detailed nature of the instrumental response function. Effects due to neglect of response-function asymmetry have been discussed in some detail (Blass and Halsey, 1981, pp. 107–112).

In Chapter 2, Jansson describes the determination of the system response function for a dispersive spectrometer system. We have made a number of such determinations using very-low-pressure samples of, for example, CO in the 5-μm region. As discussed by Jansson, one records the data and then removes the Doppler profile using deconvolution, yielding the system response function.

Our system is set up with a stepping motor drive on the grating tangent arm, and our tests show repeatability of position to better than $\pm \frac{1}{10}$ step (an average tolerance equal to 8×10^{-6} cm^{-1} at 1000 cm^{-1}). We are therefore able to scan a very weakly absorbing CO line as many as 150 times and use the average as the input data for deconvolution. In this fashion, we avoid the problems concerning the signal-to-noise ratio mentioned by Jansson in Section II.G.1 of Chapter 2.

Another useful approach is to determine the profile of the very weakly absorbing single Doppler-limited line to be used as the system response function. Instead of removing the Doppler profile by deconvolution, however, one might convolve the observed data with the theoretical Doppler profile and use the observed line directly as the effective response function for the modified data set. This technique has been tested and works well. The general approach might be especially useful where a pressure-broadened line has to be used to obtain the response function. Instead of deconvolving the theoretical Voigt profile, the data might well be convolved with the theoretical Voigt profile and the directly observed "response function" used without modification. This approach has one very positive feature, in that it does not add noise to the response function as does deconvolution (to some degree, deconvolution enhances residual noise).

V. Preparing to Deconvolve a Data Set

Preparation of data for deconvolution must begin prior to the acquisition of the first data point. Once the resolution of the system is set (in a dispersive spectrometer this is equivalent to setting the slit width), the density of data points per resolution element must be chosen as discussed in Sections III.D and III.E. There are some subtle factors that must be taken into account. For example, for continuous scanning, approximately 10 data points per resolution element are recommended (Blass, 1976a) to capture all of the information required by the data *and* the noise. On the other hand, the data-point density must be great enough to characterize the spectral lines *after* deconvolution. Because a gaussian requires only 2 data points per FWHM, 10 points per resolution element is a sufficient density in the raw data for deconvolution by a factor of up to 5. Of course, when only 2 points per FWHM remain, interpolation is necessary to render the spectrum visually useful.

Under less restrictive noise/bandwidth considerations, one might drop the density to 6 points per resolution element at the risk of some minor noise aliasing. However, deconvolution by a factor of 3 would leave only two points per FWHM of a Gaussian spectrum—a number sufficient to characterize the spectrum, although display and measurement are difficult.

In the latter event, one may use a sinc-function interpolation algorithm to increase the data-point density in the deconvolved spectrum (Willson and Edwards, 1976). A density-doubling routine is especially easy to implement. A word of warning, however—do not attempt to interpolate prior to deconvolution. For any reasonable-length interpolation filter, sufficient additional high-frequency components are added to the spectrum to destroy the validity of the deconvolution process. After deconvolution a density doubler of length 50 works well for measurement and display if 4 or 5 data points remain per FWHM. If the remaining number is 2 or 3, then a longer filter will be required for a reasonable representation (on the order of 200 to 500 points long).

Returning to the mainstream discussion of data preparation, we note that, for a 6-dB-per-octave-rolloff *RC* filter network in a lock-in amplifier, the continuous scan rate amounts to approximately one time constant per data point or 10 time constants per resolution element (Blass, 1976a). Some time is saved if only six data points are taken per resolution element. We have tried acquiring in this fashion, with no visible negative effects.

In our operations, we often establish step-scan parameters as if we were continuously scanning. Because continuous scanning requirements are more demanding, no problems arise using this approach.

Remember, also, that the signal-to-noise ratio must be high enough to accommodate the desired deconvolution (see Section IV.A). As noted, inadequate signal-to-noise ratio is the major cause of unsatisfactory deconvolution results.

A. SMOOTHING

Once the data have been recorded in machine-readable form, they may be further processed to improve the signal-to-noise characteristics of the record. Note, however, that an improvement of the signal-to-noise ratio will most often be at the expense of resolution, and thus smoothing the data record is not without problems. In Section III.C.5 of Chapter 3, Jansson comments on band limiting and polynomial filters. Examples of smoothing using polynomial filters are presented in Chapter 7.

We have carried out simulations using polynomial least-squares filters of the type described by Savitzky and Golay (1964) to determine the impact of such smoothing on apparent resolution. For quadratic filters, a filter length of one-fourth of the linewidth (at FWHM) does not seriously degrade the apparent resolution of two Gaussian lines in very close proximity.

As a rule of thumb, we generally restrict the filter length used to one-fourth of the FWHM of an observed Doppler-broadened line. Willson and Edwards (1976) have carried out numerous simulations and provide insight into the impact of smoothing on spectral data records.

In Chapter 7, the high-frequency attenuation characteristics of polynomial filters are discussed. These considerations are relevant to questions regarding the desirability of a single application of a filter of given length as opposed to multiple application of shorter filters.

Because the most serious problem arising in the deconvolution of spectra is that of noise, detailed attention to smoothing in a fashion consistent with the uniform attenuation of high-frequency noise will result in the best possible deconvolution results.

B. BASE-LINE CONSIDERATIONS

Owing to the nature of typical raw absorption data, extreme care must be used in preparing the data for deconvolution. Referring to the ordinate (data values) in Fig. 5, one sees that 0% absorptance corresponds to an arbitrary voltage at the output of the typical lock-in amplifier. A typical data record will have been converted by subtracting the lock-in output (Fig. 5) from, say, a value of 10 V, thus inverting ("tipping over") the spectral line and resulting in the appearance of a line as an emission line. However, the 0% absorptance level now corresponds to some voltage (data value) of, say, 2 V. In addition, this 0% absorptance level is generally a slowly varying function of wave number.

Anyone proceeding to deconvolve the raw, "tipped-over" data set will be disappointed (see Pliva et al., 1980; e.g., Section 3). It is easy to see why this is so when constraints are placed on deconvolution as in the Jansson method (Blass and Halsey, 1981). The most effective constraint placed on the deconvolution process in the Jansson algorithm is that 0% absorptance is a lower limit in the deconvolved estimate. That is, the deconvolved absorption spectrum is not allowed to exhibit an emission signal. If the observed data include a bias (i.e., offset) that does not represent actual absorption, then the 0% absorptance constraint will take effect at the apparent 0% absorptance level, and not at the true 0% absorptance level. This results in numerous spectral artifacts. Examples are presented in Chapter 7.

Processing the data file to remove the artificial absorption signal is relatively easy in the event that the spectrum is not too densely populated with lines and 0% absorptance levels may be found across the range of the recorded spectrum. We find that a simple point-to-point base-line subtraction method is simplest and gives the best results. This requires some accurate method to display and measure data values in the file. In our case, we use a graphic display with light pen to acquire a base-line subtraction data set. Other, less direct methods should work also. Sensitivity tests are not difficult to design and execute.

In the event that one is dealing with very dense spectra, the best method of establishing 0% absorptance levels is the inclusion of measured 0% absorptance levels periodically in the data file.

VI.   Deconvolving the Data: Constrained Deconvolution

The actual deconvolution of a data set is formally straightforward. Let $\hat{o}^{(k)}(x)$ be the kth iterative estimate of the actual spectrum $o(x)$, where x is nominally "time" viewed as a sequence-ordering variable. Further, let $i(x)$ be the actual observed spectrum that has been instrumentally convolved with the observing system response function $s(x)$. The observed data set $i(x)$ is assumed to be related to $o(x)$ by the convolution integral equation

$$i(x) = \int s(x - x')o(x')\,dx' \tag{50}$$

or, equivalently,

$$i = s \otimes o. \tag{51}$$

Let us note that $\hat{o}^{(k)}(5)$ is the amplitude of $\hat{o}^{(k)}(x)$ at $x = x_5$. Let $r^{(k)}[\hat{o}^{(k-1)}(x)]$ be a weighting function used in determining the kth estimate $\hat{o}^{(k)}(x)$ of $o(x)$. We shall have occasion to abbreviate $r^{(k)}[\hat{o}^{(k-1)}(x)]$ as $r^{(k)}[y]$.

The constrained deconvolution algorithm is then specified by

$$\hat{o}^{(k)}(x) = \hat{o}^{(k-1)}(x) + r^{(k)}[\hat{o}^{(k-1)}(x)][i(x) - \hat{o}^{(k-1)}(x) \otimes s(x)]. \quad (52)$$

The $\hat{o}^{(0)}(x)$ estimate is generally taken to be $i(x)$.

A. DECONVOLUTION ALGORITHM

Step by step, the constrained deconvolution algorithm may be stated as follows.

(1) Set $k = 1$, $\hat{o}^{(0)}(x) = i(x)$.

(2) Form

$$\hat{o}^{(k)}(x) = \hat{o}^{(k-1)}(x) + r^{(k)}[\hat{o}^{(k-1)}(x)]\Big[i(x) - \sum \hat{o}^{(k-1)}(x')s(x-x')\Big],$$

where the sum over x' is over the range of nonzero values of $s(x-x')$.

(3) Invoke termination-convergence tests. If satisfied, the current $\hat{o}^{(k)}(x)$ is the estimate of $o(x)$ and the procedure is completed. If not satisfied, set $k = k + 1$ and return to step (2).

Convergence may be monitored by forming the root-mean-square error of the current estimate as

$$e^{(k)} = \left\{ \frac{1}{N} \sum_x [i(x) - \hat{o}^{(k-1)}(x) \otimes s(x)]^2 \right\}^{1/2}. \quad (53)$$

Note that $e^{(k)}$ is not an infallible measure of convergence.

Relaxation Functions

The key to the success of constrained deconvolution is the relaxation function $r[x]$. For absorption spectra the data are scaled to lie in the range from 0 to 1. For example, Jansson (1970) used

$$r^{(k)}[y] = r^{(k)}[\hat{o}^{(k-1)}(x)] = r^{(k)}_{max}\big[1 - 2|\hat{o}^{(k-1)}(x) - \tfrac{1}{2}|\big], \quad (54)$$

where $r^{(k)}_{max}$ is a constant for the kth iteration. Blass and Halsey (1981) after Willson (1973) use

$$r^{(k)}[y] = r^{(k)}[\hat{o}^{(k-1)}(x)] = r^{(k)}_{max}\{\hat{o}^{(k-1)}(x)[1 - \hat{o}^{(k-1)}(x)]\}^p, \quad (55)$$

where p is normally set equal to 1. As Frieden (1975) points out, one may modify Eq. (54) to specify limits other than 0 and 1 by using

$$r^{(k)}[y] = r^{(k)}[\hat{o}^{(k-1)}(x)] = C\left[1 - 2(B - A)^{-1}|\hat{o}^{(k-1)}(x) - (A + B)/2|\right],$$
(56)

where A is the allowed minimum and B the allowed maximum data value.

In Chapter 7 several tests are presented using a Gaussian relaxation function such that

$$r^{(k)}[y] = r^{(k)}[\hat{o}^{(k-1)}(x)] = r_{max}^{(k)}\left\{\exp\left[-(y - 0.5)^2/0.11\right]\right\},$$
(57)

which yields $r^{(k)}[0] = r^{(k)}[1] = 0.1$ and $r^{(k)}[0.5] = 1.0$.

Another function used in Chapter 7 is described by

$$r^{(k)}[y] = 1, \qquad 0 \leq y \leq 1,$$
$$\hat{o}^{(k)}(x) = 1, \qquad y > 1,$$
$$\hat{o}^{(k)}(x) = 0, \qquad y < 0,$$
(58)

which is a form of clipping and not a relaxation function in the usual sense of the term.

B.   THE NONLINEARITY OF DECONVOLUTION

The constrained deconvolution algorithm produces estimates that cannot be obtained from the data by simple linear inverse filtering. This is most readily seen using the Blass–Halsey weight function as an example.

Using Eq. (55) for $r[y]$, we obtain for step (1) in the iterative algorithm of Section VI.A

$$\hat{o}^{(1)}(x) = i(x) + i(x)[1 - i(x)]\{i(x) - r[y] \otimes i(x)\}$$
(59)

$$= i(x) + [i(x)]^2 - [i(x)]^3 - i(x)\{r[x] \otimes i(x)\}$$
$$+ [i(x)]^2\{r[x] \otimes i(x)\}.$$
(60)

One readily sees that on an analytical level the algorithm rapidly becomes complicated. This is true even if one simply constrains the data to be nonnegative, that is,

$$r^{(k)}[y] = \hat{o}^{(k-1)}(x),$$

whence

$$\hat{o}^{(1)} = i(x) + i(x)\{i(x) - r[y] \otimes i(x)\} \tag{61}$$

$$= i(x) + [i(x)]^2 - i(x)\{r[y] \otimes i(x)\}. \tag{62}$$

Two conclusions follow readily. First, the constrained deconvolution algorithm is decidedly nonlinear in the observed data. Second, no easy analytical interpretation of the effects of a particular relaxation function may be obtained by considering the Fourier transform of $\hat{o}^{(k)}$ cast in terms of $i(x)$ and $r[i(x)]$.

In the next chapter, we present the results of tests carried out with the relaxation functions described here. Simulated spectra generated by computer as well as actual experimental spectra are used in the tests.

References

Blass, W. E. (1976a). *Appl. Spectrosc.* **30**, 287–289.

Blass, W. E. (1976b). *Appl. Spectrosc. Rev.* **11**, 57–123.

Blass, W. E., and Halsey, G. W. (1981). "Deconvolution of Absorption Spectra." Academic Press, New York.

Blass, W. E., and Nielsen, A. H. (1974). *In* "Methods of Experimental Physics" (D. Williams, ed.), Vol. 3A, 2nd Ed., (Chapt. 2.2) pp. 126–192. Academic Press, New York.

Boyd, W. J., Jennings, D. E., Blass, W. E., and Gailar, N. M. (1974). *Rev. Sci. Instrum.* **45**, 1286–1288.

Davenport, W. B., Jr., and Root, W. L. (1958). "An Introduction to the Theory of Random Signals and Noise." McGraw-Hill, New York.

Frieden, B. R. (1975). *In* "Picture Processing and Digital Filtering" (T. S. Huang, ed.), p. 177. Springer-Verlag, Berlin and New York.

James, J. F., and Sternberg, R. S. (1969). "The Design of Optical Spectrometers." Chapman & Hall, London.

Jansson, P. A. (1970). *J. Opt. Soc. Am.* **60**, 184.

Jennings, D. E. (1974). Ph.D. Dissertation, University of Tennessee, Knoxville.

Kruse, P. W., McGlauchlin, L. D., and McQuistan, R. B. (1962). "Elements of Infrared Technology: Generation, Transmission, and Detection." Wiley, New York.

McCubbin, T. K., Jr. (1961). *J. Opt. Soc. Am.* **51**, 887.

Pliva, J., Pine, A. S., and Willson, P. D. (1980). *Appl. Opt.* **19**, 1833.

Savitzky, A., and Golay, M. J. E. (1964). *Anal. Chem.* **36**, 1627.

Stewart, J. E. (1970). "Infrared Spectroscopy." Dekker, New York.

Todd, T., and McCubbin, T. K. (1977). *Appl. Spectrosc.* **31**, 326.

Townes, C. H., and Schalow, A. (1955). "Microwave Spectroscopy." McGraw-Hill, New York.

Treffers, R. R. (1977). *Appl. Opt.* **16**, 3103–3106.

Wadsworth, F. O. (1894). *Philos. Mag.* **38**, 337–351.

Willson, P. D. (1973). Ph.D. Thesis, Michigan State University, East Lansing.

Willson, P. D., and Edwards, T. H. (1976). *Appl. Spectrosc. Rev.* **12**, 1.

Chapter 7 | Deconvolution Examples

Paul Benjamin Crilly

Department of Electrical Engineering, The University of Tennessee
Knoxville, Tennessee

William E. Blass
George W. Halsey

Department of Physics and Astronomy, The University of Tennessee
Knoxville, Tennessee

List of Symbols

b	constraint operator
CSD	constrained signal-space deconvolution
E_{RMS}	root-mean-square error based on $o(x) - \hat{o}^{(k)}(x)$
FTIR	Fourier transform infrared spectrum
FWHM	full width at half maximum

$i, o, \hat{o}$	vectors of $i(x)$, $o(x)$, $\hat{o}(x)$, where x is a discrete variable
$\otimes$	denotes convolution operation
$i, i(x)$	observed spectrum, or data
$n, n(x)$	additive noise
$o, o(x)$	true spectrum, not "distorted" by observation
$\hat{o}^{(k)}(x)$	kth estimate of $o(x)$
$\hat{o}, \hat{o}(x)$	estimate of $o(x)$
$p, p(x)$	positive constraint operator
PSD	power spectral density
$r_o, r_{\max}^{(k)}$	constant coefficient of relaxation function $r[o^{(k)}(x)]$
$r[\hat{o}^{(k)}(x)]$	relaxation function used in constrained deconvolution
RMSE	root-mean-square error defined in Eq. (53) of Chapter 6
s	matrix form of $s(x)$, where x is a discrete variable
$s, s(x)$	response or spread function
$\text{sinc}(x)$	$(\sin \pi x)/(\pi x)$
$\text{sinc}^2(x)$	$(\sin^2 \pi x)/(\pi x)^2$
SNR	signal-to-noise ratio
TDL	tunable diode laser
x	sequence-ordering variable, such as time, wave number, or index

I. Introduction

Since 1973 we have been exploring deconvolution, first as a fascinating novelty and recently as a valuable tool. Deconvolution was originally viewed with a large measure of skepticism. However, over the years, in the process of testing and evaluating the practical limits of deconvolution, we have gained a certain measure of confidence in the results. Therefore, we can improve the quality of the measurement process.

Being relatively simple to use, iterative deconvolution methods have advantages over other techniques, including that no *a priori* knowledge of the data's statistics is required, easy incorporation of constraints into the algorithm, and that iterations can cease when errors exceed a given bound.

This chapter evaluates various iterative deconvolution algorithms that are commonly used to restore degraded chromatographic or spectroscopic data. The evaluation criteria will include root-mean-square (RMS) errors, relative errors in peak size, and rates of convergence. The iterative algorithms to be evaluated include Van Cittert's method, Van Cittert's method with constraint operators, relaxation-based methods, and Gold's ratio method. The presentation also includes some enhancements that improve convergence properties. Examples are chosen from gas chromatographic peak data and interferometric, Raman infrared, grating tunablediode, and γ-ray spectra.

II.   Examining the Deconvolution Process

A.   SYSTEM MODELING

Mathematically speaking, deconvolution refers to the method or methods used to solve the convolution integral equation

$$i(x) = \int s(x - x')o(x')\, dx' + n(x) \tag{1a}$$

or

$$\hat{o}^{(k+1)}(x) = \hat{o}^{(k)}(x) + b\left[i(x) - s(x) \otimes \hat{o}^{(k)}(x)\right]. \tag{1b}$$

where i is the measured spectrum, s the system spread function or impulse response function, o is the ideal spectrum free of any instrumental effects, and n is additive noise. With spectroscopic or chromatographic data, the effect of Eq. (1) is to degrade peak resolution to the extent that two or more peaks may be observed as one peak. This is illustrated in Fig. 1; spectrum a shows the ideal peak, b and c show the effects of distortion and noise.

B.   ITERATIVE METHODS

The iterative methods used in this chapter are as follows:

   1. Van Cittert's:

$$\hat{o}^{(k+1)}(x) = \hat{o}^{(k)}(x) + b\left[i(x) - s(x) \otimes \hat{o}^{(k)}(x)\right]. \tag{2}$$

   2. Van Cittert's with positive constraints:

$$\hat{o}^{(k+1)}(x) = p(x)\,\hat{o}^{(k)}(x) + b\left[i(x) - s(x) \otimes p(x)\,\hat{o}^{(k)}(x)\right]. \tag{3}$$

with

$$p(x) = \begin{cases} 1, & \hat{o}^{(k)}(x) \geq 0, \\ 0, & \hat{o}^{(k)}(x) < 0. \end{cases}$$

3. Jansson's:

$$\hat{o}^{(k+1)}(x) = \hat{o}^{(k)}(x) + r\left[\hat{o}^{(k)}(x)\right]\left[i(x) - s(x) \otimes \hat{o}^{(k)}(x)\right], \quad (4a)$$

with

$$r\left[\hat{o}^{(k)}(x)\right] = r_0\left[1 - 2\left|\hat{o}^{(k)}(x) - \tfrac{1}{2}\right|\right]. \quad (4b)$$

4. Gold's ratio:

$$\hat{o}^{(k+1)}(x) = \hat{o}^{(k)}(x) \cdot i(x)\bigg/\left[s(x) \otimes \hat{o}^{(k)}(x)\right]. \quad (5)$$

III. Simulations

A. EXAMPLE DEFINITIONS

To evaluate the preceding iterative methods, data sets that simulate ideal and overlapped spectroscopic peaks were generated at various signal-to-noise ratios and are shown in Fig. 1. The overlapped peaks were decon-

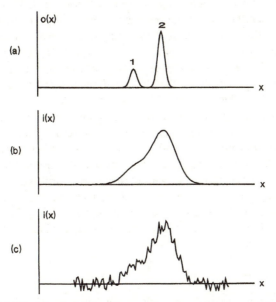

Figure 1   (a) Ideal signal, (b) distorted version with resolution of 0.55 and signal-to-noise ratio of 100:1, (c) distorted version with resolution of 0.55 and signal-to-noise ratio of 4:1 (Crilly, 1991b, © IEEE; reprinted with permission).

volved using each of the above four methods and the estimates evaluated with respect to RMS error and rate of convergence. Though RMS error provides a measure of the overall quality of the estimate, it may not directly relate to the goal of the process, which is to correlate peak size (area or height) with some physical parameter that characterizes the sample. Therefore, recovery of peak areas and their variances are also considered.

It should be noted that for the evaluations in Section III of this chapter, RMS error will be defined as

$$E_{\text{RMS}} = \sqrt{\sum_x [o(x) - \hat{o}(x)]^2}, \tag{6}$$

where $\hat{o}(x)$ is the final estimate of $o(x)$ and the summation is made over all the sample values of $o(x)$ and $\hat{o}(x)$. The discrete variable x is the index value of elements in the data set.

Because we know the true signal $o(x)$, and for the reasons presented by Eqs. (2)–(6) of Chapter 5, we choose the metric in above Eq. (6) versus the one of Eq. (53) in Chapter 6 that uses $[i(x) - s(x) \otimes \hat{o}^{(k)}(x)]$. Note that though Eq. (6) is especially valuable for simulation, Eq. (53) of Chapter 6 is needed when the goal of the process is the estimation of $o(x)$.

Figure 1a shows function $o(x)$ consisting of two Gaussian peaks whose area percentages are 25% (peak 1) and 75% (peak 2). The unequal peak areas were chosen because a given error in measured area of the smaller peak has a larger effect on the calculated relative error than the same error would have had it occurred in the larger peak.

The peaks of Figs. 1b and 1c were generated by convolving function $o(x)$ of Fig. 1a with a Gaussian-shaped impulse response function so that the resultant function, $i(x)$, had a resolution of 0.55 and signal-to-noise ratios of 100:1 and 4:1. Resolution is defined as the distance between two peaks divided by their average width (Snyder and Kirkland, 1979). Signal-to-noise ratio is defined as the ratio of the signal height to the peak-to-peak noise width. In order to provide a fair statistical basis for the merits of a given method, 10 independent data sets were generated for each of the two noise conditions.

B. RESULTS

The first test was to evaluate each of the four methods with respect to RMS error and rate of convergence. The data of Fig. 1b were deconvolved without any prefiltering; the RMS errors versus iteration are plotted in

Fig. 2. The data of Figs. 1b and 1c were prefiltered using cross-correlation and autocorrelation filtering, where the following substitutions are made in Eqs. (2)–(5) prior to deconvolution:

$$i(x) \otimes s(-x) \to i(x), \tag{7a}$$

$$s(x) \otimes s(-x) \to s(x). \tag{7b}$$

The error results are plotted in Figs. 3 and 4 as a function of iteration number.

To give a visual idea of the estimate quality for a given method, the results of deconvolving the data of Fig. 1c with the prefiltering of Eq. 7 are plotted in Figs. 5–8.

Each of the 10 data sets were deconvolved, the peak areas quantitated, and then averaged. The relative errors and measurement variances are tabulated in Tables I and II.

C. CONCLUSIONS

As the tables and figures indicate, Jansson's method generally gave the best results. This was expected because this technique constrains the estimates within their physical limits, but does so in such a way that

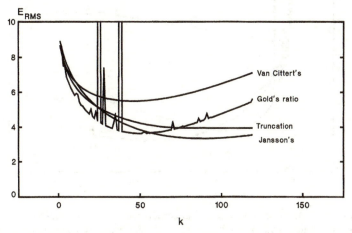

Figure 2 RMS error comparisons and convergence rates for various algorithms. Observed signal had a resolution of 0.55 and signal-to-noise ratio of 100:1. No prefiltering was used (Crilly, 1991b, © IEEE; reprinted with permission).

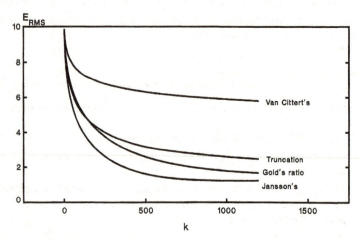

Figure 3 RMS error comparisons and convergence rates for various algorithms. Observed signal had a resolution of 0.55 and signal-to-noise ratio of 100:1. Auto/cross-correlation prefiltering was used (Crilly, 1991b, © IEEE; reprinted with permission).

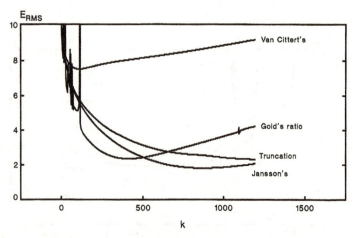

Figure 4 RMS error comparisons and convergence rates for various algorithms. Observed signal had a resolution of 0.55 and signal-to-noise ratio of 4:1. Auto/cross-correlation prefiltering was used (Crilly, 1991b, © IEEE; reprinted with permission).

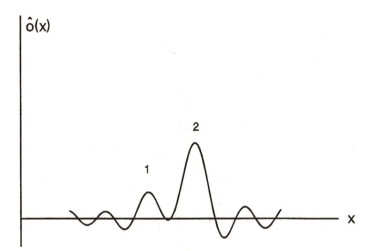

Figure 5 Signal of Fig. 1c deconvolved using Van Cittert's method and auto/cross-correlation prefiltering (Crilly, 1991b, © IEEE; reprinted with permission).

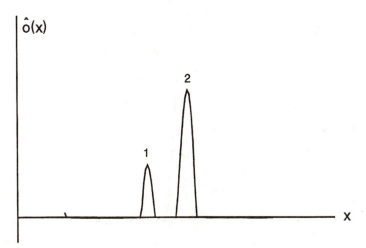

Figure 6 Signal of Fig. 1c deconvolved using truncation operator deconvolution method and auto/cross-correlation prefiltering (Crilly, 1991b, © IEEE; reprinted with permission).

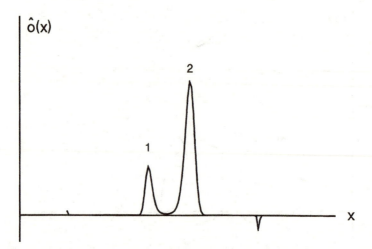

Figure 7 Signal of Fig. 1c deconvolved using Gold's ratio method and auto/cross-correlation prefiltering (Crilly, 1991b, © IEEE; reprinted with permission).

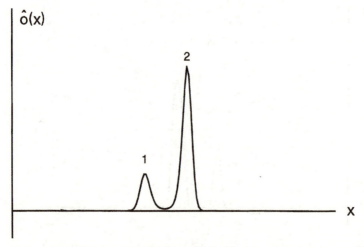

Figure 8 Signal of Fig. 1c deconvolved using Jansson's method and auto/cross-correlation prefiltering (Crilly, 1991b, © IEEE; reprinted with permission).

Table I   **Relative Error Comparisons for Various Iterative Deconvolution Methods**[a]

| SNR of original data | Relative error for peak 1 (%)[b] | | |
	Truncation operator Eq. (3)	Jansson's Eq. (4)	Gold's ratio Eq. (5)
100:1	−3.7	−0.9	−1.8
100:1[c]	−4.9	−0.5	−1.2
4:1[c]	−5.1	−2.5	−1.3

[a] From Crilly, 1991b, © IEEE; reprinted with permission.
[b] Values based on average of 10 measurements.
[c] Data prefiltered prior to deconvolution.

information is not truncated. For example, a truncated negative value appearing in an early iteration does not necessarily force a null value for all subsequent iterations. It would be unreasonable to expect only physically realizable values arising early in the process when the solution $\hat{o}^{(k)}(x)$ is still a rough approximation.

Gold's ratio with cross-correlation prefiltering gave surprisingly reasonable results because this method does not incorporate any constraints. And as noted in Chapter 4, this method is used primarily for data for which the noise distribution is Poisson and therefore does not have negative values. However, as shown in Figs. 2 and 4, it has erratic convergence properties if prefiltering is not used or the data are extremely noisy. This due to the

Table II   **Peak Size Variances for Various Iterative Deconvolution Methods**[a]

| SNR of original data | Standard deviation for peak 1 (%)[b] | | |
	Truncation operator Eq. (3)	Jansson's Eq. (4)	Gold's ratio Eq. (5)
100:1	0.9	1.3	0.5
100:1[c]	0.3	0.2	0.2
4:1[c]	9.3	9.6	4.5

[a] From Crilly, 1991b, © IEEE; reprinted with permission.
[b] Values based on average of 10 measurements.
[c] Data prefiltered prior to deconvolution.

division in Eq. (5) whereby finite noise values in $i(x)$ are divided by nearly zero values in expression $[s(x) \otimes \hat{o}^{(k)}(x)]$. Heasley *et al.* (1984) also did a comparison between Jansson's and Gold's ratio methods to restore Uranus and Neptune planet images. Their results state Jansson's method generally produced sharper images compared to Gold's ratio method.

The results of Figs. 7 and 8 show that Jansson's and Gold's ratio methods provided estimates that looked virtually identical to the original peaks in Fig. 1a. On the other hand, the results of the algorithm of Eq. (2), which used a truncation operator to suppress spurious results, showed considerable distortion in the base-line region. The estimate did not have the rounded Gaussian characteristics of the original signal.

As expected of a linear unconstrained technique, Van Cittert's method did not provide satisfactory results. In fact, the results were such that peak areas could not be properly quantitated.

IV. Deconvolution Examples

We conclude this chapter by presenting several examples of deconvolution of real data. The examples include high-resolution grating spectra, tunable-diode-laser (TDL) spectra, a Fourier transform infrared spectrum (FTIR), laser Raman spectra, high-resolution γ-ray spectrum, and gas chromatographic peaks. Unless otherwise noted, 100 iterations of Jansson's method were used for all cases.

A. INTERFEROMETRIC SPECTRA

The authors have witnessed many debates concerning the deconvolution of Fourier transform spectra. Some insist that it cannot work; others insist that it does, but that it should not; and others just assume that it works. In Fig. 9, an unapodized spectrum [response function $(\sin \pi x)/\pi x = \text{sinc}(x)$] is shown in trace b. For such a spectrum there will be sidelobes and "negative absorption" if the natural linewidths are narrower than the full width of the sinc-shaped response function. These are seen in Fig. 9, where the linewidth is three points and the response function width eight points. Here the phrase "instrument response function" may have a slightly different definition, but the meaning is clear. For such a response function, the direct deconvolution methods fall short.

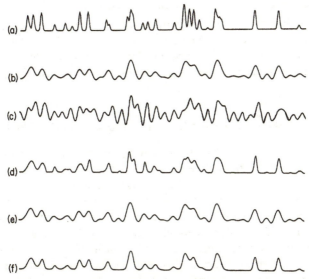

Figure 9   Deconvolving $s(x) = \text{sinc}(x)$ and $s(x) = \text{sinc}^2(x)$ convolved data. Trace a is the original spectrum $o(x)$, trace b the result of convolving with an eight-point sinc function, and trace c the result of unconstrained deconvoltuion using the same sinc function for just 10 iterations. Trace d is the result of 100 iterations using "zero clipping." Trace e is the original spectrum convolved with an $8\sqrt{2}$-point sinc-squared function, trace f the result of deconvolving trace e with the same sinc-squared function for 100 iterations using the Jansson-type relaxation function $r^{(k)}[o^{(k-1)}(x)]$.

The negative sidelobes of the unapodized spectrum render the constraints defined earlier useless. Trace c of Fig. 9 shows the result of unconstrained deconvolution. After only five iterations the algorithm begins to diverge and, as seen in Fig. 9, the spectrum is worse for our trouble. Trace d shows the result of zero clipping. Although there is some improvement and convergence does occur, the resulting spectrum appears real but is not a faithful representation of the original spectrum shown in trace a, particularly where several lines overlap in the convolved spectrum. Trace e represents the same spectrum with triangular apodization used to generate a response function $s(x) = (\sin^2\pi x)/\pi^2 x^2 = \text{sinc}^2(x)$. Although some resolution is lost, the negative lobes are eliminated and the positive sidelobes greatly reduced. Such a spectrum can be deconvolved with the Jansson weighting technique even in the presence of the sidelobes (which are now all nonnegative). The result is shown in trace f. This spectrum shows resolution equal to or slightly better than the unapodized spectrum shown in trace b and the sidelobes have been almost completely eliminated. For a very crowded spectrum this would appear to be a useful technique, particularly if there is interest in the weaker lines of the spectrum.

B. PRESSURE-BROADENED INFRARED AND RAMAN SPECTRA

In all the previous simulations we have assumed a narrow Gaussian line profile that results from Doppler broadening. This is often not the case, owing to the presence of pressure broadening that results in a Lorentzian line profile at sufficiently high pressures. This often means an unavoidable loss of resolution. To observe a spectrum when the transition intensities are low, it is often necessary to increase the sample gas pressure. In this case the pressure broadening can be sufficient to mask important spectral features. Conversely, when the pressure is low enough to allow these important spectral features to be resolved, they may be too weak to observe at achievable signal-to-noise ratios. To some extent this problem can be alleviated by using deconvolution to remove pressure-broadening effects (as well as instrumental effects). This is demonstrated for two cases: a simulated absorption spectrum and a simulated Raman spectrum.

For an absorption spectrum the effects of pressure broadening can be approximated by convolving the spectrum with a Lorentzian of the proper width. (An accurate representation would require us to carry out the convolution on the absorption coefficient spectrum.) Thus a spectrum with an eight-point Lorentzian line profile convolved with an eight-point Gaussian response function can be deconvolved with a Voigt profile having Gaussian and Lorentzian components each eight points in width. This is shown in Fig. 10. Trace a represents an infrared absorption spectrum with a Gaussian line profile three points wide. Trace b is the same spectrum generated using an eight-point Lorentzian for the line profile. Trace c is the convolution of trace b with an eight-point Gaussian, and trace d is the result of deconvolving trace c with the appropriate Voigt profile. As can be seen in examining Fig. 10, most of the pressure-broadening effects have been eliminated, resulting in Gaussianlike spectral lines with a full width at half maximum (FWHM) of from four to five points. The relative intensities and positions are well reproduced except for the feature marked "x" in trace d, which should be partially resolved at the apparent resolution.

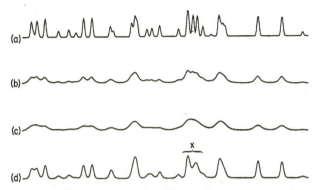

Figure 10 Removing pressure-broadening effects with deconvolution. Trace a is a simulated absorption spectrum using a three-point Gaussian line shape. Trace b is the same infrared spectrum except that an eight-point Lorentzian was used for the line shape. Trace c is the result of convolving trace b with an eight-point Gaussian. Trace d is the result of deconvolving trace c with a Voigt profile having Lorentzian and Gaussian widths of eight points. Feature marked "x" should be better resolved at the apparent deconvolved resolution.

This scheme was also used to test pressure-broadening removal in a Raman spectrum. Here, no approximations are needed and any pressure-broadening effects can be considered as part of the instrument response function if the pressure broadening is the same for all lines in the spectrum. The results are shown in Fig. 11. The recovery of the original spectrum is quite good and the resulting line profile is approximately a Gaussian with a FWHM of less than four points for most lines. Though the recovery is better in the Raman case, the features marked "x" in trace d of Fig. 11 should be better resolved if the resolution were as good as indicated by the widths of single lines.

C. *GRATING SPECTRA*

For several years now we have routinely deconvolved spectra recorded on the grating spectrometer at the University of Tennessee at Knoxville. Two examples are presented here. The first is part of the ν_4 band of CD_3F recorded in thirteenth order using a 31.6/mm grating. The spectrum was recorded once with a resolution of 0.010 cm^{-1} in a scan that covered most of the ν_4 band. The spectrum was smoothed and base-line corrected prior to deconvolution, and a section of this data set (RQ$_4$ region) is shown in

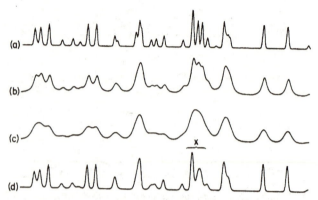

Figure 11 Replication of Fig. 10 for Raman-type data. The constraint used was for negative suppression only. All plots are scaled so that the maximum signal is 1 inch. This gives the *appearance* of nonconservation of area.

Fig. 12 as trace a. The data set was deconvolved using the relaxation method described by Eq. (4b). The resulting spectrum was then interpolated to double the data-point density, and the RQ_4 region is shown in trace b of Fig. 12. A calculated spectrum, using molecular parameters from a previous lower resolution run ($\Delta\bar{\nu} = 0.028$ cm^{-1}), is shown in trace c using a resolution corresponding to the Doppler width ($\Delta\bar{\nu} = 0.004$ cm^{-1}). The analysis of ν_4 using this deconvolved data set was successful and in excellent agreement with other high-quality results for CD_3F.

The second example of deconvolved grating spectra is part of the ν_3 band of $^{28}SiH_4$, and is shown in Fig. 13. Trace a was recorded at a pressure of 1 Torr of natural-abundance SiH_4 at a resolution of 0.020 cm^{-1} in twelth order of the 31.6/mm grating. This spectrum was recorded to study the ν_3 spectrum of the less abundant isotopes ($^{29}SiH_4$, 4.70%; $^{30}SiH_4$; 3.09%) and was later deconvolved. Owing to the isotopic shift in the ν_3 band, the interaction between ν_3 and ν_1 splits the J manifolds of $^{29}SiH_4$ and $^{30}SiH_4$ much more than those of $^{28}SiH_4$. The J manifolds of $^{28}SiH_4$ were then rescanned 12 times at higher resolution (0.012 cm^{-1}). These scans were averaged prior to further processing. The signal-to-noise ratio was approximately 60:1 for the individual scans and approximately 200:1 in the averaged spectrum. The averaged spectrum was not smoothed further, and the base-line-corrected spectrum for P_6 is shown in trace b of Fig. 13. A calculated spectrum, based on a preliminary analysis of ν_1 and ν_3, is

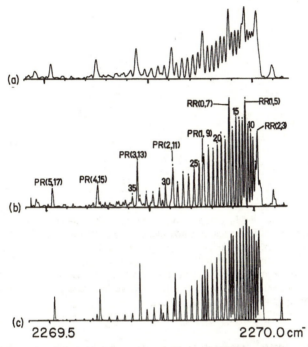

Figure 12 Q branch of the RQ_4 subband of ν_4 of CD_3F recorded on a grating spectrometer. Trace a is the raw data after smoothing and base-line subtraction. The resolution based on the widths of single lines is 0.010 cm^{-1}. Trace b is the result of deconvolving this data set with a Gaussian of FWHM = 0.010 cm^{-1}. The resolution is 0.0045 cm^{-1}. Trace c is a calculated spectrum based on the analysis of an earlier spectrum with a resolution of 0.028 cm^{-1}. The resolution in trace b is 0.0045 cm^{-1} and in trace c is 0.0040 cm^{-1}.

shown in trace c using a resolution of 0.004 cm^{-1}, corresponding to the Doppler width. The deconvolved spectrum is shown in trace d of Fig. 13 and exhibits a linewidth of from 0.003 to 0.004 cm^{-1}. The instrument function used was a Gaussian 0.012 cm^{-1} wide.

D. TUNABLE-DIODE-LASER SPECTRA

Tunable diode lasers have in recent years pushed the limits of resolution well beyond that achievable using conventional means. For the purposes of infrared molecular spectroscopy, the limiting factor is the Doppler broad-

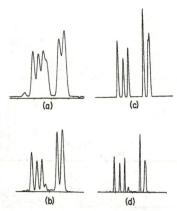

Figure 13 P_6 of $^{28}SiH_4$ recorded on a grating spectrometer. Trace a is from a continuous scan of natural-abundance silane at a resolution of 0.020 cm^{-1}. Trace b is the same region recorded separately at a resolution of 0.012 cm^{-1} and is the average of 12 scans. Trace c is P_6 calculated at a resolution of 0.004 cm^{-1}, trace d the result of deconvolving the data shown in trace b. The resulting resolution is between 0.003 and 0.004 cm^{-1}, and the relative intensities and positions are in good agreement with the calculated spectrum.

ening of spectral lines rather than the resolution when using a TDL spectrometer. Sophisticated techniques have been used to overcome the Doppler limit problem, such as cooling the sample under consideration and sampling through gas jets. Deconvolution presents a simpler way to limit the effects of Doppler broadening without investing in complicated and expensive instrumentation. The ν_9 band of ethane provides an interesting example.

The spectrum of ethane exhibits a splitting in all spectral lines at very high resolution. This effect is due to the torsional motion of the molecule that splits the energy levels. If the splittings in the upper and lower states of a transition differ, a splitting of the spectral line associated with that transition will occur. For the ν_9 band this splitting is approximately equal to the Doppler width at room temperature. The splitting increases with J and is enhanced in RQ_0 owing to the combined effects of a Coriolis interaction and l-doubling. For high values of J this splitting is resolved, but for lower values of J the splittings are unresolved even when using a tunable diode laser.

Figures 14 and 15 demonstrate the success of deconvolution beyond the Doppler limit. Trace a in both figures is the undeconvolved data and

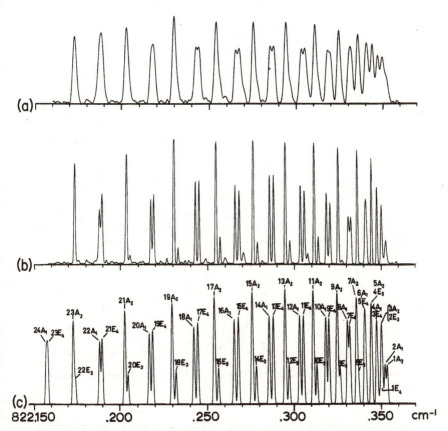

Figure 14 Tunable-diode-laser spectrum of RQ_0 of ν_9 of ethane. Trace a is the average of 250,000 scans and exhibits linewidths of 0.0022 cm^{-1} (the Doppler width is 0.0018 cm^{-1}). Trace b results from the deconvolution of the data in trace a using a Gaussian with a FWHM of 0.0022 cm^{-1} as a response function. Trace c is the Q branch calculated using a model that includes torsional splitting effects $\Delta\nu = 1.95$ mk. Trace c is calculated for $\Delta\nu = 0.00075$ cm^{-1}, which is less than one-half the 300 K Doppler width.

represents the result of signal averaging over a large number of scans to produce a signal-to-noise ratio of approximately 300:1. The spectra were base-line corrected using a 0% absorption corresponding to the apparent base line between lines and a 100% absorption level was assumed, which gave reasonable relative intensities for the range of J values observed. The spectra were then interpolated to a linear frequency scale of 0.00025

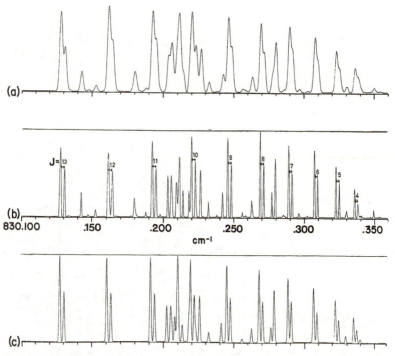

Figure 15   Tunable-diode-laser spectrum of RQ_3 of ν_9 of ethane. All details are the same as in Fig. 14.

cm^{-1} per data point using a 3-inch germanium étalon producing a fringe spacing of 0.0161 cm^{-1}. The linewidths were nearly constant across the scans and approximately 0.0022 cm^{-1}. The spectra were then deconvolved using a Gaussian response function having a FWHM of 0.0022 cm^{-1}. The results are shown in trace b of both figures. Trace c is a calculated spectrum (using constants obtained from fitting the entire ν_9 band) using linewidths of 0.00075 cm^{-1}, less than half the Doppler width. The agreement is very good, limited in this case by the linearization of the diode-laser scans.

E.   INTERFEROMETRIC ABSORPTION SPECTRA

Figures 16 and 17 show the results of using deconvolution to remove the sidelobes present in a Fourier transform spectrum. These two spectral sections are taken from the ν_2 absorption band of ammonia at about 848

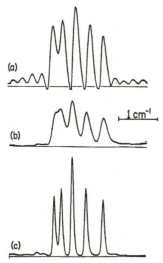

Figure 16 Fourier transform spectrum of ν_2 of ammonia. Trace a is a section of the infrared absorption spectrum of ammonia recorded on a Digilab Fourier transform spectrometer at a nominal resolution of 0.125 cm^{-1}. In this section of the spectrum near 848 cm^{-1} the sidelobes of the sinc response function partially cancel, but the spectrum exhibits "negative" absorption and some sidelobes. Trace b is the same section of the ammonia spectrum using triangular apodization to produce a sinc-squared transfer function. Trace c is the deconvolution of the sinc-squared data using a Jansson-type weight constraint.

and 828 cm^{-1}, respectively. The unapodized spectrum exhibits a resolution of 0.125 cm^{-1}, and is shown in trace a. The spacing of the lines in Fig. 16 is such that the sidelobes, when added together, partially cancel, minimizing their effect. Sidelobes and the apparent negative absorption between the lines are both still present. The spacing of the lines in Fig. 17 is such that the sidelobes add constructively, accentuating their effect and producing spurious absorption lines between the true features. Trace b shows the same spectrum using triangular apodization. Here the sidelobes are greatly reduced and the negative absorption features are removed but the resolution is reduced. A single well-isolated line between these two sections was chosen for the instrument profile. The deconvolved spectrum presented in trace c still shows very weak sidelobes, but no negative absorption points. The resolution lost in the apodization has been recovered, the apparent linewidths of the deconvolved spectrum being about two-thirds of those in the unapodized spectrum.

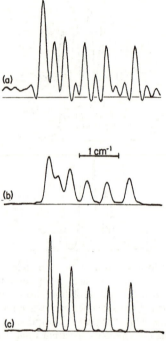

Figure 17   Same as Fig. 16 except $\bar{\nu} \approx 828$ cm^{-1}. Sidelobes add constructively in the unapodized spectrum and produce very large sidelobes. The strongest feature is two unresolved lines.

F.   PURE ROTATIONAL LASER RAMAN SPECTRA

Trace a of Fig. 18 represents part of the pure rotational laser Raman spectrum of O_2. This example is a single scan of the S_3 transition $J = 3$ to $J = 5$ in the ground state. This transition is split into three components because of the permanent dipole moment of the oxygen molecule in the ground state. This scan was recorded as part of a series of scans in a study of pressure broadening in oxygen. The pressure was 1.0 atm and induced a pressure broadening of approximately 0.1 cm^{-1}. In addition to pressure broadening (Lorentzian), the profile was affected by Doppler broadening (0.042 cm^{-1}, Gaussian) and the instrument function (0.035 cm^{-1}, quasi-Gaussian). These effects combine to mask the splittings almost completely. The spectrum was recorded with 0.0051 cm^{-1} per data point, using 300

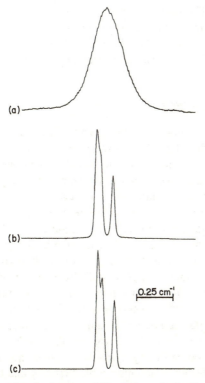

Figure 18 Pure rotational spectrum of O_2. Trace a is the S_3 transition recorded at a pressure of 1.0 atm. Trace b is the result of deconvolving the S_3 profile with a Voigt profile to remove most of the pressure broadening, Doppler broadening, and instrument effects. Trace c was calculated using a 0.035-cm^{-1} Gaussian profile and calculated spin splittings. The traces are scaled to the same height.

points to scan each line in the pure rotational spectrum. The peak in S_3 is approximately 2400 counts and there is a nearly constant background of 100 counts.

Prior to deconvolution, the background was subtracted and the data were smoothed with a 15-point quadratic least-squares polynomial followed by a 19-point quartic least-squares polynomial. The data were then scaled from 0 to 1. The S_3 profile was deconvolved using a relaxation function of the form

$$r^{(k)}[\hat{o}^{(k-1)}(x)] = \hat{o}^{(k-1)}(x) \exp[-\hat{o}^{(k-1)}(x)] \qquad (8)$$

for 75 iterations. The root-mean-square error was reduced by a factor of 5 and the resolution of the three components of S_3 is in fairly good agreement with the theory. The deconvolved spectrum is shown in trace b of Fig. 18, and a calculated spectrum with a 0.035-cm^{-1} Gaussian profile is shown in trace c.

G. GAMMA-RAY SPECTRA

Figure 19 shows the γ-ray emission of Tm^{159} from 120 to 250 keV using a Tm^{159} source produced in a beam experiment. The response function here is in many ways similar to that of the tunable-diode-laser spectrum shown in Section D. That is, the response function is partially due to Doppler broadening and is not constant over even a short section of the spectrum. Here the FWHM ranges from 7.0 points at the low-energy (left) end of the section shown to about 7.8 points at the high-energy (right) end, each data point corresponding to 0.25 keV. The sample under consideration is produced in such a way that there is a sizable background signal from many weak unresolvable γ-ray peaks and also background from the Compton scattering edges for higher energy γ rays. The average background over this region is about 19,000 counts per channel and the highest peak exhibits about 25,000 counts per channel above background. Trace a of Fig. 19 represents the base-line-corrected data. Prior to deconvoltuion these data are smoothed using a seven-point quadratic and a nine-point quartic Savitzky–Golay smoothing profile (Savitzky and Golay, 1964). The response function used was a Gaussian with a FWHM of seven points. The weighting scheme used was of the form of Eq. (8), which seems to maintain the relative intensities better than a weighting scheme with only a lower limit constraint. The deconvolved result is shown in trace b.

H. GAS CHROMATOGRAPHIC PEAKS

In gas chromatographic (GC) analysis, peak resolution is proportional to the square root of the column length. However, analysis time is directly proportional to column length and thus, all else being equal, if we want to double resolution, we need to increase the column length by a factor of 4. Obviously, resolution can be increased by improving the instrument in some other way, but this may involve other tradeoffs. Using deconvolution to achieve peak resolution may be the only desirable alternative. Deconvolution can be a useful tool not only to improve the instrument's resolution, but also to produce significantly faster GC analysis (Crilly, 1987a–c; 1992).

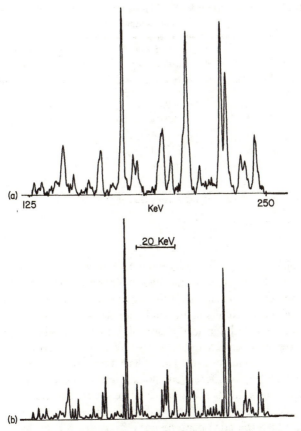

Figure 19 Gamma-ray spectrum of Tm^{159}. Trace a is the recorded spectrum after the continuum has been subtracted. Trace b is the result of smoothing and deconvolving this spectrum with a Gaussian profile having a width corresponding to the narrowest features observed in the spectrum.

In the GC analysis of ethyl benzene and *m*-xylene by Crilly (1987b, c; 1992) achieving minimal acceptable peak resolution required a column that was 50 m long and 0.320 mm in diameter, and took 16 minutes. The same analysis was done using a 10-m length and 0.530-mm-diameter capillary column and the resultant peak resolution (as previously defined in Section III.A. of this chapter) was an unacceptable 0.42. When deconvolution was subsequently performed, the resolution increased to 2. The total

Table III   Relative Errors for Peaks Resolved via Deconvolution versus a
Long Column[a]

Sample number[b]	Peak	Relative error with a long column	Relative error with a short column and deconvolution
1	A	−0.04	+0.08
	B	+0.04	−0.07
2	C	−1.84	−1.25
	D	+0.39	+0.36
3	E	+2.01	−4.68
	F	−0.20	+0.46

[a] From Crilly, 1987b, c; 1992; © 1992, IEEE; reprinted with permission. Values based on average of 20 measurements.
[b] For peaks shown in Fig. 20.

analysis time including computation was 1.5 minutes. This included the 160 iterations needed to perform the deconvolution on an IBM PS/70 personal computer. The results are shown in Table III and Fig. 20. The plots of Fig. 20 show the results before and after deconvolution.

It should be noted that in the case of gas chromatographic peaks, the raw observed peak heights and areas (before and after deconvolution) may not reflect the sample mixture. This is because chromatographic detectors respond differently to different compounds. For example, if equal amounts of *m*-xylene and ethyl benzene are analyzed with a thermal conductivity detector, the detector's bias is such that the *m*-xylene peak will be about 10% larger than the ethyl benzene peak. To overcome this bias, response factors are calculated for each compound and then the peak quantities are adjusted.

A good approximation to the impulse response function can be a single peak having a retention time similar to that of the sample being analyzed. In this case, we used *m*-xylene.

Table III shows the error comparison between conventional GC analysis where a long column was used to achieve acceptable resolution versus deconvolution. As Table III shows, peak quantitation using deconvolution gave results similar to those obtained from a long column. In other tests (Crilly, 1987b, c; 1992) deconvolution gave orders of magnitude improvement in quantitation accuracy of severely overlapped peaks as compared to the widely used graphical techniques of shoulder quantitation and perpen-

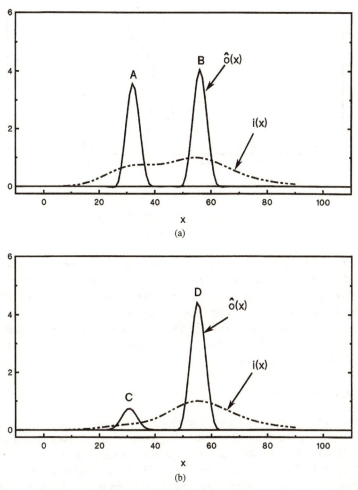

Figure 20 Gas chromatographic peaks of ethyl benzene and *m*-xylene. Broken line plots are the original chromatograms, solid line plots are chromatograms deconvolved using Jansson's method. (a) 50/50 mixture, (b) 20/80 mixture, (c) 10/90 mixture (Crilly, 1992, © IEEE; reprinted with permission).

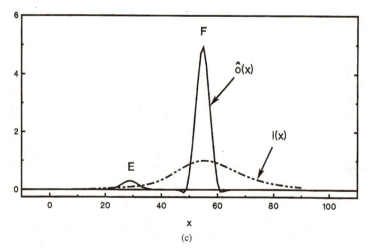

Figure 20   (*Continued*)

dicular drop (Crilly, 1987c). Schure (1991) provides another example of constrained deconvolution to improve the resolution of GC peak data.

A comparable analysis using the same overlapped peak data with Gold's ratio deconvolution method was done by Crilly (1991a). In that study, the relative errors resulting from Gold's ratio method were twice that of Jansson's. As already reported in this chapter, Gold's method had no advantage over Jansson's.

I.   CONCLUSIONS

From several years of testing and using deconvolution techniques, we have developed a positive attitude toward constrained signal-space deconvolution (CSD). Undoubtedly, our favorable attitude stems from the extensive testing of CSD woven into actual practical use of it. Given that proper attention is paid to sampling rates and signal-to-noise ratio in the information band, CSD is capable of and will consistently produce high-quality results. There is no question that CSD can be misused; however, so can all observing instruments. For example, a spectrometer may be scanned at a rate several times too high for the signal bandwidth. The result is a set of corrupted observations. On the other hand, if one uses CSD on data for which the SNR is 3:1 in the information band and 20:1 overall, the failure of CSD is not surprising. One can conclude that CSD properly used, where the response function is well characterized and the SNR appropriate, will result in a factor of 2 to 3 in resolution enhancement and zero artifact generation.

Remember also that, as pointed out by Blass and Halsey (1981) and Crilly (1987a–c; 1992), deconvolution can be used to achieve resolution otherwise unavailable or to obtain a specific required resolution using hardware having a factor of 2 to 3 less in resolving power. To the dedicated experimental scientist this may sound like heresy. Even today, when an instrument is lacking in resolution, the reply is often "get a better instrument, or improve your practice."

Today, as in 1984 upon publication of the first edition of this book, the question is still, "Would you rather use deconvolution or build a more highly resolving instrument?" Generally, this is not a well-posed question! However, the answer, if both time and resources are unlimited, is to build a better instrument. If not, then deconvolution is often the preferred course of action.

References

Blass, W. E., and Halsey, G. W. (1981). "Deconvolution of Absorption Spectra." Academic Press, New York.

Crilly, P. B. (1987a). *J. Chemom.* **1**, 79–80.

Crilly, P. B. (1987b). *J. Chemom.* **1**, 175–183.

Crilly, P. B. (1987c). Ph.D. Thesis, New Mexico State University, Las Cruces.

Crilly, P. B. (1991a). *J. Chemom.* **5**, 85–95.

Crilly, P. B. (1991b). *IEEE Trans. Instrum. Meas.* **40**, 558–562.

Crilly, P. B. (1992). *IEEE Trans. Ind. Electron.* **39**, 20–24.

Heasley, J. N., Pilsher, C. B., Howell, R. R., and Caldwell, J. J. (1984). *ICARUS* **57**, 432–442.

Kawata, S., and Ichioka, Y. (1980a). *J. Opt. Soc. Am.* **70**, 762–768.

Kawata, S., and Ichioka, Y. (1980b). *J. Opt. Soc. Am.* **70**, 768–772.

Savitzky, A., and Golay, M. J. E. (1964). *Anal. Chem.* **36**, 1627–1639.

Schure, M. R. (1991). *J. Chromatogr.* **550**, 51–69.

Snyder, L. R., and Kirkland, J. J. (1979). "Introduction to Modern Liquid Chromatography." Wiley, New York.

Chapter 8 | Application to Electron Spectroscopy for Chemical Analysis

Robert D. Davies
Peter A. Jansson

College of Optical Sciences, University of Arizona, Tucson

List of Symbols

A_1, A_2	relative intensities of the x-ray $K\alpha$ doublet
ESCA	electron spectroscopy for chemical analysis
f	fraction of electrons scattered inelastically
$H(x)$	1 when $x > 0$; 0 when $x \leq 0$; additional background in Chapter 1
$H_-(x)$	$H(-x) = 1 - H(x)$
HWHM	half-width at half maximum
$(i)_j$	data $i(x)$ sampled with energy spacing Δx

i_l	lth data element, the element at the high-energy end of the scan range
i_L	number of electrons uniformly scattered from energies above limit x_L
$i(x), i$	observed ESCA spectrum; the data
$i_S(x), i_S$	observed spectrum contribution due to inelastically scattered electrons
l	number of data elements in a digitized scan
$(\hat{o}_B)_j$	discrete version of estimate of function $o_B(x)$ sampled with energy spacing Δx
$\hat{o}_i^{(k)}$	kth estimate of the true spectrum as described in Chapter 1, here identified as estimate of $o_R(x)$
$o(x), o$	spectrum devoid of broadening due to natural and instrumental effects
$o_B(x), o_B, i_R$	ESCA spectrum as observed less inelastic-scattering broadening
$o_R(x), o_R$	spectrum showing both natural and energy analyzer broadening
$s_B(x), s_B$	$\delta(x) + s_S$, the combined base-line spread function due to electrons inelastically scattered in sample plus those unscattered
$s_E(x), s_E$	spread function due to electrical filtering
$s_M(x), s_M$	spread function due to energy analyzer broadening
$s_N(x), s_N$	spread function due to natural broadening of ESCA spectral line
$s_R(x), s_R$	$s_X(x) \otimes s_E(x)$
$s_S(x), s_S$	spread function due to inelastic scattering
$s_X(x), s_X$	spread function due to broadening by exciting x-ray line
x	electron kinetic energy
x_1, x_2	relative energies of the x-ray Kα doublet components
x_L	upper energy limit of a spectrometer scan
$\delta(x), \delta$	Dirac δ function or impulse
Δx_X	half-width at half maximum (HWHM) of each of the x-ray Kα doublet components
κ	relaxation function
κ_0	constant in relaxation function

I. Introduction

The pioneering efforts of K. Siegbahn and associates in the 1950s led to the development of a new type of spectroscopy uniquely suited to the study of surfaces (Siegbahn *et al.*, 1967). The technique, in Siegbahn's own nomenclature, is that of electron spectroscopy for chemical analysis (ESCA), which can reveal information about both chemical bonding and elemental composition within about 50 Å of the surface of a sample. It is the x-ray-excited variety of electron spectroscopy; other modes of excita-

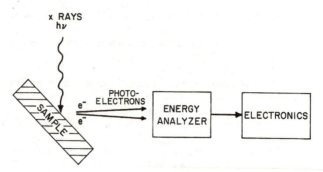

Figure 1 ESCA spectrometer.

tion exist (Herglotz, 1982). ESCA has proved valuable in practical studies of material properties, catalysis, and adhesion, as well as fundamental investigations into the nature of surface chemical bonds. Herglotz and Suchan (1975) have reviewed ESCA methods, instruments, results, and limitations for workers in surface research. A series edited by Brundle and Baker (1977, 1978, 1979, 1981) is also helpful to workers becoming familiar with this field. A more recent review has been prepared by Riviere (1993).

In a typical ESCA experimental, a sample is placed in an evacuated chamber and bombarded with x rays, as illustrated schematically in Fig. 1. The x rays dislodge inner-shell electrons, which leave the sample with kinetic energy determined by the difference of the incident x-ray photon energy and the binding energy of the electron before the interaction. The energy analysis of these escaping electrons gives the binding-energy spectrum of the electrons belonging to the various atoms that make up the sample surface.

In addition to the principal influence of the atomic energy levels, the binding energy of inner-shell electrons is affected to a lesser degree by the outer electrons that participate in chemical bonding. The various atomic lines in the spectrum therefore appear slightly shifted in energy. The degree of shift depends on the nature of the binding of the atom from which the electron was ejected. This shift is typically small, often not larger than the breadth of the exciting x-ray line, energy analyzer broadening, and other effects.

Detailed analysis of the chemical shifts demands higher resolution than a basic ESCA spectrometer can provide. The needed resolution improvement may be achieved experimentally by monochromatizing the x rays and/or by high-resolution energy analysis of the electrons. Either way, because of the reduced slit widths required, penalties are paid in decreased signal-to-noise ratio and longer observation time. Furthermore, the problem of sample charging (Section IV.D) is typically more severe when monochromatized x rays are used (Brundle and Baker, 1981).

An elegant instrumental technique for partially circumventing the signal-to-noise versus resolution trade-off has been found (Siegbahn *et al.*, 1972). In this technique, x rays are dispersed in energy across the sample. The ejected electrons pass into an energy analyzer the dispersion of which exactly cancels the dispersion of the x rays. The result is the acquisition of a high-resolution spectrum. The added cost of the required equipment, however, combined with the sample-charging disadvantage and the increasing availability of low-cost computation, serves to provide a keen motivation for the development of a computational means of improving resolution.

All known methods for computational resolution improvement, however, involve signal-to-noise versus resolution trade-offs. Clean data having good signal-to-noise ratios are required if useful resolution improvement is to be obtained. The computational approach was facilitated for us by the availability of a spectrometer (Lee, 1973) that provides a high signal-to-noise ratio while retaining moderate to good resolution.

The work reported in this chapter was conducted in 1971, at the Engineering Physics Laboratory of E. I. du Pont de Nemours and Company. This research was performed as part of the ESCA instrument development program active at that time, and was presented by Jansson and Davies (1974). An example of a deconvolved polyester spectrum obtained in the early work appeared in the review article by Herglotz and Suchan (1975).

II. Distortions Present in ESCA Spectra

The resolution of overlapping spectral peaks depends on their separations, intensities, and widths. Whereas separation and intensity are predominantly functions of the sample, peak width is strongly influenced by the instrument's design. The observed line is a convolution of the natural line, a function characteristic of inelastically scattered electrons that produces a skewed base line, and the instrument function. The instrument function is, in turn, the convolution of the x-ray excitation line shape, the broadening

Table I   **Spread-Function Component Definitions**

$s_S(x)$	spread function due to inelastic scattering
$s_N(x)$	spread function due to natural broadening of spectral line observed
$s_X(x)$	spread function due to broadening by exciting x-ray line
$s_B(x)$	$\delta + s_S$, the combined base-line spread function due to electrons inelastically scattered in sample plus those unscattered
$s_M(x)$	spread function due to energy analyzer broadening
$s_E(x)$	spread function due to electrical filtering

inherent in the electron energy analyzer, and the effect of electrical filtering. This description is summarized in Table I.

$$i = s_B \otimes \left| \begin{array}{c} s_X \otimes s_E \otimes s_M \\ \leftarrow \text{instrument} \rightarrow \\ \text{function} \end{array} \right| \otimes s_N \otimes o \right|$$

$$\longleftarrow i_R \longrightarrow$$
$$\longleftarrow o_B \longrightarrow$$

Two contributions to broadening can be thoroughly and easily characterized—the excitation line and electrical filtering. They are therefore good candidates for deconvolution. Although a third contribution—inelastic scattering—is a function of sample composition and morphology, it can be well approximated by a simple model. This model serves as the basis of an effective procedure to correct the resulting uneven base line. Spectral resolution can be significantly improved by mathematically removing the broadening and distortion produced by these three effects.

A. *ELECTRON INELASTIC SCATTERING*

Many photoelectrons are not able to emerge from the sample without suffering energy loss by inelastic collisions. At a given kinetic energy, in addition to electrons from an atomic energy-level transition at that energy, one observes some electrons from higher-energy transitions that have partially lost their kinetic energy because of inelastic scattering. The general shape of these energy-loss curves is well known. Tougaard (1988) has reviewed the field and formulated an accurate physical model that explains these data and is useful for corrections. In this work we employed a much simpler model that is instructive, and was reported previously (Jansson and Davies, 1974). It was found adequate to prepare the data for the primary task of removing the excitation-line broadening.

The simplifying assumption is that scattered electrons are uniformly redistributed to lower energies, so that the background flux of electrons at a given energy is proportional to the integral of the flux at higher energies. This background correction models the scattering contribution at energy x as

$$i_S(x) = f\int_x^\infty o_B(x')\, dx', \tag{1}$$

where $o_B(x')$ is the number of unscattered electrons at energy x', and f is a fraction of the electrons scattered uniformly (in this approximation) to lower energy. We further observe that this integral may be converted to a convolution by introducing the Heaviside step function in abscissa-reversed form:

$$i_S(x) = f\int_{-\infty}^\infty H_-(x - x')o_B(x')\, dx', \tag{2}$$

where $H_-(x) = H(-x) = 1 - H(x)$. When expressed as a convolution, the number of unscattered electrons at a given energy appears as a simple Dirac δ function. For the total contribution at x we then have

$$i(x) = o_B(x) + i_S(x) = \int_{-\infty}^\infty [\delta(x - x') + fH_-(x - x')]o_B(x')\, dx'. \tag{3}$$

The kernel of this equation may be identified by the characteristic base-line response function

$$s_B(x) = \delta(x) + fH_-(x), \tag{4}$$

and Eq. (3) may be abbreviated

$$i = s_B \otimes o_B. \tag{5}$$

B. X-RAY EXCITATION BROADENING

The Lorentzian shape of x-ray emission lines is well founded in quantum theory and has been substantiated experimentally (Hoyt, 1932). Siegbahn *et al.* (1967) discuss the aluminum anode x-ray source as applied to ESCA. Beatham and Orchard (1976) list doublet separations and half-widths derived from the literature and optimized by computer simulation. Källne and Åberg (1975) and Senemaud (1968) also provide values.

We may write the spread-function contribution due to the x-ray source as the following Lorentzian doublet:

$$s_X(x) = \sum_{j=1}^{2} \frac{A_j}{1 + \left[(x - x_j)/\Delta x_X\right]^2}. \tag{6}$$

In this expression the A_j are the relative intensities and the x_j the relative energy coordinates of the components. The quantity Δx_X is the component half-width at half maximum (HWHM).

C. OTHER BROADENING

1. Electron Energy Analyzer

The precise nature of broadening due to the energy analyzer depends on analyzer design and may not be easy to compute, even in the ideal case. Measurement may also be difficult. The determination of the broadening function $s_M(x)$ may be attempted with the aid of narrow spectral lines or spectral lines of known shape, or by supplying the analyzer with electrons from a thermionic source (Lee, 1973).

The energy analyzer used in the present investigation is a unique high-étendue unit developed by Lee (1973) that facilitates the high signal-to-noise ratio desirable for deconvolution. The effect of broadening due to the energy analyzer was minor when compared with the x-ray excitation broadening affecting the present experiments.

2. Electrical Filters

As discussed in Chapter 1, any linear analog filter network may be used. Its performance may always be described by convolution with a filter function, here called s_E. Many modern digital filters may be thus described. The instrument employed in the present work used a simple single-stage *RC* filter of the type analyzed in Section II.D of Chapter 2.

3. Natural Line

The principal inherent broadening of the ESCA line is related to lifetime effects in a manner similar to that discussed in Section I.A of Chapter 2.

III. Correction of Distortions

In this investigation, only electron inelastic scattering and x-ray line broadening were chosen for substantial correction. There were two principal reasons: first, these broadening mechanisms account for the largest part of the distortion, and, second, their contributions are easily determined or approximated.

A. INELASTIC SCATTERING BASE LINE

Because the function given by Eq. (4) neither obliterates nor strongly suppresses the high Fourier frequencies in the data, we would expect a linear method to perform relatively well. A simple iterative approach based on the direct method of Section I of Chapter 3 does, in fact, prove effective.

From Eqs. (1) and (3) we may write the total contribution of the observed electron flux as

$$i(x) = o_B(x) + f \int_x^\infty o_B(x') \, dx'. \tag{7}$$

For data taken over a limited scan range it is convenient to break up the integral:

$$i(x) = o_B(x) + i_L + f \int_x^{x_L} o_B(x') \, dx'. \tag{8}$$

In this equation x_L is the high-energy scan limit, and we have defined the number of electrons uniformly scattered from energy above x_L as

$$i_L = f \int_{x_L}^\infty o_B(x') \, dx'. \tag{9}$$

Solving for the number of unscattered electrons and converting the resulting equation to a discrete form suitable for use with digitized data, we obtain the spectrum as it would appear in the absence of scattering:

$$(o_B)_j = \begin{cases} (i)_j - i_L - f\Delta x \sum\limits_{k=j+1}^{l} (o_B)_k, & j = l-1, l-2, \ldots, 0, \\ (i)_j - i_L, & j = l, \end{cases} \tag{10}$$

where indices j and k denote the jth and kth kinetic energy intervals of width Δx, respectively, and l is the index corresponding to the energy of the upper scan limit x_L.

Equation (10) may be used as a recurrence formula to obtain an estimate of $(o_B)_j$ that depends on having previously obtained estimates of $(o_B)_{j'}$ for $j' = j + 1, \ldots, l$. The quantity f must be chosen correctly for this procedure to work. One possible method of determining f begins with choosing a section of spectrum that begins and ends in a "base-line" region devoid of spectral structure, so that $(o_B)_l = (o_B)_0 = 0$. An assumed value of f is used to compute the estimate $(\hat{o}_B)_j$ based on Eq. (10). The last value, $(\hat{o}_B)_0$, will be either underestimated or overestimated, depending on the initial assumption for f. This information may be used in a simple Newton's method update of f and repeated until convergence is achieved [$(\hat{o}_B)_0 \approx 0$].

For $j = l$, we have $(\hat{o}_B)_l = (i)_l - i_L$. For scans started in the "base-line" region we find $(i)_l = i_L$. It may sometimes be convenient to subtract the value $(i)_l$ from the entire data set before further processing so that the data may be treated as if $i_L = 0$.

It is also possible (and instructive) to express Eq. (10) in a nonrecurrence form containing the data values $(i)_k$. After some algebra, we obtain

$$(o_B)_j = (i)_j - i_L - f\Delta x \sum_{k=j+1}^{l} [(i)_k - i_L](1 - f\Delta x)^{k-j-1}, \tag{11}$$

$$j = 1, \ldots, l-1.$$

The quantity f may be determined and Eq. (11) applied in the same manner as for Eq. (10). This time, of course, the $(\hat{o}_B)_j$ may be computed in either ascending or descending sequence. With both Eqs. (10) and (11) it may be desirable to use the average of a group of points to fix values for the end points $(i)_0$ and $(i)_l$. Only three to five iterations proved to be necessary in our applications (Jansson and Davies, 1974).

A widely used method described by Shirley (1972) is similar, but it employs instead a summation over the observed electron flux $(i)_k$, which includes contributions from electrons already scattered. Our own iterative method is based on the assumption of uniform scattering of a fixed fraction of the electrons that would be observed in the complete absence of scattering. Our formula reduces to the Shirley formula when the quantity $|(1 - f\Delta x)^l - 1|$ is small. This occurs when $f\Delta x$ is small.

Other workers have carried out Van Cittert deconvolution of the x-ray line contribution without treating inelastic scattering (Wertheim, 1975; Wertheim and Hüfner, 1975), or they have used Van Cittert's method to remove the scattering contribution (Madden and Houston, 1976). Wertheim and Hüfner (1975) have called into question the use of the "traditional" background correction methods that are based on the assumption of electron scattering. They present supporting arguments and data. Prior removal of the base line is, however, necessary to take full advantage of the physical-realizability constraints imposed by the method that we have used to remove the x-ray broadening.

We recognize that the present simple model for inelastic scattering does not account for certain details such as structure in the energy-loss spectrum due to plasmons (Davis and Lagally, 1981) and numerous other considerations (Tougaard, 1988). Nevertheless, its simplicity and effectiveness bear out its practice as a useful approximation.

B. X-RAY EXCITATION BROADENING

A review of deconvolution methods applied to ESCA (Carley and Joyner, 1979) shows that Van Cittert's method has played a big role. Because the Lorentzian nature of the broadening does not completely obliterate the high Fourier frequencies as does the sinc-squared spreading encountered in optical spectroscopy (its transform is the band-limiting rect function), useful restorations are indeed possible through use of such linear methods. Rendina and Larson (1975), for example, have used a multiple filter approach. Additional detail is given in Section IV.E of Chapter 3.

Constraining the solution to be positive can nevertheless provide the added benefits of reduced sensitivity to noise and improved resolution. Gold's iterative ratio method (Chapter 1, Section IV.A), for example, has been used successfully by a number of workers, including MacNeil and Dixon (1977) and Delwiche *et al.* (1980). MacNeil and Dixon have compared it with the standard Van Cittert method, which is linear.

Other constrained methods have also been applied. Beatham and Orchard (1976) experimented with Biraud's method but experienced only limited success. Vasquez *et al.* (1981) found that maximum entropy is capable of yielding excellent results on simulated ESCA data. The authors, who used Burg's method, cite its freedom from need for trial-and-error optimization. They did, however, have to develop methods of dealing with problems of instability and lack of an order-selecting criterion.

For the present work, we chose the constrained method described by Jansson (1968) and Jansson *et al.* (1968, 1970). See also Section VI of Chapter 4 and supporting material in Chapter 3. This method has also been applied to ESCA spectra by McLachlan *et al.* (1974). In our adaptation (Jansson and Davies, 1974) the procedure was identical to that used in the original application to infrared spectra except that the data were pre-smoothed three times instead of once, and the variable relaxation factor was modified to accommodate the lack of an upper bound. Referring to Eqs. (16) and (17) of Section VI.B of Chapter 4, we set $\kappa = 2\hat{o}_n^{(k)}\kappa_0$ for $\hat{o}_n^{(k)} \le \frac{1}{2}$ and $\kappa = \kappa_0 \exp[\frac{1}{2} - \hat{o}_n^{(k)}]$ for $\hat{o}^{(k)} > \frac{1}{2}$. This function is seen to apply the positivity constraint in a manner similar to that previously employed but eliminates the upper bound in favor of an exponential falloff. We also experimented with $\kappa = \kappa_0$ for $\hat{o}_n^{(k)} > \frac{1}{2}$, and found it to be equally effective. As in the infrared application, *only 10 iterations were needed.*

C. INSTRUMENTAL CONSIDERATIONS

We applied this method to the removal of broadening by $s_X(x)$, which describes the x-ray line broadening. In generating $s_X(x)$, we were guided by the Kα doublet intensity, separation, and width data of Senemaud (1968). By using component half-widths somewhat larger than the literature values, we partially compensated for the energy analyzer broadening $s_M(x)$ as well. In the present experiment, we took the aluminum Kα doublet components to be separated by 0.430 eV and to have a half-width of 0.35 eV each. The components were given a 2:1 intensity ratio and scaled to normalize the area under the combined peak. Figure 2 shows the resulting widened $s_X(x)$.

We decided to use the same procedure employed in the earlier infrared studies, retaining the original parameters, that is, sample rate, number of passes, type of smoothing, and so on. For the procedure to work properly in this form, it was critical that the data be scaled to lie between 0 and 1

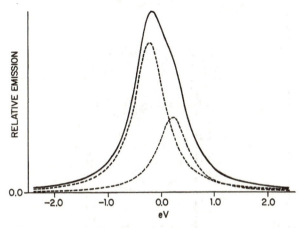

Figure 2 Aluminum x-ray Kα doublet, widened to compensate partially for energy analyzer broadening. Dotted curves, components; solid curve, their sum.

and that sample spacing relative to $s_X(x)$ be approximately the same as that used in the original infrared studies. Here we used 52 points per electron volt.

As noted in the introduction, resolution correction requires data of superior quality. In the present study we were fortunate to have access to a spectrometer that provided the needed high counting rate without resolution degradation. Two design features combined to provide this capability: a high-intensity aluminum Kα x-ray source and an energy analyzer having uniquely high étendue. The source has been described by Davies and Herglotz (1969) and the analyzer by Lee (1973).

IV. Applications

A. SILVER DOUBLET

The silver $3d_{3/2}-3d_{5/2}$ doublet at approximately 367- and 373-eV binding energy was chosen for an initial demonstration of the method. Figure 3 compares raw data obtained from our spectrometer with the result after base-line correction and deconvolution. It reveals a nearly threefold improvement in resolution. There is a conspicuous absence of the noise and spurious peaks that are typical of most methods. The slight peak asymme-

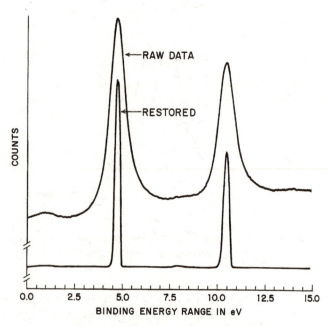

Figure 3   Silver $3d_{3/2}$–$3d_{5/2}$ doublet.

try is a consequence of the point-successive nature of the iteration scheme. The resulting linewidth represents the remaining broadening due to the natural line plus the finite bandpass of the electron energy analyzer, the small amount of electrical filtering employed, and the function used to presmooth the data.

B. ARTIFICIAL DOUBLETS

We observed the silver doublet again, this time with a 27-Hz square-wave modulation applied to the retarding potential of the energy analyzer. The spectrometer thus was made to alternate rapidly between two different scan ranges, the energy separation between ranges being dictated by the peak-to-peak amplitude of the square wave. Our control over the duty cycle of the square wave was used to adjust the relative intensities of the two spectra being superimposed. The concept applied here may be understood by noting that the observation time spent in each of two states of the square wave is proportional to the number of electrons counted in that state. The energy separation of the superimposed spectra was controlled by variation of the square-wave amplitude. By this means, artificial doublets were created for the purpose of testing deconvolution. The data recorded were otherwise exactly as they would be rendered in any normal spectrometer scan.

Figure 4 shows both data and resulting deconvolution when the silver doublet is observed with a 1-V peak-to-peak square-wave modulation having 2:1 duty-cycle ratio. Each of the doublet components is itself doubled. This component doubling is apparent in the raw data only as a modest broadening but becomes quite obvious after deconvolution. Figure 5 shows a similar run employing a 0.75-V square wave.

C. SILICON

Unaided by deconvolution, only instruments employing monochromatized x rays are able to resolve the $2p_{1/2}$ and $2p_{3/2}$ lines in the spectrum of silicon. Figure 6a displays our raw data before deconvolution; there is no

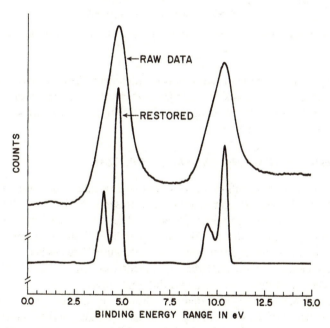

Figure 4 Artificial doublets: silver spectrum modulated by a 1-V peak-to-peak square wave.

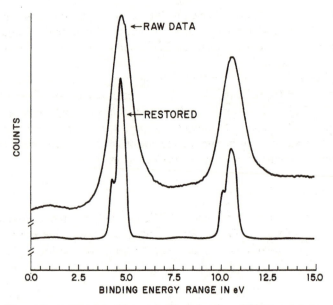

Figure 5 Artificial doublets as in Fig. 4, but for a 0.75-V modulation.

obvious evidence of the presence of two components. Yet, the deconvolved spectrum in Fig. 6b shows both components clearly. They are separated by less than 0.6 eV. The resolution obtained compares favorably with that obtained by other methods.

D. SAMPLE CHARGING IN POLYESTER FILM

The carbon $1s_{1/2}$ transition was observed in a sample of polyester film at near 1195 eV. The raw data indicate the presence of the three different carbon sites in the polyester molecule. The deconvolved data reveal additional spectral structure (Fig. 7). The additional structure appears in the form of a threefold repetition of the same curve. We easily reproduced the raw data, but each time the experiment was run, the deconvolution produced a threefold repetition of a different curve.

These results are readily explained (Brundle and Baker, 1981) as being a consequence of the nonuniform electrostatic charging of the sample. The sample loses electrons continuously during the observation process. Owing to the nonconducting nature of the sample, the resulting charge buildup is

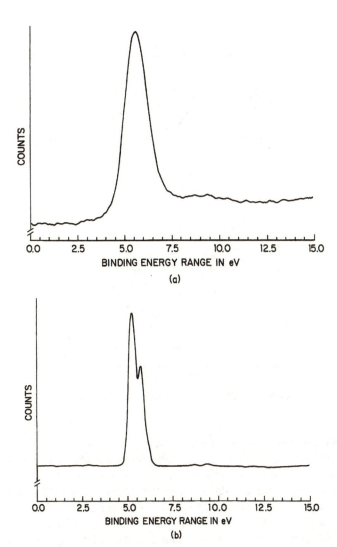

Figure 6 The $2p_{1/2}$ and $2p_{3/2}$ lines of silicon: (a) raw data; (b) result of deconvolution.

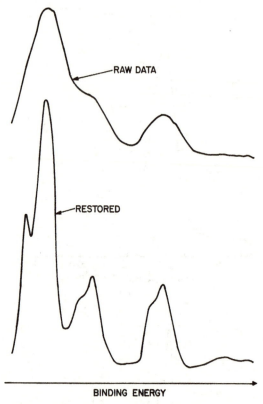

BINDING ENERGY

Figure 7 Carbon $1s_{1/2}$ transition in polyester film.

spatially distributed in a nonuniform manner. Electrons leaving the sample from different areas of the surface move through slightly different potential fields. The observed repeated curve can thus be explained as the potential energy distribution of the surface of the sample. Photoelectrons from each of the characteristic carbon sites are subjected to identical energy shifts. The raw data are therefore seen to result from the convolution of the desired spectrum $o(x)$ with the potential energy distribution.

The foregoing observations suggest that further deconvolution may be possible. Specifically, we could attempt to remove sample-charging effects by using an observed isolated line as a measure of the potential distribution on the surface. Like all computer processing discussed in this chapter, this correction would be applied to data acquired in the normal way and would involve no alteration of the experimental apparatus or sample conditions, as would the electron flood gun method of minimizing such charging effects (Huchital and McKeon, 1972; Wagner *et al.*, 1979).

E. RELATED RESEARCH

Other researchers too (Anderson, 1990; Sprenger and Anderson, 1991) have judged the present ESCA deconvolution method superior to alternatives, which were found to produce artifacts. In his work on synthetic and silicon-wafer spectra, Anderson found that "the results are comparable or better than performed with monochromatised high-resolution XPS ... ." Sprenger and Anderson least-squares fitted gaussians to ESCA spectra of sintered-ceramics, already partially resolved by this method of deconvolution. The a aluminum:silicon molar composition of 44:56 thus obtained was close to the theoretical value of 41:59, even though the data for the four spectral lines deconvolved were completely overlapped, appearing as a single peak with only the barest trace of a shoulder.

V. Concluding Remarks

In the present work, we have shown that a well-known simplified model of inelastic scattering can be described as a convolution and have applied an iterative base-line correction method based on it. The method proved effective in preprocessing spectra before deconvolution of x-ray line broadening. A deconvolution method constrained to yield physically realizable solutions was effective in removing this broadening. We tested the method by modulating the potential energy of conducting samples to create artificial doublets. Deconvolution of spectra from nonconducting samples explicitly revealed the shape of the potential-energy distribution caused by the electrostatic charging and suggested the possibility of removing charging-induced broadening by further deconvolution.

Acknowledgments

The authors thank their colleague J. D. Lee, for highly valued advice and contributions to stimulating discussions, and their former colleague H. K. Herglotz, for his guidance as technical leader of the ESCA program at the

DuPont Engineering Physics Laboratory and for the impetus he gave to the research reported here. The authors are especially indebted to their close collaborator in the present work, H. L. Grams, for computer programming and for valuable insights. H. L. Suchan is thanked for his substantial engineering contribution in the design of the apparatus. The authors are also grateful to their colleagues for valuable criticisms and suggestions concerning the manuscript for this chapter.

References

Anderson, O. (1990). *Vacuum* **41**, 1700–1702.

Beatham, N., and Orchard, A. F. (1976). *J. Electron Spectrosc. Relat. Phenom.* **9**, 129–148.

Brundle, C.R., and Baker, A. D., eds. (1977, 1978, 1979, 1981). "Electron Spectroscopy: Theory, Techniques and Applications," Vols. 1–4. Academic Press, New York.

Carley, A. F., and Joyner, R. W. (1979). *J. Electron Spectrosc. Relat. Phenom.* **16**, 1–23.

Davies, R. D., and Herglotz, H. K. (1969). *Adv. X-Ray Anal.* **12**, 496–505.

Davis, G. D., and Lagally, M. G. (1981). *J. Vac. Sci. Technol.* **18**, 727–731.

Delwiche, J., Hubin-Franskin, M. J., Caprace, G., and Natalis, P. (1980). *J. Electron Spectrosc. Relat. Phenom.* **21**, 205–218.

Herglotz, H. K. (1982). *In* "Surface Treatments for Improved Performance and Properties" (J. J. Burke and V. Weiss, eds.), pp. 19–49. Plenum, New York.

Herglotz, H. K., and Suchan, H. L. (1975). *Adv. Colloid Interface Sci.* **5**, 79–103.

Hoyt, A. (1932). *Phys. Rev.* **40**, 477–483.

Huchital, D. A., and McKeon, R. T. (1972). *Appl. Phys. Lett.* **20**, 158–159.

Jansson, P. A. (1968). Ph.D. Dissertation, Florida State Univ., Tallahassee.

Jansson, P. A., and Davies, R. D. (1974). *J. Opt. Soc. Am.* **64**, 1372.

Jansson, P. A., Hunt, R. H., and Plyler, E. K. (1968). *J. Opt. Soc. Am.* **58**, 1665–1666.

Jansson, P. A., Hunt, R. H., and Plyler, E. K. (1970). *J. Opt. Soc. Am.* **60**, 596–599.

Källne, E., and Åberg, T. (1975). *X-Ray Spectrom.* **4**, 26–27.

Lee, J. D. (1973). *Rev. Sci. Instrum.* **44**, 893–898.

McLachlan, A. D., Jenkin, J. G., Liesegang, J., and Leckey, R. C. G. (1974). *J. Electron Spectrosc. Relat. Phenom.* **3**, 207–216.

MacNeil, K. A. G., and Dixon, R. N. (1977). *J. Electron Spectrosc. Relat. Phenom.* **11**, 315–331.

Madden, H. H., and Houston, J. E. (1976). *J. Appl. Phys.* **47**, 3071–3082.

Rendina, J., and Larson, P. (1975). Paper Number 66, The Pittsburgh Conference on Analytical Chemistry and Applied Spectroscopy, Cleveland Convention Center, Cleveland, Ohio, March 3–7 (abstract only).

Rivière, J. C. (1993). *Materials Science and Technology* **9**, 365–377.

Senemaud, C. (1968). Thèsis, docteur ès Sciences Physiques, l'Université de Paris.

Shirley, D. A. (1972). *Phys. Rev.* **B5**, 4709–4714.

Siegbahn, K., Nordling, C., Fahlman, A., Nordberg, R., Hamrin, K., Hedman, J., Johansson, G., Bergmark, T., Karlsson, S., Lindgren, I., and Lindberg, B. (1967). "ESCA-Atomic, Molecular and Solid State Structure Studied by Means of Electron Spectroscopy." Almquist and Wiksells, Uppsala.

Siegbahn, K., Hammond, D., Fellner-Feldegg, H., and Barnett, E. F. (1972). *Science* **176**, 245–252.

Sprenger, D., and Anderson, O. (1991). *Fresenius J. Anal. Chem.* **341**, 116–120.

Tougaard, S. (1988). *SIA, Surf. Interface Anal.* **11**, 453–472.

Vasquez, R. P., Klein, J. D., Barton, J. J., and Grunthaner, F. J. (1981). *J. Electron Spectrosc. Relat. Phenom.* **23**, 63–81.

Wagner, C. D., Riggs, W. M., Davis, L. E., Moulder, J. F., and Muilenberg, G. E., eds. (1979). "Handbook of X-Ray Photoelectron Spectroscopy." Perkin-Elmer, Eden Prairie, Minnesota.

Wertheim, G. K. (1975). *J. Electron Spectrosc. Relat. Phenom.* **6**, 239–251.

Wertheim, G. K., and Hüfner, S. (1975). *Phys. Rev. Lett.* **35**, 53–56.

Chapter 9 | Deconvolution in Optical Microscopy

Jason R. Swedlow

Department of Cellular and Molecular Pharmacology
The University of California, San Francisco
San Francisco, California

John W. Sedat

Department of Biochemistry and Biophysics
The University of California, San Francisco
San Francisco, California

David A. Agard

Howard Hughes Medical Institute and
Department of Biochemistry and Biophysics
The University of California, San Francisco
San Francisco, California

List of Symbols

B	upper bound for the sum of the squares of the noise, ε, in POCS restoration
CCD	charge-coupled device
d_f	focal distance of objective lens
FFT	fast Fourier transform
$i(x, y)$	two-dimensional recorded image
$i(x, y, z)$	three-dimensional recorded image
$\hat{i}^{(k)}(x, y, z)$	estimate of recorded image; convolution of the point-spread function and the estimate of the object at iteration k
$i_j(x, y)$	two-dimensional image at focal plane j in a three-dimensional image stack
$i_p(x, y)$	projection image generated by summing a three-dimensional image along z
$I(u, v)$	Fourier transform of two-dimensional recorded image
$I(u, v, w)$	Fourier transform of three-dimensional recorded image
j	index for a series of focal planes
k	iteration number during restoration
$M^{(k)}(u, v, w)$	Fourier transform of $\mu^{(k)}(x, y, z)$; result of inverse filtering the lack of closure of the current estimate during restoration
NA	numerical aperture of objective lens
$o(x, y)$	two-dimensional object
$o(x, y, z)$	three-dimensional object
$\hat{o}(x, y, z)$	calculated estimate of $o(x, y, z)$
$\hat{o}^{(k)}(x, y, z)$	estimate of $o(x, y, z)$ at iteration k
$O(u, v)$	Fourier transform of two-dimensional object
$O(u, v, w)$	Fourier transform of three-dimensional object
OTF	optical transfer function
POCS	projection onto convex sets
PSF	point-spread function
r	Rayleigh criterion
$s(x, y)$	two-dimensional point-spread function
$s(x, y, z)$	three-dimensional point-spread function
u, v, w	three-dimensional coordinates in Fourier space
x, y, z	three-dimensional coordinates in real space
$\otimes$	convolution operation
α	acceptance angle of objective lens
β	adjustable parameter for restoration by regularization
Δz	distance from the in-focus plane
ε	noise estimate for POCS restoration

η	refractive index of the medium between the objective lens and the sample
κ	noise estimate for Weiner filter
λ	wavelength of light used for imaging
$\mu(x, y, z)$, $\mu^{(k)}(x, y, z)$	discrepancy between the current estimate of the object and the real object; discrepancy at iteration k during restoration
$\tau(u, v)$	two-dimensional optical transfer function; Fourier transform of two-dimensional PSF
$\tau(u, v, w)$	three-dimensional optical transfer function; Fourier transform of three-dimensional point-spread function

I.   Introduction

The goal of modern cell biology is an understanding of the molecular interactions that cause a cell to feed, grow, respire, and replicate. In almost all cases, the synthesis, localization, and disposal of many thousands of proteins, lipids, nucleic acids, sugars, and ions are tightly regulated. Beyond a chemical description of these compounds (e.g., the amino acid sequence of a protein) we require a knowledge of each compound's function and, perhaps most importantly, a description of where and when in the life of the cell each function is carried out. Epifluorescence microscopy is one of the most powerful techniques for localizing molecules of all types within cells. Biologically active molecules can be specifically labeled with a fluorescent "tag" by a number of different techniques. The localization of this tag is then recorded with a microscope equipped to detect the fluorophore.

Unfortunately, most cells are only micrometers in size and their constituents are significantly smaller, often close to or below the resolution limit of the optical microscope. This resolution limit and the image blurring discussed in the following section interfere with image interpretation and are therefore significant problems in modern cell biology. This situation is exacerbated by the fact that cells are three-dimensional structures and that the nominal resolution of the microscope is at least twofold worse along the optical axis than in the image plane. This problem is a classic inversion problem and therefore can be addressed by deconvolution methods. Here we describe the nature of microscopic image blurring and methods we and others have used to restore digital microscopic images. Because this chapter represents an application of deconvolution methods to a class of important biological problems, we also discuss critical technical issues such as data collection, preprocessing, and optical transfer function (OTF) generation so that the novice may effectively use these algorithms.

II. Image Blurring in Optical Microscopy

A. TWO-DIMENSIONAL IMAGING OF A TWO-DIMENSIONAL SPECIMEN

The following discussion is a summary of work by Castleman (1979) and Agard (1984). We first consider image formation from a two-dimensional sample located precisely at the plane of focus of the objective lens. If we apply a three-dimensional orthogonal coordinate system, where x and y are in the focal plane of the specimen and z coincides with the optical axis of the microscope, then a simple magnification transformation is all that is necessary to relate the sample coordinates to x' and y', the orthogonal axes in the recorded image. For this two-dimensional case, the fluorophore distribution in the focal plane of the specimen $o(x, y)$ is blurred by the microscope smearing or point-spread function (PSF), $s(x, y)$, to generate the observed image $i(x, y)$. The function $s(x, y)$ includes effects due to the finite circular aperture of the objective lens, the wavelengths of the excitation and emission light, and various optical aberrations. To a remarkably good approximation, the imaging system can be described as a simple convolution operation (in this case two-dimensional) (Bracewell, 1986):

$$i(x, y) = \int_{-\infty}^{\infty} \int_{-\infty}^{\infty} s(x - x', y - y')o(x', y') \, dx' \, dy', \qquad (1)$$

or more compactly,

$$i(x, y) = s(x, y) \otimes o(x, y), \qquad (2)$$

where the $\otimes$ symbol indicates a convolution operation. The Fourier transform of the PSF $s(x, y)$ yields the optical transfer function (OTF) $\tau(u,v)$. The OTF describes the alteration of specimen spatial frequencies by the microscope (effectively all of which occurs in the objective lens). It is often convenient to transform the convolution of Eq. (2) into Fourier space. The convolution theorem (see, for example, Chapter 1, this volume) allows us to determine $i(x, y)$ from its Fourier transform $I(u, v)$, by multiplication of the Fourier transforms of $o(x, y)$ and $s(x, y)$:

$$I(u, v) = \tau(u, v) \cdot O(u, v). \tag{3}$$

The combination of Eq. (3) along with the fast Fourier transform (FFT) generally results in substantial improvements in computational efficiency when calculating convolutions.

For an ideal lens having a circular pupil function, $\tau(u, v) = \tau(q)$, where $q = \sqrt{u^2 + v^2}$,

$$\tau(q) = \frac{1}{\pi}(2\beta - \sin 2\beta) \, \text{jinc}\left[\frac{8\pi w}{\lambda}\left(1 - \frac{|q|}{f_c}\right)\frac{q}{f_c}\right], \tag{4}$$

$$w = -d_f - \Delta z \cos \alpha + (d_f^2 + 2d_f\Delta z + \Delta z^2 \cos^2\alpha)^{1/2},$$

$$\beta = \cos^{-1}\left(\frac{q}{f_c}\right),$$

$$f_c = \frac{2\eta \sin \alpha}{\lambda},$$

$$\text{jinc}(x) = 2\frac{J_1(x)}{x}.$$

In this formulation, α is the acceptance angle of the circular aperature of the objective lens, d_f is the focal distance of the lens, λ is the wavelength of the light used in the image, η is the refractive index of the medium between the lens and the sample, and Δz is the deviation from the in-focus focal plane (see later) (Castleman, 1979). The product $\eta \sin \alpha$ defines the numerical aperature (NA), a quantity that determines the range of frequencies collected by the objective lens. This expression is sufficiently accurate for small amounts of defocus, $\Delta z \leq 2\lambda$.

Equation (4) fully expresses the resolving capability of a lens. However, a simpler measure of resolution is provided by the Rayleigh criterion, which specifies that two infinitely small point sources emitting incoherent radiation can be considered "resolved" from one another if separated by r, the distance from the center of one point to the first zero of the Bessel function in Eq. (4):

$$r = \frac{0.61\lambda}{\eta \sin \alpha}. \tag{5}$$

As λ increases or α decreases, the limit of resolution in the image decreases. For almost all applications in optical microscopy, λ ranges from 400 to 700 nm and the highest NA attainable is 1.4 ($\alpha = 67.5°$), limiting the Rayleigh resolution of $s(x, y)$ to approximately 200 nm. In fact, many cellular structures of biological interest are significantly smaller than 200 nm, so images of these structures are substantially blurred.

B. THREE-DIMENSIONAL IMAGING OF A TWO-DIMENSIONAL SPECIMEN

Three-dimensional imaging in the light microscope is accomplished through the practice of optical sectioning. Typically, a high-resolution computer-controlled stepper motor is attached to the focus control, and the focal plane of the objective lens is systematically moved in discrete steps through the sample. The result is a set of two-dimensional images recorded at different focal planes: $i_j(x, y)$, or more specifically: $i(x, y, z)$.

We can extend the previous two-dimensional analysis further to include the effects of out-of-focus lens aberrations. That is, when the focal plane of the lens deviates from the location of a two-dimensional sample by a distance Δz, the finite depth-of-field of the objective lens results in the sample being further blurred:

$$i(x, y, \Delta z) = s(x, y, \Delta z) \otimes o(x, y). \tag{6}$$

One-dimensional slices through the radially symmetric OTFs for an ideal lens are plotted for different values of defocus in Fig. 1 [these can be calculated using Eq. (4)]. Note the dramatic reduction in high-frequency information with increasing defocus.

C. THREE-DIMENSIONAL IMAGING OF A THREE-DIMENSIONAL SPECIMEN

For a thick specimen $o(x, y, z)$, $i(x, y, z)$ at a given focal plane z_j is generated by summing the images from N focal planes. Each of these images is the fluorophore distribution in that focal plane convolved with the appropriate PSF:

$$i(x, y, z_j) = \sum_{k=1}^{N} s(x, y, z_j - z_k) \cdot o(x, y, z_k) \, \Delta z. \tag{7}$$

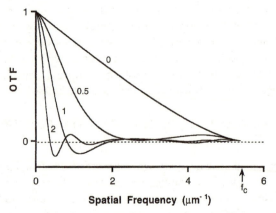

Figure 1 The two-dimensional OTF. Plots of the two-dimensional image plane OTF at various levels of defocus for a theoretical 60 × /NA1.4 objective. The OTF is radially symmetric, so a single one-dimensional slice is used. OTFs at 0, 0.5, 1, and 2 μm defocus are shown. Most of high-frequency content of OTF associated with the Rayleigh criterion is missing from optical planes at large defocus. The position of the critical frequency, f_c, is shown.

In the limit of $N \rightarrow \infty$ and $\Delta z \rightarrow 0$, this becomes the continuous definite integral over t, the thickness of the specimen:

$$i(x, y, z) = \int_0^t s(x, y, z - z') \cdot o(x, y, z') \, dz'. \tag{8}$$

It is apparent that Eqs. (7) and (8) are just convolutions, as in Eqs. (1) and (2). The three-dimensional image is the convolution of the specimen with the microscope's three-dimensional PSF. It follows that if $s(x, y, z)$ is known, the resolution of recorded three-dimensional images can then be restored by any of a number of inversion techniques. In our hands, constrained, iterative deconvolution has proved the most powerful and reliable.

Figure 2 shows two views of an empirical PSF for a high-NA lens. A fluorescent bead smaller than the resolution limit is blurred to a diameter of roughly 0.2 μm in the image plane (Fig. 2A). However, along the optical axis, significant amounts of fluorescence flux are distributed more than 2 μm away from the bead (Fig. 2A and B). In thick samples, an image of two point sources separated by the Rayleigh limit [Eq. (5)] may also contain light from other focal planes. In this case, the frequency response

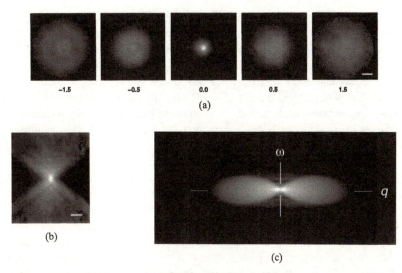

Figure 2 Three-dimensional PSF and OTF. The PSF, a three-dimensional image of a subresolution bead, and the OTF, the FFT of the PSF, show the relative difference in information content along the optical axis and in the image plane. (a) A series of optical sections at the indicated defocus from a PSF measured with a $100 \times$ /NA1.4 lens and a 0.12-μm bead. Scale bar = 1 μm. (b) A single image plane showing an orthogonal view through the center of the PSF shown in A. The optical axis of the microscope coincides with the vertical axis of the image; the in-focus plane coincides with the horizontal axis of the image. Scale bar = 1 μm. (c) OTF of the PSF shown in A and B. To generate the OTF, the FFT of the PSF has been radially and axially averaged (see text).

of the lens is sufficient to resolve the two sources, but out-of-focus blurring prevents an observer from discriminating them. Thus, in practice, the limit of resolution may be significantly poorer than that given by the Rayleigh criterion. This property means that *image blurring in microscopy is an inherently three-dimensional problem and that optimal resolution can only be attained if this blurring is corrected or eliminated.*

D. GENERATING THREE-DIMENSIONAL OPTICAL TRANSFER FUNCTIONS

In order to generate $\hat{o}(x, y, z)$, a good estimate of the specimen, we need an accurate description of the PSF and its associated OTF. As we have seen, the best restorations will be obtained when the full three-dimensional image distribution is included in the calculation. The PSF $s(x,y,z)$ can be calculated for an ideal objective lens given a knowledge of λ and the lens NA [Eq. (4)] (Goodman, 1968; Sheppard and Wilson, 1982; Agard, 1984). However, because of optical imperfections inherent to all real lenses and path length errors introduced during sample mounting (Scalettar *et al.*, 1996; Hiraoka *et al.*, 1990b; Gibson and Lanni, 1991), we have had greatest success with OTFs derived from empirically measured PSFs using fluorescent beads less than $\lambda/4$ (0.05–0.1 μm). The resulting three-dimensional OTF can be used for calculation of $\hat{o}(x, y, z)$ (see later). Because of signal-to-noise limitations in the observed PSF, it is generally advisable to average the OTF in the image plane (Hiraoka *et al.*, 1990b). This radial averaging derives from the circular symmetry of the PSF (Goodman, 1968) and assumes spatial invariance in the image. These criteria are usually met in high-NA lenses of good quality. It is also possible to average the OTF about the in-focus plane, along the optical axis. This assumes that all wave fronts in the optical path are spherical, thereby producing a PSF that is mirror symmetric about the in-focus plane. In some images, this condition is not met and the images are said to contain spherical aberration. If the spherical aberration is small, the OTF can be symmetrized along the optical axis, producing a real OTF. Otherwise, the axial asymmetries are best left, producing a complex OTF. We have found that we can correct spherical aberration using radially averaged OTFs from PSFs containing appropriate amounts of spherical aberration (Scalettar *et al.*, 1996).

III. Data for Image Deconvolution

Perhaps the most important factor in generating good image restorations is the quality of the starting data and the accuracy of the OTF. Noise in the image must be minimized while detector sensitivity is maximized. For accurate image processing, the detector must have good geometric accuracy. In addition, the detector elements must have as wide a dynamic range as possible and respond linearly to impinging photons over the whole dynamic range. We and others have found that cooled, slow-scan, scientific-grade charge-coupled device (CCD) cameras fit all of these requirements (Hiraoka *et al.*, 1987). These cameras generate images with a maximum dynamic range of 2^{16} different grey levels and are exquisitely linear over the whole range. Most importantly, the noise generated by the camera is usually significantly smaller than the Poisson noise inherent in photon detection. The cost of CCDs has dropped significantly in the last few years, making them available to the dedicated microscopist. A discussion of the CCD architecture and function is beyond the scope of this chapter, but the interested reader is referred to any of a number of reports (Hiraoka *et al.*, 1987; Aikens *et al.*, 1989).

The other aspect of data quality necessary for good image restoration is the pixel sampling rate. This is determined in each two-dimensional image by the effective detector pixel size in image space. For a three-dimensional image, the sampling rate of the z dimension is determined by the focus interval between optical sections. As with any signal measurement, the minimum sampling rate is given by the Shannon sampling theorem as twice the highest frequency component of the image (Bracewell, 1986). For an objective lens with the highest NA (1.4) the pixel size should therefore be ≤ 0.1 μm in the image plane and ≤ 0.35 μm along the optical axis—we typically use 0.1–0.2 μm. Image deconvolution is quite sensitive to the proper sampling of all frequency components. For example, in our own studies, three-dimensional images recorded with a high-NA lens with a focus interval of greater than 0.5 μm yield only small improvements in resolution and contrast after deconvolution, although this can probably be improved through interpolation of the observed data (see below).

In practice, there are a number of systematic data errors that must be minimized or removed prior to deconvolution. A significant source of error is caused by variation in the position of the Hg arc used for sample illumination—as the arc wanders the illumination pattern changes, resulting in illumination differences between optical sections. Installation of a fiber optic scrambler between the Hg arc lamp and the back focal plane of the objective produces spatially homogeneous light and effectively removes any spatial illumination gradients (Kam *et al.*, 1993). The measured number of photons at each pixel in the CCD is also affected by characteristic variations in the gain and offset for a given pixel, which can be accurately calibrated from a series of dark images and uniform bright images. Variations in shutter timing and in the output from the Hg arc lamp can be corrected by direct measurement of the total number of illumination photons impinging on the sample. Finally, photobleaching of the fluorophore used to stain the samples for fluorescent microscopy can cause a loss of flux emanating from the region of the sample being imaged. This can be corrected by determining the rate of decay of flux in the image. A detailed discussion of the correction CCD data errors has been published (Chen *et al.*, 1995).

IV.   Constrained, Iterative Deconvolution of Microscopic Images

A.   METHODS

Unfortunately, the OTF from an optical microscopic shows that the region of support approximates a solid of revolution formed from a figure-eight (Fig. 2C). It is quite clear that significant information about the optical axis is missing. In addition, the magnitude of the signal ranges widely within the observable region. Together, these problems are sufficiently severe as to render linear restoration methods inappropriate. For some time we have been examining numerous nonlinear, positivity-constrained iterative approaches for recovering $\hat{o}(x, y, z)$ from the measured $i(x, y, z)$. One particular set of methods, derived from the pioneering work of Van Cittert, Jansson (Jansson *et al.*, 1970), and Gold (1964), seems to have the optimal combination of convergence rates and stability in the presence of noise to be applicable to a wide variety of biological specimens. This approach has allowed us and others to make important and novel observations.

The general philosophy is to seek a positively constrained solution bounded in z, $\hat{o}(x, y, z)$, that when convolved with the PSF equals the observed data:

$$i(x, y, z) = s(x, y, z) \otimes \hat{o}(x, y, z). \qquad (9)$$

In reality, the presence of noise reduces the equality to a best approximation. The solution is developed in an iterative manner, and we have investigated several different update schemes for generating each successive estimate $\hat{o}^{(k+1)}(x, y, z)$ based on $i(x, y, z)$ and the previous estimate $\hat{o}^{(k)}(x, y, z)$. In all cases, we begin the first iteration with $o^{(0)}(x, y, z) = i(x, y, z)$. Convergence is assayed by the calculation of the error estimate, $R^{(k)}$, at every cycle, k.

$$R^{(k)} = \frac{\sum\limits_{x, y, z} |i(x, y, z) - \hat{i}^{(k)}(x, y, z)|}{\sum\limits_{x, y, z} i(x, y, z)}, \qquad (10)$$

where

$$\hat{\imath}^{(k)}(x, y, z) = s(x, y, z) \otimes \hat{o}^{(k)}(x, y, z).$$

It should be noted that $R^{(k)}$ provides only a measure of algorithm convergence and that other factors (e.g., image contrast, frequency content) should also be used to assess the quality of restoration.

1. Algebraic Update

Our initial efforts were based on the Van Cittert update scheme originally proposed for spectroscopic analysis [Frieden (1975); see Chapters 3 and 4, this volume] with clipping of the nonphysical negative part of the solution:

$$\hat{\imath}^{(k)}(x, y, z) = s(x, y, z) \otimes \hat{o}^{(k)}(x, y, z), \tag{11}$$

$$\hat{o}^{(k+1)}(x, y, z) = \hat{o}^{(k)}(x, y, z) + [i(x, y, z) - \hat{\imath}^{(k)}(x, y, z)];$$

$$\text{if} \quad \hat{o}^{(k+1)}(x, y, z) < 0, \quad \text{then set} \quad \hat{o}^{(k+1)}(x, y, z) = 0;$$

$$k \to k + 1.$$

We tried numerous variations of relaxation functions of the type employed by Jansson (Chapter 4) to enforce bounds. Though in many forms of spectroscopy such limits are knowable, in image restoration only the lower bound is generally well characterized. We have found that eliminating them entirely and enforcing the zero lower bound by truncation seems to produce the best results for our multidimensional optical images. This approach is extremely robust, especially in the presence of noise. However, its convergence rate is rather slow.

2. Geometric Update

An alternative update method that gives faster convergence than Eq. (11), is based on the ratio of $i(x, y, z)$ and $\hat{\imath}^{(k)}(x, y, z)$, as suggested by Gold (1964):

$$\hat{\imath}^{(k)}(x, y, z) = s(x, y, z) \otimes \hat{o}^{(k)}(x, y, z), \tag{12}$$

$$\hat{o}^{(k+1)}(x, y, z) = \hat{o}^{(k)}(x, y, z) \frac{i(x, y, z)}{\hat{\imath}^{(k)}(x, y, z)};$$

$$\text{if} \quad \hat{o}^{(k+1)}(x, y, z) < 0, \quad \text{then set} \quad \hat{o}^{(k+1)}(x, y, z) = 0;$$

$$k \to k + 1.$$

With our data, this method converges to a stable solution within 15–25 iterations. We have continued processing for up to 1000 iterations with only limited improvement in image contrast or resolution. The method is stable using data with signal-to-noise ratios from 1.5:1 to greater than 10:1. Examples of the quality of restoration obtained with this technique are shown in Figs. 3 and 4.

3. Regularized Algebraic Update

A simple analysis of the algebraic update scheme indicates the basis for a possible improvement through regularization of the update. After any cycle of Eq. (11) the discrepancy between the current estimate and the ideal, converged solution can be called $\mu(x, y, z)$:

$$i(x, y, z) = s(x, y, z) \otimes [\hat{o}^{(k)}(x, y, z) + \mu(x, y, z)]. \tag{13}$$

By rearranging and substituting in $\hat{i}^{(k)}(x, y, z)$ we get

$$s(x, y, z) \otimes \mu(x, y, z) = i(x, y, z) - i^{(k)}(x, y, z). \tag{14}$$

From this, it is clear that $\mu(x, y, z)$ can be recovered by inverse filtering the lack of closure. This is best accomplished in Fourier space using a Wiener filter approach to compensate for noise and the zeros in the OTF:

$$M^{(k)}(u, v, w) = \frac{I(u, v, w) - \hat{I}^{(k)}(u, v, w)}{\tau(u, v, w) + \kappa}, \tag{15}$$

where $I(u, v, w)$ and $\hat{I}(u, v, w)$ are the Fourier transforms of $i(x, y, z)$ and $\hat{i}(x, y, z)$, respectively, $M^{(k)}(u, v, w)$ is the Fourier transform of $\mu^{(k)}(x, y, z)$, and κ is an adjustable filter parameter. Once determined, $M^{(k)}(u, v, w)$ is Fourier transformed and

$$\hat{o}^{(k+1)}(x, y, z) = \hat{o}^{(k)}(x, y, z) + \mu^{(k)}(x, y, z),$$

and one then proceeds as in Eq. (11).

In practice, this approach is extremely fast, yet after many cycles becomes unstable in the presence of noise. The optimal compromise between speed and stability is to use the regularized update for the first three cycles and then to continue with either the simple algebraic scheme of Eq. (11) or the geometric update scheme of Eq. (12).

B. *NOISE SUPPRESSION AND IMAGE APODIZATION*

Repeated application of Eqs. (11), (12), and (15) restores lost information but also causes the addition of high-frequency noise to $\hat{o}^{(k)}(x, y, z)$. To

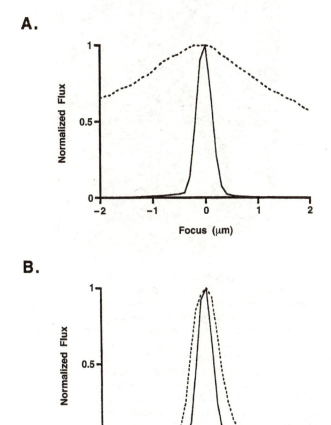

Figure 3 Restoration of PSF by iterative deconvolution. (A) The integrated flux along the optical axis of a PSF measured with a Nikon 100 × /NA1.4 lens. The integrated flux is the sum of the intensity in each optical section and shows the distribution of light in the three-dimensional volume. (B) The flux along a line parallel to the optical axis through the center of the bead. Iterative deconvolution produces a marked enhancement in resolution along the optical axis. All plots have been normalized to the peak flux in the image (at the center of the bead). The plots show flux before (—) and after (---) restoration with the OTF calculated from the PSF.

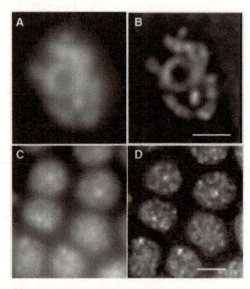

Figure 4 Use of iterative deconvolution on three-dimensional images of chromosomes. (A, B) Optical section from three-dimensional image of fixed *Drosophila* embryonic mitotic chromosomes. Image was recorded with a 100 × /NA1.4 lens. Chromosomes were visualized by treating chemically fixed embryos with a dye (4',6-diamidino-2-phenylindole) that becomes fluorescent on binding the DNA. (C, D) Optical section from time-lapse, three-dimensional image of living *Drosophila* embryonic mitotic chromosomes during metaphase. Chromosomes were visualized *in vivo* by injecting the embryo with fluorescently labeled histones H2A and H2B, proteins that are integral components of chromatin. Image was recorded with a 60 × /NA1.4 lens. (A, C) Before restoration by iterative deconvolution with appropriate OTF. (B, D) After restoration by iterative deconvolution with appropriate OTF. Scale bar = 5 μm.

minimize this, the guess is periodically smoothed by a Fourier space Gaussian-shaped low-pass filter that limits the resolution extrapolation of frequencies beyond the resolution limit of the objective lens. We typically apply this filter after every five iterations and to the final $\delta^{(k)}(x, y, z)$. Given the diverse structures observed in biological microscopy, we have not attempted any further constraints on the data.

A second issue arises from the exact position of the object in the three-dimensional image stack. Some images contain objects that abut or continue beyond one or more of the edges of the three-dimensional image volume that was collected. Because of the cyclical nature of the FFT, the resulting discontinuities in the image lead to inconsistencies with the mechanism of image formation described by Eq. (2). The net result is that large ripples are found to contaminate the deconvolved image. Owing to the fact that the PSF stretches over a long distance along the optical axis, this problem is quite severe in the z direction. As a consequence, careful attention to image apodization is very important.

We have used two approaches to deal with this problem (Agard *et al.*, 1989). For edge differences within the image plane, a small number of edge pixels are remapped to interpolate smoothly one edge of the image to its opposite. Alternatively, the edges can be rolled down to an average or background value with a Gaussian weighting function. For boundary differences between the top and bottom sections, the image stack is padded with extra sections and the flux distribution in the top section is smoothly rolled to that in the bottom section. In our hands, this technique works better than simple zero padding and is critical to successful image restoration: because of the limited spatial information along the z axis that is passed by the objective lens (see Fig. 2C), discontinuities between optical sections will tend to dominate a wide range of frequencies along z. This is especially important in the regularization scheme proposed in Eq. (15). We typically expand the z axis 1.5- to 2-fold by filling with a weighted pixel-by-pixel linear combination of the intensities in the top and bottom sections.

To minimize the computational and memory impact of this significant expansion along z, the expansion is done only during the filter operation in the inner loop. The three-dimensional Fourier filter (or inverse filter) is performed by first calculating the two-dimensional FFTs for each plane of the estimated or observed data only over the known region along z. The Fourier transform is then completed by expanding each z line by linear interpolation between the first and last values, followed by the one-dimensional FFT. The data are then filtered by multiplication with the OTF, reverse transformed in the z dimension, and only the relevant regions along z are stored in preparation for the inverse two-dimensional FFTs. The net result is no increase in memory requirements and only a small increase in computation time. Generally, because radix-2 FFTs are more efficient, expanding z to the next power of two provides a computationally efficient way of processing three-dimensional image data sets with an arbitrary number of sections. In summary, these apodization methods stabilize the algorithm and allow it to be used for a wide variety of image data.

C. MULTIWAVELENGTH IMAGES

One of the most powerful applications of fluorescence microscopy is the localization of two or more cellular components in the same specimen. This is accomplished by labeling different probes specific for different components with different fluorophores. A microscope equipped with appropriate bandpass interference filters can selectively image each of these fluorophores while excluding the others. Superimposition of the resulting images shows the relative localization of the individual components.

This type of experiment presents a problem for iterative deconvolution or any other restoration technique. As shown in Eqs. (4) and (5), the OTF and resolution limit of a given lens are strong functions of λ. This implies that a different OTF should be used for each λ used in an experiment. The collection of PSFs for all wavelengths and lenses used in a laboratory can be time-consuming and difficult: we have not been able to obtain sufficiently bright and photostable beads to generate good PSFs at all of the wavelengths used in our experiments. However, as shown in Eq. (4), the OTF is a function of the ratio of the image sampling and critical frequencies of the objective lens. We have found that to a first approximation, an OTF measured at a given λ can be used to generate OTFs at any new λ by simply scaling the sampling frequency of the known OTF by a ratio of the critical frequencies at the known and unknown wavelengths.

By contrast, our efforts to interpolate OTFs *between* different lenses of similar magnification and numerical aperture have yielded inferior results. Therefore, we always generate at least one PSF for each lens. To date, we have not found it necessary to generate a new PSF for each sample, e.g., by adding fluorescent beads to our samples (Fay *et al.*, 1989; Loew *et al.*, 1993) or calculating the PSF directly from the sample itself (Liu and Holmes, 1993). In fact, our recent results suggest that the most significant sample-dependent effect found in microscopic images is spherical aberration, as discussed above. We have found that we can substantially correct the effects of spherical aberration in images from real samples using similarly aberrated PSFs generated from single fluorescent beads (Scalettar *et al.*, 1996).

D. TIME-LAPSE IMAGES

Because the light microscope can be used to record images of living specimens, there is increasing interest in studying the localization of one or more cellular components *in vivo* by recording time-lapse three-dimensional movies. We have successfully conducted such experiments and restored images by processing each individual three-dimensional image in a time-lapse sequence as described in Eqs. (11)–(15) (Fig. 4). However, in some cases, physical storage space may be limited, or in a time-lapse experiment, the events being studied may occur too quickly to allow collection of a sequence of three-dimensional data sets. In this case, one can sacrifice the performance of three-dimensional images and collect single optical sections. A compromise is achieved by generating a projection image $i_p(x, y)$,

$$i_p(x, y) = \left[\sum_z i(x, y, z) \right] / N. \tag{16}$$

The summation can be done directly on the CCD by waiting until the end of the optical section stack to read out data from the chip. Alternatively, it can be done computationally using the individually observed optical sections. In this case, it is possible to generate a set of quasi-stereo paired images for each set of optical sections. In either case, it is then possible to process these images using a two-dimensional deconvolution. This approach is based on the Fourier projection theorem (Bracewell, 1986), which states that the Fourier transform of a projected image of a three-dimensional object is just a two-dimensional slice at the same projection angle through the three-dimensional transform. For untilted data,

$$I(u, v, 0) = O(u, v, 0) \cdot \tau(u, v, 0). \tag{17}$$

This implies that the two-dimensional projection image i_p can be deconvolved using the $w = 0$ central section through the three-dimensional OTF, $\tau(u, v, w)$. This technique, termed synthetic projection microscopy (Agard et al., 1989), is extremely rapid and has generated quite satisfactory results (Hiraoka et al., 1989). However, when possible, the full three-dimensional image and OTF should be used because these provide the most information about the specimen (Swedlow et al., 1993).

V. Other Methods

A. REGULARIZATION

An alternative method of generating $\hat{o}(x, y, z)$ is to minimize the difference

$$\sum_x \sum_y \sum_z |i(x, y, z) - \hat{o}(x, y, z) \otimes s(x, y, z)|^2. \tag{18}$$

Because of noise present in the data, least-squares and other direct methods of minimization fail to give a satisfactory restoration. Carrington, Fay, and co-workers have used a regularized form of Eq. (18) by adding a term that tends to smooth the solution and enforces nonnegativity (Carrington, 1990):

$$\sum_x \sum_y \sum_z |i(x,y,z) - \hat{o}(x,y,z) \otimes s(x,y,z)|^2$$

$$+ \beta \int_x \int_y \int_z |\hat{o}(x,y,z)|^2 \, dx \, dy \, dz, \qquad \hat{o}(x,y,z) \geq 0,$$

where β is an empirically chosen weighting. Minimization is achieved using the conjugate gradient method. This approach has been used for a number of different types of images and has proved quite useful. One published comparison found that regularization provided better resolution and less noise than iterative deconvolution (Fay *et al.*, 1989). However, the deconvolution scheme used in this test did not use the apodization or noise suppression schemes previously discussed. A careful comparison between regularization and iterative deconvolution methods has not yet been done, but would certainly prove informative.

B. POCS METHODS

We have developed a method of using projection onto convex sets (POCS) for restoration of three-dimensional data (Koshy *et al.*, 1990). In this scheme, we use an implementation of the Eq. (8) that includes an explicit noise estimate, ε:

$$i(x,y,z) = o(x,y,z) \otimes s(x,y,z) + \varepsilon.$$

If B is an estimate of the upper bound for the sum of the squares of the noise, then we constrain $\hat{o}(x,y,z)$ such that the norm

$$\|\hat{o}(x,y,z) \otimes s(x,y,z) - i(x,y,z)\|^2 \leq B. \tag{19}$$

The set of all $\hat{o}(x,y,z)$ that satisfy this constraint forms a closed convex set. A second convex set is given by the nonnegativity constraint. Alternate projection of these sets leads to an accurate estimate for $\hat{o}(x,y,z)$. In our hands, this method has produced the highest resolution restorations, providing that the correct OTF is used. The method appears to be somewhat sensitive to mismatches between the measured OTF and the sample. We are currently working to understand this problem and fully implement this approach.

C. CONFOCAL MICROSCOPY

All of the methods described to this point depend on the collection of both in-focus and out-of-focus information by a digital camera followed by image restoration by various means. However, it is also possible to minimize the light that reaches the detector so that only those photons from the region near the in-focus plane will be detected. This principle is the basis of the design of the confocal microscope, where the sample is scanned with a diffraction-limited spot from a focused laser beam, and a pinhole placed in the optical path is used to reject out-of-focus light. A complete discussion of the many implementations of this type of microscopy is beyond the scope of this chapter—the interested reader is referred to the textbook by Pawley (1995).

In principle, the highest resolution image would be obtained by three-dimensional iterative deconvolution of confocal microscopic images. In fact, this does appear to be the case (Shaw and Rawlins, 1991). However, because many of the photons in the confocal microscope never reach the detector, the light efficiency and sensitivity of these microscopes appear to be dramatically less than a conventional, or "wide-field," microscope. Many of our experiments involve the introduction of fluorescent probes into living cells, making the cells extremely photosensitive. Light efficiency is therefore a major concern in such experiments. For this reason, we have chosen to continue to use the wide-field microscope and image restoration by iterative deconvolution.

VI. Uses of Constrained, Iterative Deconvolution in Cell Biology

We and others have used constrained, iterative deconvolution for image restoration in a number of different types of experiments and specimens. In our own laboratory, we are studying the structure and dynamics of nuclei and chromosomes in the embryo of *Drosophila melanogaster*. These structures are small relative to the resolution limit of the light microscope —the spherical nuclei are approximately 5 μm in diameter and the rod-shaped chromosomes are 1–12 μm long, but 0.3 μm wide. Interpretation of images of these structures requires the high resolution afforded by three-dimensional constrained, iterative deconvolution. The method was used to map the paths and relative positions of all eight chromosomes in single nuclei (Hiraoka *et al.*, 1990a). We have also recorded time-lapse images of chromosome assembly in living *D. melanogaster* embryos and shown that this process initiates at specific sites on the nuclear periphery (Hiraoka *et al.*, 1989). In addition, we have followed the distribution of DNA topoisomerase II, an important chromosomal enzyme, as chromosomes pass through the events of cell division in a living embryo (Swedlow *et al.*, 1993). This study identified specific populations of the enzyme that transit on and off chromosomes at specific times.

A number of other groups have used deconvolution for images in other types of experiments. The detailed localization of various proteins and RNAs involved in RNA metabolism in the yeast *Saccharomyces cerevisiae* and the pea *Pisum sativum* has been described (Highett *et al.*, 1993; Wilson *et al.*, 1994; Beven *et al.*, 1995). High-resolution microscopic images were used for the determination of the exact positions of many of the nuclear and nucleolar components in yeast and plants and suggested a model for the function of these molecules. A second example is a series of papers describing the localization of the protein encoded by the pro-tooncogene c-*src* in a number of different cells. This protein is present in all cells; though its normal function is unknown, cells bearing mutated forms of this protein become transformed, i.e., cancerous. The resolution afforded by conventional fluorescence microscopy was insufficient to allow determination of the localization of the normal protein. Deconvolution of three-dimensional images showed that the protein is concentrated in the microtubule organizing center, apparently associated with membrane vesicles (Kaplan *et al.*, 1992). Mutations that cause the protein to transform cells change the localization of the protein to regions of the cell involved in adhesion of the cell to surfaces (Kaplan *et al.*, 1994). A third example of the application of deconvolution in cell biology is the localization of the product of the *germ cell-less* gene, a protein involved in the formation of the germ cells (i.e., those cells that will develop into sperm or eggs) in the embryo of *D. melanogaster* (Jongens *et al.*, 1994). Deconvolved three-dimensional images showed that this protein was localized to the nuclear periphery of the germ cells and colocalized with components of the nuclear envelope. Along with other data, this finding suggested that this protein might function by regulating the transport of other molecules involved in the development of the germ cells into the nucleus.

VII. Future Directions

The application of iterative deconvolution to digital microscopy has proved quite powerful and has led to a number of new insights not possible with conventional microscopy. However, we anticipate a number of modifications that would further improve the quality of restoration, accelerate the convergence, and make it applicable to a wider range of images. We believe the following enhancements will be most useful.

A. LIGHT SCATTERING

Many of the most interesting biological questions require imaging through relatively thick tissues (> 10 μm) with refractive indices significantly higher than the surrounding medium (usually water or glycerol). In addition, tissues are heterogeneous materials containing many irregular small volumes of varying refractive index. When imaging through such a material, light scattering becomes a serious problem. To date, we have ignored this issue and assumed that the PSF is, to a first approximation, unaffected by light scattering. This is reasonable as a first approximation, but is clearly not exact. Many corrections to atmospheric scattering effects have been developed in astronomy (Thompson and Castle, 1992) and application of these techniques to biological microscopy may prove quite useful. The success of these methods in the heterogeneous environment of biological tissue remains to be determined.

B. COHERENT IMAGES

We have focused our attention on restoration of fluorescence images generated using incoherent light. A number of other imaging regimes wherein coherent or semicoherent radiation is used to generate contrast are widely used in biological microscopy. In general, these methods use phase-shift differences generated as light passes through cellular components with differing refractive indices [for descriptions of these methods, see Inoué (1986)]. Our current implementation of the deconvolution methodology is not appropriate for these imaging schemes, but it is possible to imagine straightforward modifications that would allow restoration of such images. For example, the presence of a phase component in the coherent image must be included. Also, in some of these schemes, the assumption of radial symmetry of the PSF is not valid owing to the presence of differentially polarized radiation generated by a dichroic material. However, the PSF should be mirror symmetric about a plane perpendicular to the polarization plane. The transfer function of a coherent imaging system has been described (Sheppard and Mao, 1989), opening the way to the application of restoration methods to such images.

C. MULTIGRID APPROACHES

When collecting and processing images from a microscope, there is always a compromise between image size and image sampling—finer sampling generates more pixels. An increase in image size almost always translates into increased processing time (other factors such as storage space and file transfer time are also affected). This raises the question: What sampling rate is sufficient to collect the complete frequency range available and/or allow accurate restoration of an image?

Typically, a sampling rate twice that of the resolution of the signal recording system is sufficient to record all detectable frequencies from a sample. This is the Shannon limit (Bracewell, 1986). However, for accurate restoration of microscopic image data, we have found that the sampling rate should be three- to fivefold higher than the highest measurable frequency. This ensures that the imaging of small objects is continuous over the sampling grid and that the restoration is not limited by the digital sampling. To test this, we have compared the quality of restoration of three-dimensional images of *Drosophila* chromosomes collected with a $60 \times$ /NA1.4 lens (resolution along the optical sectioning axis is approximately 0.7 μm), where the optical sectioning interval was 0.1, 0.2, or 0.1 μm derived from the 0.2 μm image by linear interpolation. Surprisingly, restoration of both the 0.1 μm measured and the 0.1 μm calculated images gave small but significant improvements in image definition and contrast when compared to the 0.2 μm image. Very little difference was noted between the restored 0.1 μm measured and calculated images. However, these small improvements come with a penalty of a twofold increase in memory requirements and computational time.

An expansion of this type of manipulation may provide a method to accelerate the restoration of microscopic images, and, if necessary, the accuracy of restoration. In this case, a three-dimensional image is recorded with a sampling rate sufficient to record all spatial frequencies collected by the objective lens. At the start of the iterations, the sampling rate in all three dimensions is decreased by a factor of 2, 3, or 4 by averaging or subsampling (decreasing image size by factors of 8, 27, or 64); this gives a dramatic reduction in processing time. After 10 iterations, $ô(x, y, z)$ is reinterpolated back to the original sampling rate and the processing is continued for the remaining 10 iterations using the original, fully sampled $i(x, y, z)$. In our preliminary experiments, this multigrid scheme provides a significant decrease in processing time while preserving most of the image contrast and resolution available by processing the full image. This is especially helpful for immediate processing of three-dimensional images as they are collected from the microscope. It is important to note that the amounts of interpolation used in the image plane and along the optical axis can be adjusted independently and vary with the required resolution and processing efficiency.

VIII. Conclusions

The combination of iterative deconvolution and digital microscopy has been a major advance in modern cell biology and has provided a number of new and significant insights. The method is reasonably fast and, very importantly for use by nonexperts, it is quite robust. With the recent availability of fast computer workstations, the technique is now being used by a growing number of groups to obtain high-resolution images in a variety of experimental settings.

Acknowledgments

We thank Peter J. Shaw for his contributions to the early parts of this work. Work in the authors' laboratories has been supported by the Howard Hughes Medical Institute (J. W. S. and D. A. A.) and by grants from the National Institutes of Health to J. W. Sedat (GM-25101) and D. A. Agard (GM-31627). J. R. S. is a Damon Runyon–Walter Winchell Postdoctoral Fellow.

References

Agard, D. A. (1984). *Annu. Rev. Biophys. Bioeng.* **13**, 191–219.
Agard, D. A., Hiaroka, Y., Shaw, P., and Sedat, J. W. (1989). *Methods Cell Biol.* **30**, 353–377.

308   **Jason R. Swedlow** *et al.*

Aikens, R. S., Agard, D. A., and Sedat, J. W. (1989). *Methods Cell Biol.* **29**, 291–313.

Beven, A. F., Simpson, G. G., Brown, J. W., and Shaw, P. J. (1995). *J. Cell Sci.* **108**, 509–518.

Bracewell, R. N. (1986). "The Fourier Transform and Its Applications." McGraw-Hill, New York.

Carrington, W. (1990). *Proc. Soc. Photo-Opt. Instrum. Eng.* **1205**, 72–83.

Castleman, K. R. (1979). "Digital Image Processing." Prentice-Hall, Englewood Cliffs, New Jersey.

Chen, H., Swedlow, J. R., Grote, M., Sedat, J. W., and Agard, D. A. (1995). *In* "Handbook of Biological Confocal Microscopy" (J. Pawley, ed.), 2nd Ed., pp. 197–210, Plenum, New York.

Fay, F. S., Carrington, W., and Fogarty, K. E. (1989). *J. Microsc.* **153**, 133–149.

Frieden, B. R. (1975). *Top. Appl. Phys.* **3**, 177–248.

Gibson, S. F., and Lanni, F. (1991). *J. Opt. Soc. Am.* **8**, 1601–1613.

Gold, R. (1964). Report No. ANL-6984. Argonne National Laboratory, Chicago.

Goodman, J. W. (1968). "Introduction to Fourier Optics." McGraw-Hill, San Francisco.

Highett, M. I., Beven, A. F., and Shaw, P. J. (1993). *J. Cell Sci.* **105**, 1151–1158.

Hiraoka, Y., Sedat, J., and Agard, D. (1987). *Science* **238**, 36–41.

Hiraoka, Y., Minden, J. S., Swedlow, J. R., Sedat, J. W., and Agard, D. A. (1989). *Nature* (*London*) **342**, 293–296.

Hiraoka, Y., Agard, D. A., and Sedat, J. W. (1990a). *J. Cell Biol.* **111**, 2815–2828.

Hiraoka, Y., Sedat, J. W., and Agard, D. A. (1990b). *Biophys. J.* **57**, 325–333.

Inoué, S. (1986). "Video Microscopy." Plenum, London.

Jansson, P. A., Hunt, R. H., and Plyler, E. K. (1970). *J. Opt. Soc. Am.* **60**, 596–599.

Jongens, T. A., Ackerman, L. D., Swedlow, J. R., Jan, L. Y., and Jan. Y. H. (1994). *Genes Dev.* **8**, 2123–2136.

Kam, Z., Jones, M. O., Chen, H., Agard, D. A., and Sedat, J. W. (1993). *Bioimaging* **1**, 71–81.

Kaplan, K. B., Swedlow, J. R., Varmus, H. E., and Morgan, D. O. (1992). *J. Cell Biol.* **118**, 321–333.

Kaplan, K. B., Bibbins, K. B., Swedlow, J. R., Amaud, M., Morgan, D. O., and Varmus, H. E. (1994). *EMBO J.* **13**, 4745–4756.

Koshy, M., Agard, D. A., and Sedat, J. W. (1990). *Proc. Soc. Photo-Opt. Instrum. Eng.* **1205**, 64–71.

Liu, Y.-H., and Holmes, T. J. (1993). *MSA Bull.* **23**, 189–198.

Loew, L. M., Tuft, R. A., Carrington, W., and Fay, F. S. (1993). *Biophys. J.* **65**, 2396–2407.

Pawley, J. (1995). "Handbook of Confocal Microscopy," 2nd Ed. Plenum, New York.

Scalettar, B. A., Swedlow, J. R., Sedat, J. W., and Agard, D. A. (1996). *J. Microsc.* **182**, 50–60.

Shaw, P. J., and Rawlins, D. J. (1991). *J. Microsc.* **163**, 151–165.

Sheppard, C. J. R., and Wilson, T. (1982). *Proc. R. Soc. London Ser. A* **379**, 145–158.

Sheppard, C. J. R., and Mao, X. Q. (1989). *J. Opt. Soc. Am.* **6**, 1260–1269.

Swedlow, J. R., Sedat, J. W., and Agard, D. A. (1993). *Cell* **73**, 97–108.

Thompson, L. A., and Castle, R. M. (1992). *Opt. Lett.* **17**, 1485–1487.

Wilson, S. M., Datar, K. V., Paddy, M. R., Swedlow, J. R., and Swanson, M. S. (1994). *J. Cell Biol.* **127**, 1173–1184.

Chapter 10 | Deconvolution of Hubble Space Telescope Images and Spectra

Robert J. Hanisch
Richard L. White
Ronald L. Gilliland

Space Telescope Science Institute
Baltimore, Maryland

List of Symbols

A	constant (in relaxation function)
$\mathring{A}$	angstrom (10^{-10} meter)
$\mathscr{A}$	amplitude of wave front
b	constant

B	constant (in relaxation function)
$C(x, x')$	complete data (counts) in pixel x from location x'
CCD	charge-coupled device
EM	expectation maximization
f, g	functions
F, F^{-1}	forward and inverse Fourier transform operators
FFT	fast Fourier transform
FOC	Faint Object Camera
FOS	Faint Object Spectrograph
GHRS	Goddard High Resolution Spectrograph
HSP	High Speed Photometer
HST	Hubble Space Telescope
i	the data (blurred image)
$\hat{i}$	estimate of the data
$(\tilde{i})$	estimate of the data (oversampled)
j	square root of -1
k	iteration counter
$\mathscr{L}$	likelihood function
LSA	large science aperature
LSF	line spread function
m, n	indices
MEM	maximum entropy method
N	number of elements in a vector or array
NASA	National Aeronautics and Space Administration
o	object intensity
$\hat{o}$	estimate of object intensity
O	Fourier transform of object o
OTA	optical telescope assembly
p_1, p_2	independent variables
$\mathscr{P}$	probability function
PSF	point-spread function
r	relaxation function
RL	Richardson–Lucy (algorithm)
RMS	root mean square
s	point-spread function (blurring function)
SSA	small science aperture
t	threshold (multiple of standard deviation)
u, v	independent variables (frequency domain)
u, U	unit vector and its Fourier transform
w, W	weight vector and its Fourier transform
$\mathscr{W}$	complex pupil function
WFPC	Wide-Field/Planetary Camera
x, x', y	independent variables (spatial domain)
α	correction factor
β	conjugate gradient multiplier
γ	constant

ϵ	additive noise
λ	wavelength
λ, μ, ν	indices
ϕ	phase of wave front
Φ	objective function
ρ	ratio of two images
σ	standard deviation
τ	Fourier transform of point-spread function
$\otimes$	convolution operator
$*$	complex conjugate

I. Introduction

In April 1990 the Hubble Space Telescope (HST), a joint program of the National Aeronautics and Space Administration (NASA) and the European Space Agency, was boosted into orbit by NASA's Space Shuttle. The HST was the first of NASA's "Great Observatories," a program of four major space-based observatories spanning the wavelength domain from infrared through γ-rays. Owing to its position high above the blurring and obscuring effects of the earth's atmosphere, HST promised to deliver unprecedented clarity and sensitivity in imaging and spectroscopy in the visible and ultraviolet parts of the spectrum.

Shortly after its launch, however, the HST was diagnosed as suffering from spherical aberration in its 2.4-m primary mirror. The aberration is now understood to have resulted from the use of a faulty measuring device in testing the mirror figure during its grinding and polishing by the Perkin–Elmer Corporation. The mirror was polished to an excellent surface, but unfortunately it was the wrong shape. Burrows *et al.* (1991) were the first to measure the aberration, finding an error of -0.43 waves root mean square (RMS) at $\lambda 633$ nm, corresponding to a 4.6-μm optical path length error for marginal rays at paraxial focus. The error in the surface increases in proportion to the fourth power of the distance from the center of the mirror, resulting in a marginal focus (optimal focus for light incident at the edge of the mirror) that is 43 mm from paraxial focus. The aberration caused the point-spread function (PSF) to cover a region over 5 seconds of arc in diameter, with only about 15% of the light in a sharp core with radius 0.1 second of arc. Had the telescope performed at the design specification, 70–80% of the light would have been concentrated within the core. The core of the PSF was formed by the inner region of the mirror, and the halo (composed of diffraction rings and tendrils) originated from aberrated rays from the outer region of the mirror and from obscurations in the optical telescope assembly (OTA). An example PSF for the aberrated HST is shown in Fig. 1.

The initial reaction by the astronomical community, NASA, and, indeed, by Congress was that the Hubble Space Telescope was a technical

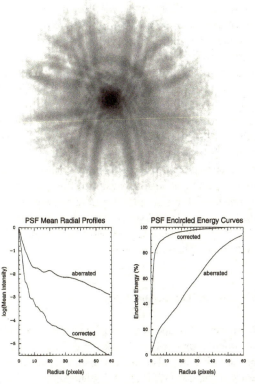

Figure 1 An example of the point-spread function (PSF) for the aberrated Hubble Space Telescope (shown for a bandpass centered at 547 nm), where the grey scale covers the range from 0 to 25% of the peak intensity and the intensity is scaled logarithmically. Corresponding radial profile and encircled energy plots are shown, and are compared to the performance after corrective optics were installed on the December 1993 Servicing Mission.

and financial disaster. However, after some months of discussions, study, and analysis, we found that certain scientific programs could still be carried out, and a number of imaging and spectroscopic experiments could be facilitated through the use of deconvolution (or restoration) of HST images and spectra. It is important to realize that the aberrated HST images retained most of the spatial resolution of an unaberrated system. This was evident in the sharp core of the PSF. However, with only a small fraction of the light concentrated in the core, dynamic range was severely compromised and sensitivity was lost. The bright, extended halo made it difficult to distinguish objects in crowded fields, and impossible to carry out a number of key scientific programs that required high dynamic range (such as the search for galaxies underlying quasars and the detection of planets around other stars by direct imaging).

In the summer of 1990 a number of astronomers and image processing scientists began adapting known deconvolution methods to HST data. Like all inverse problems, HST image restoration was nontrivial. The primary difficulties can be grouped into several categories: problems with the data, problems with the PSF, and problems with algorithms. In this chapter we discuss each of these aspects of HST image restoration, paying particular attention to the deconvolution algorithms that have proved most beneficial. A number of review papers are available that provide good overviews of the HST image restoration problem and variety of approaches, for example, Adorf *et al.* (1993), Gilliland (1991), Hanisch (1991, 1994), Hanisch and White (1993, 1994), Hanisch and Mo (1993), Lucy (1992), Murtagh (1993), and White and Hanisch (1991).

II.   Hubble Space Telescope Data

The Hubble Space Telescope was launched with a complement of five scientific instruments, plus a Fine Guidance System that is used not only for spacecraft pointing but also for astrometric observations. The Wide-Field/Planetary Camera (WFPC) had eight 800 × 800 pixel charge-coupled device (CCD) detectors arranged in two 2 × 2 arrays, and the Faint Object Camera (FOC) uses image-intensified photocathode detectors. The Goddard High Resolution Spectrograph (GHRS), the Faint Object Spectrograph (FOS), and the High Speed Photometer (HSP) completed the complement of instruments. The performance of the HSP was severely compromised by the spherical aberration problem due to the combination of reduced throughput in its small apertures and complications in spacecraft pointing and stability. Both cameras and both spectrographs were used extensively, however, and image and spectral deconvolution efforts were undertaken.

The eight WFPC CCDs constituted two partially independent cameras —the Wide-Field Camera, with a total field of view of 2.6 minutes of arc, and the Planetary Camera, with a total field of view of 1.1 minutes of arc. The latter was designed, as its name implies, to be optimized for imaging of the planets and provided reasonably well-sampled data. The Wide-Field Camera's larger field of view, but with the same number of pixels, means that throughout its range of wavelength sensitivity the data are undersampled. Wide-Field Camera pixels are 0.1×0.1 second of arc, and Planetary Camera pixels are 0.043×0.043 second of arc. The FOC, on the other hand, is well-sampled, and in its high-resolution mode of operation (at a focal ratio $f/96$) has 0.022×0.022 second of arc pixels. At $f/96$, however, the FOC field of view is only 11×11 seconds of arc.

The dominant sources of noise for the WFPC are photon noise (or shot noise, with a Poisson distribution), and additive (Gaussian) read-out noise originating in the analog-to-digital (A/D) converters on the CCDs. Cosmic ray contamination, data transmission errors, and other flaws in the detectors contribute to the effective noise, but in most cases it is possible to identify and mask errant pixels. Weak cosmic ray hits can be difficult to distinguish from faint stars, owing to the combination of undersampling and the sharp core of the PSF. Unambiguous identification of cosmic ray hits requires the use of multiple exposures, where deviant pixels can be identified by comparison with the mean value and expected standard deviation for each pixel in the set of observations. Bright cosmic rays can be identified in a single frame via a similar comparison between the suspect pixel value and pixels in the immediate vicinity, although this method is impractical when cosmic rays hit close to the core of the image of a bright star. The FOC is subject only to photon noise but suffers from low dynamic range, low count rates, and strong nonlinearities. Data from both cameras were geometrically distorted, with FOC being the more severe.

The majority of the image restoration efforts were aimed at WFPC images, because this turned out to be the more heavily subscribed instrument and was much more linear in photometric response than the FOC. However, WFPC image restoration was further complicated by a strong space-variance in the PSF (described further below).

The HST spectrographs are similar in design. They both have quiet, photon-counting detectors. For most purposes they can both be treated as essentially ideal detectors that are entirely limited by the statistics of photon counting. Several different focal plane apertures are available for the spectrographs. The astronomer has a choice between using a small aperture that passes little of the light but produces spectra with resolution undegraded by the spherical aberration, or using a large aperture that passes most of the light but leads to degraded spectral resolution. In the latter case one can attempt to restore higher resolution through deconvolution methods. For modest resolution this approach can be successful; however, if one needs the highest resolution possible it is better to use the small apertures and pay the penalty of lost light (see Section VI). The spectrographs provide an excellent proving ground for image processing methods. Not only are numerical experiments on one-dimensional spectra much more computationally tractable than experiments on large images, but it is also possible to determine the "ground truth" for blurry, large-aperture spectra by recording spectra of the same objects through the small apertures.

III. The Hubble Space Telescope Point-Spread Function

The HST point-spread function presents a number of challenges for image restoration. First, in the WFPC the PSF is strongly space-variant owing to the strong gradient in phase across the aperture introduced by the spherical aberration and the internal obstructions within the WFPC optical system.

All instrumental response functions are further complicated by time variability of the PSF. Prior to the 1993 HST Servicing Mission, the power-producing solar panels tended to expand and contract significantly as the spacecraft moved in and out of the earth's shadow. This was expected, but the mechanism that was designed to compensate for the effect did not work properly, and the solar panels would suddenly "snap" into position after building up too much stress. The sudden motion induced a vibration into the entire spacecraft that could not always be compensated by the pointing control servo system. The resulting jitter introduced into HST images added to the image blurring in a time- and exposure-dependent way. Data from the Fine Guidance Sensors needed to be employed in order to construct a jitter function, and the jitter function was then convolved with the PSF predicted from an optical model of the telescope and cameras to obtain a best estimate to the actual PSF.

Another source of time variability to the PSF is an overall thermal expansion and contraction of the spacecraft. Again, this was an expected phenomenon, though not at the magnitude observed. There is a net variation in the effective focal length of the telescope of ± 2–3 μm, which is correlated with orbital period and spacecraft attitude. This periodic focus variation is superimposed on a long-term focus change resulting from desorption, or outgassing, from the telescope support structure. As the telescope structure shrinks, focus is monitored and the HST secondary mirror is moved to compensate. Both the short- and long-term time variability problems limit the utility of model PSFs. To date the best image restoration results have relied on empirically determined PSFs (e.g., from well-exposed observations of stars, ideally in the same field of view and quite close to the object of interest so that space- and time-variability problems can be ignored).

With a PSF that is independent of atmospheric blurring (astronomical "seeing," which is usually the dominant feature in the PSFs of ground-based optical telescopes), the color or wavelength dependence of the PSF is significant. PSFs must be modeled or observed for each different filter or spectral range. When an object is observed through a broad-band spectral filter, the PSF is actually dependent on the spectral energy distribution of the object being observed! This latter effect, however, is more subtle than all of our other uncertainties about the details of the PSF and has, thus far, been ignored in HST image restorations.

The combination of the spatial and temporal variations in the PSF made it difficult to compute model PSFs, because one could not fully predict all properties of the optical system. It was similarly difficult to utilize observed PSFs for deconvolution purposes, because one could not guarantee the presence of a good PSF star in one's field of view. Even with a PSF star (i.e., well-isolated, with good exposure) one would introduce errors owing to the space variance of the response function. Observers might try to observe a PSF star in the same field location as their study object, but this then introduced the problems of both time variability and color dependence. Thus, the fundamental limitation to HST image restoration, both prior to and following the First Servicing Mission, is in determining accurate PSF models.

IV. The Richardson–Lucy Algorithm and Adaptations

Prior to HST's spherical aberration problem, astronomers used deconvolution methods most extensively in the area of radio interferometry. The most frequently used deconvolution method in radio astronomy is the CLEAN algorithm (Högbom, 1974; Schwarz, 1978; Clark, 1980; Cornwell, 1983; Schwab, 1984), and a number of investigators attempted to adapt CLEAN to the HST imaging case. These adaptations were partly successful (Keel, 1991, Busko, 1994), but as Cornwell (1994) pointed out, CLEAN is intrinsically unstable for direct imaging data such as that from HST.

Numerous experiments were carried out using linear and nonlinear restoration methods, with two algorithms emerging as de facto standards: the Richardson–Lucy (RL) method (Richardson, 1972; Lucy, 1974) with extensions by Snyder (1991), White (1994a), Lucy (1994), and Lucy and Hook (1992), and the maximum entropy method (Narayan and Nityananda, 1986; Gull and Skilling, 1991; Cornwell and Evans, 1985; Sault, 1990), with implementations by Weir (1991) and Wu (1994a). Almost all algorithms were used in a nonspatially variant mode, using small subfields of HST images in order to avoid PSF mismatch problems. For deconvolution of spectroscopic data the Jansson–Van Cittert algorithm (see Chapter 4) proved to be the most beneficial.

A. THE BASIC RICHARDSON–LUCY ALGORITHM

Following the notation used throughout this book,* we assume that $o(x)$ is the intensity of the sky at position x, and $s(x, x')$ is the point-spread function of the telescope, camera, and detector being used to image the sky. For simplicity and with no loss of generality, we conduct our discussion here as if there were only one spatial independent variable. Generalization to two orthogonal spatial coordinates is straightforward. In the absence of noise, a camera with infinitely small pixels would produce a blurred image

$$i(x) = \int o(x')s(x, x')\, dx'. \tag{1}$$

---

*The present notation corresponds to Lucy's (1974) notation as follows: $i \rightarrow \phi$, $n \rightarrow \tilde{\phi}$, $o \rightarrow \psi$, and $s \rightarrow P$.

We shall refer to the range of x as "data space" and the range of x' as "image space." In most astronomical applications image space and data space are very similar, in that $i(x)$ looks rather similar to the unblurred image $o(x')$; however, this is not necessarily the case, and it is important to maintain a clear distinction between the two. As a concrete example, if the data are collected by scanning a long, narrow slit across the sky in various directions (as was done by the IRAS satellite), the raw data hardly resemble an image at all. In such cases, operations such as taking the difference $o(x') - i(x)$ are really quite senseless.

Many discussions of image formation at this point introduce an additive noise term $\epsilon(x)$ that corrupts $i(x)$; however, this is not correct when the noise is a result of counting statistics. In that case, $\bar{i}(x)$ is the number of counts that would be measured at x if we were to average a large number of independent observations. The observed data $i(x)$ are distributed about the mean $\bar{i}(x)$ with a Poisson distribution, so the probability of getting $i(x)$ counts in a pixel when the mean is $\bar{i}(x)$ is

$$\mathscr{P}(i|\bar{i}) = e^{-\bar{i}}\bar{i}^{\,i}/i!. \tag{2}$$

The likelihood of an entire set of data values is the product of the probabilities of each value, so the log likelihood is

$$\ln \mathscr{L} = \int \left[i(x)\ln \bar{i}(x) - \bar{i}(x) - \ln i(x)! \right] dx. \tag{3}$$

The goal of image restoration is to recover the image $o(x)$ from the (noisy) data $i(x)$, given the equations of image formation, Eq. (1), and of Poisson likelihood, Eq. 3. The maximum-likelihood solution, $\hat{o}$, maximizes $\ln \mathscr{L}$. The maximum-likelihood solution may not be the "best" solution to the image restoration problem, because there is not universal agreement on the meaning of "best." However, the maximum-likelihood solution is of interest because it can be shown to be *unbiased*: if we compute $\hat{o}$ from many independent sets of data $i(x)$, most properties of the mean o converge to their true values. Consequently, the maximum-likelihood image has linear photometric properties: one can, in principle, measure the intensity of sources on the restored image with unbiased results. The maximum-likelihood image is, however, very unsuitable as an estimate of o in some cases, because noise in the data can be amplified to an objectionable level. We will return to this point later as we deal with adaptations to the method to suppress noise amplification.

At any rate, how do we find the maximum-likelihood solution to Eqs. (1) and (3)? Take the derivative of $\ln \mathscr{L}$ with respect to o:

$$\frac{d \ln \mathscr{L}}{do(x')} = \int \left[\frac{i(x)}{\hat{i}(x)} - 1 \right] s(x, x')\, dx = 0; \qquad (4)$$

but the RL iteration is

$$\hat{o}^{(k+1)}(x') = \hat{o}^{(k)}(x') \frac{\int [i(x)/\hat{i}^{(k)}(x)] s(x, x')\, dx}{\int s(x, x')\, dx}, \qquad (5)$$

where $\hat{o}^{(k)}$ is the kth estimate of o and $\hat{i}^{(k)}$ is computed from $\hat{o}^{(k)}$ using Eq. (1):

$$\hat{i}^{(k)}(x) = \int \hat{o}^{(k)}(x') s(x, x')\, dx'. \qquad (6)$$

We now see immediately from Eq. (4) that once we reach the maximum-likelihood solution, the ratio of the two integrals in Eq. (5) is unity and $\hat{o}^{(k+1)} = \hat{o}^{(k)}$; in other words, if the iteration converges, it must converge to the maximum-likelihood solution. If an object exists such that the blurred image $\hat{i}(x)$ is identical to the data $i(x)$, this is obviously a solution. However, in realistic cases the noise in the data precludes the existence of any object that fits the data perfectly. This iteration finds the object that best fits the data subject to the nonnegativity constraint.

Note that the integrals in Eq. (5) are over x, not x'. Thus, they are *not* equivalent to blurring the ratio $i/\hat{i}$ using the PSF. In fact, the integral over x in Eq. (5) is a correlation of the ratio and the PSF. Also note that Lucy (1974) couched his derivation of this iteration in terms of s being a probability density function and so required $\int s(x, x')\, dx = 1$. This is not necessary, and in fact it is helpful to retain the denominator explicitly. Both of these points are discussed further in Section IV.B.

Several properties of the RL iteration are worth pointing out. First, if s, $\hat{o}^{(k)}$, and n are nonnegative, then $\hat{o}^{(k+1)}$ must also be nonnegative. This is a necessary feature of any solution involving Poisson statistics, because one obviously cannot have a negative number of counts in the data, nor can one have a negative mean count rate. If both i and $\hat{i}$ are zero, their ratio should be replaced with unity. If $\hat{i}$ is zero and i is nonzero then their ratio becomes infinite; however, that can only happen if one starts the iteration with an $\hat{i}^{(0)}$ that is already zero over the entire region spanned by the PSF (which is a mistake, because the multiplicative correction factor for $\hat{o}$ can never change the value of initially zero values).

How do we know that the RL iteration will converge? The empirical answer is that it *does* converge: if we start with an initial guess $\hat{i}^{(0)}(x')$ that is very far from the correct answer, a modest number of iterations (typically 20–30) brings the computed $\hat{i}(x)$ into good agreement with $i(x)$. The likelihood as computed from Eq. (3) always increases.

Perhaps a more satisfying answer to the convergence question is that the RL iteration has been proven to converge, at least in the weak sense that iterations never cause the likelihood to decrease. Shepp and Vardi (1982) derived the RL iteration as a special case of the expectation-maximization (EM) algorithm of Dempster *et al.* (1977). Briefly, the EM algorithm is a method for finding the maximum likelihood solution for a broad class of problems. Imagine that the data actually in hand are just part of a "complete" data set that potentially has a great deal more information. For Shepp and Vardi's derivation of the RL iteration, the incomplete data are the counts $i(x)$ in each pixel, whereas the complete data are the counts $C(x, x')$ in pixel x that came from location x' on the sky. Thus, $i(x) = \int C(x, x')\, dx'$. The EM iteration consists of two steps: first, given the current estimate $\hat{o}^{(k)}$, compute the most likely values for $C(x, x')$ given $i(x)$. With a little thought we find that

$$
\begin{aligned}
C(x, x') &= \frac{i(x)\hat{o}^{(k)}(x')s(x, x')}{\int \hat{o}^{(k)}(x')s(x, x')\, dx'} \\
&= \frac{i(x)}{\hat{i}(x)}\hat{o}^{(k)}(x')s(x, x'),
\end{aligned}
\tag{7}
$$

i.e., we assign $i(x)$ to the x' pixels in proportion to the contribution of each x' value to x. The second part of the iteration is to take the complete data and compute the $\hat{o}(x')$ that gives the maximum likelihood for C. With the clever choice of $C(x, x')$, this is easy:

$$
\hat{o}^{(k+1)}(x') = \frac{\int C(x, x')\, dx}{\int s(x, x')\, dx}.
\tag{8}
$$

When we substitute $C(x, x')$ from Eq. (7), we see that this iteration is identical to the RL iteration.

Dempster *et al.* (1977) proved that the EM algorithm converges to the maximum-likelihood solution [Wu (1983) corrected an error in their proof]; we can therefore conclude that the RL iteration must converge. The Shepp and Vardi derivation of the RL iteration using the EM algorithm is interesting in that it demonstrates how the concept of "complete data" can be used to turn a hard maximum-likelihood problem into two easy problems. The simple derivation based on the derivative of $\ln \mathscr{L}$ is more enlightening, however, because it shows the close connection between the RL iteration and the likelihood and image formation equations.

B.   DISCRETE FORM OF EQUATIONS

For practical computation the continuous equations (6) and (5) are usually converted to discrete equations:

$$\hat{\imath}_m = \sum_n \hat{o}_n s_{mn}, \tag{9}$$

and

$$\hat{o}_n^{(k+1)} = \hat{o}_n^{(k)} \frac{\displaystyle\sum_m (i_m/\hat{\imath}_m^{(k)}) s_{mn}}{\displaystyle\sum_m s_{mn}}. \tag{10}$$

Here the PSF s_{mn} should be interpreted as the continuous $s(x, x')$ integrated over pixels in the x space, i.e.,

$$s_{mn} = \int_{x_m}^{x_{m+1}} s(x, x_n') \, dx, \tag{11}$$

where x_m and x_{m+1} are the boundaries of bin m. This is the natural normalization of s if the PSF is determined by observations of point sources through the same camera; if s is computed from an optical model, it may be necessary to do the integration over pixels.

This is not the only way to convert the continuous equations to a discrete form, though it is the most common approach. Rather than tabulating $\hat{o}(x')$ at a grid of x' coordinates, it might be preferable to express $\hat{o}$ as a sum of some basis functions and to solve for the coefficients of those functions. One must be careful about the choice of basis functions when adopting this approach, because neither the basis functions nor their coefficients may become negative if the RL iteration is to work properly. For example, suppose the desired solution is the best linear function of x'. Let $\hat{o}(x') = ax' + b(1 - x')$, where x' has been normalized so that $0 \leq x' \leq 1$. With $a \geq 0$ and $b \geq 0$, this form for $\hat{o}$ includes all nonnegative linear functions. If we substitute $\hat{o}$ into Eqs. (6) and (5) we can find an iteration for updating a and b. This example is probably not useful in practice, but it demonstrates that there is more than one way to map the continuous equations into discrete equations.

Both the continuous equations and the discrete equations have been written here as functions of a single variable, but they are easily generalized to functions of two (or more) variables: simply interpret m and n as ordered pairs, and take sums over all values of (m_1, m_2) or (n_1, n_2).

C. SPATIALLY INVARIANT PSFs

If the PSF is spatially invariant, then $s_{mn} = s_{m-n}$ and the sums in Eqs. (9) and (10) are convolutions. In this case the computations can be speeded up by using fast Fourier transforms (FFTs). Equations (9) and (10) become

$$\hat{\imath}^{(k)} = s \otimes \hat{o}^{(k)} \tag{12}$$

and

$$\hat{o}^{(k+1)} = \hat{o}^{(k)} \frac{(i/\hat{\imath}^{(k)})s(-m)}{s(-m) \otimes u}, \tag{13}$$

where u is a vector of ones. Note that the FFT of $s(-m)$ is just the complex conjugate of the transform of s. Thus, the RL iteration with a spatially invariant PSF becomes

$$\hat{\imath}^{(k)} = F^{-1}(\tau O^{(k)}), \tag{14}$$

$$\hat{o}^{(k+1)} = \hat{o}^{(k)} \frac{F^{-1}[F(i/\hat{\imath}^{(k)})\tau^*]}{F^{-1}(\tau U)}, \tag{15}$$

where $F(\ )$ represents the FFT, $F^{-1}(\ )$ is the inverse FFT,* is the complex conjugate operator, and upper case symbols correspond to the FFTs of the same variable shown in lower case. Note that U will simply be a delta function as long as there is no missing or bad data (but see Section IV.D).

Although RL iterations using FFTs are much faster than iterations using direct sums, there is no requirement in the RL algorithm that the PSF be spatially invariant. The sums in Eqs. (9) and (10) can for a general PSF just be viewed as matrix multiplications, and the PSF can vary arbitrarily with x and x'. This is important for the HST image restoration problem, because the PSF in the Wide-Field/Planetary Camera does vary with position, as has been described earlier. Owing to the magnitude of the computational problem, however, fully space-variant image restorations have not been attempted. Partially space-variant restorations have been implemented on parallel processors (e.g., Cobb *et al.*, 1993; Boden *et al.*,

1995) in which the image is segmented into (overlapping) subimages, each of which is deconvolved separately using a space-invariant PSF appropriate to that area of the detector (Trussell and Hunt, 1978a, b).

D. BAD PIXELS AND IMAGE BOUNDARIES

Now we come on the first complication: suppose there are one or more bad pixels in the data. This is a common problem in both ground-based astronomical images and HST data: CCDs are subject to cosmic ray hits, which effectively eradicate any information about the intensity of the affected pixels, and both CCDs and photon-counting detectors are subject to saturation.

A common approach to this problem is to interpolate values for the affected pixels using neighboring pixels; however, this is not really correct. Obviously any features of the restored image that rely on the invented values inserted into bad pixels are suspect, and there really is no way to know whether the substituted values are causing trouble. Moreover, interpolating data values in this or other circumstances corrupts the Poisson statistics of the data and is always dangerous.

The correct solution to this problem is to change our definition of the data grid to exclude the affected pixels. For example, suppose we have 10 data pixels numbered 1–10, and pixel 3 is bad for some reason. The real data grid then consists of pixels $1, 2, 4, \ldots, 10$—because we have no information from pixel 3, it might as well not exist. We simply omit the $m = 3$ row of the s_{mn} matrix in Eqs. (9) and (10) to get the correct RL iteration.

But how do we handle missing rows if the PSF is spatially invariant so that we are using Eqs. (14) and (15) to perform the iteration? The solution is to replace s_{mn} with $w_m s_{mn}$, where $w_m = 0$ at bad pixels and $w_m = 1$ at good pixels. Then the iteration in Eqs. (14) and (15) becomes

$$\hat{\imath}^{(k)} = F^{-1}(\tau O^{(k)}), \tag{16}$$

$$\hat{o}^{(k+1)} = \hat{o}^{(k)} \frac{F^{-1}[F(wi/\hat{\imath}^{(k)})\tau^*]}{F^{-1}(W\tau)}. \tag{17}$$

This iteration is exactly equivalent to ignoring the pixels where $w = 0$. Thus, it is easily possible to remove bad pixels from the data without inventing data and without compromising the efficiency of the iteration.

This is an important idea that will recur below: when possible, we should try to understand quirks of the data as being the result of quirks in the PSF matrix s. If a simple mapping of some problem onto s can be identified, then the required modification to the RL iteration becomes obvious.

What special treatment is necessary near the edges of the image? There are two aspects to this question: one must avoid the wrap-around problems that can occur when using FFTs to do the convolutions, and one should allow the image $\hat{o}$ to be of broader spatial extent (i.e., comparable to the extent of the PSF) than the observed data—this allows the algorithm to account for photons in the image that originate in sources located outside the region measured directly by the detector.

The solution to the FFT problem is simple and well-known: an array being convolved with a vector of length N should be padded with at least N zeros before the FFT is computed, and then the extra N elements should be stripped off after the inverse FFT. This ensures that even though convolutions computed using an FFT are implicitly periodic, the result is a nonperiodic convolution as desired.

The second problem is a bit more subtle. Consider an observed image that has a bright star just off the edge of the detector. This problem is exactly equivalent to having bad pixels, and is handled using a mask w in the same way. Thus, the iteration is allowed to provide a complete estimate of the object $\hat{o}$ at all pixels, despite the fact that some pixels in the observed data $\hat{i}$ are contaminated or, in fact, unobserved. Figure 2

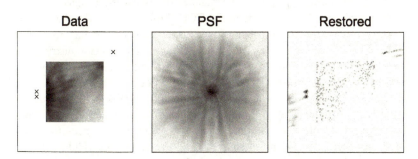

| Data | PSF | Restored |

Figure 2 An example of an image with bright stars near the image boundary, which are reasonably well restored even though only a fraction of the stars' images were actually observed. The star positions for this simulation are shown as crosses in the image at left. The restored image at right has almost all of the signal confined to the central pixel for each star, and the location of each star agrees with the original model. Contaminated or otherwise unobserved pixels are restored through the use of an appropriate weighting function [see Eqs. (16) and (17)].

shows an example of successfully restoring an image with stars located beyond the original image boundaries.

E. READ-OUT NOISE AND FLAT-FIELD CORRECTIONS

CCD detectors have two primary components of noise—photon counting noise with a Poisson distribution, which is accommodated implicitly in the derivation of the RL iteration, and read-out noise with a Gaussian distribution, which is not properly accommodated. The presence of read-out noise means that pixels with low levels of incident photons can properly have negative data values. These will cause the RL iteration to fail. The brute force approach is to simply impose a floor on the data values in the input image, but this tends to bias the result.

The proper approach to handling read-out noise was developed by Snyder (1991) and Snyder *et al.* (1993). The observed data $i(x)$ are actually the sum of a number of Poisson-distributed variables and the Gaussian-distributed read-out noise:

$$i(x) = \left[i_{obj}(x) + i_{ext\text{-}bkgd}(x) + i_{int\text{-}bkgd}(x) + i_{dark}(x) + i_{bias}(x) \right] + \epsilon(x),$$

$$(18)$$

where $i_{obj}(x)$ is the number of counts resulting from radiation from the source object, $i_{ext\text{-}bkgd}(x)$ is the number of counts from external background radiation, $i_{int\text{-}bkgd}(x)$ is the number of counts from internal background radiation (e.g., intrinsic to the CCD chip), $i_{dark}(x)$ is the number of counts from thermoelectrons, and $i_{bias}(x)$ is the number of counts from the bias applied to the CCD. All of these terms are Poisson-distributed random variables. The term $\epsilon(x)$ represents the Gaussian-distributed read-out noise of mean μ and variance σ^2. For $\mu = \sigma^2$ and if σ^2 is large, $\epsilon(x)$ approximates a Poisson-distributed variable with mean σ^2 (Feller, 1968). The validity of this approximation can be assured by adding an appropriate offset to the data. Thus, we accommodate read-out noise and all other potential noise sources (while retaining the validity of the maximum-likelihood derivation of the RL iteration) by modifying the iteration to

$$\hat{o}^{(k+1)} = \hat{o}^{(k)} \frac{F^{-1}\left\{ F\left[w\left(\dfrac{i + \sigma^2 + i_{bkgd}}{\hat{i}^{(k)} + \sigma^2 + i_{bkgd}} \right) \right] \tau^* \right\}}{F^{-1}(W\tau)}.$$

$$(19)$$

As a practical matter, if an observed data value $i(x)$ is so negative that the sum $i + \sigma^2 + i_{\text{bkgd}}$ remains negative, the pixel is masked out by setting the corresponding element in the weight array w to zero.

An explicit correction for the variation in the detector response from pixel to pixel, the so-called flat-field correction, is accommodated by extending the definition of the weight array. Instead of giving weights of unity for good pixels, the weights are adjusted according to the amount of the flat-field correction. Assuming the data values have been corrected for the flat-field variations, pixels with less than perfect sensitivity will have been scaled up (typical corrections are of order 10% or less). In order to avoid giving undue significance to the counts in these pixels, however, they are assigned reduced weights equal to the inverse of the flat-field correction factor. Snyder (1991) approached the problem slightly differently, assuming that the data are not flat-field corrected and including both the correction and the weighting in the iteration.

F. SUBSAMPLING

The results from the RL algorithm can be improved, especially in the case of undersampled data, by computing the restoration on a more finely sampled grid. This is possible if one has knowledge of the PSF on finer scales, e.g., via optical modeling techniques or via observations taken at subpixel offsets. Lucy and Baade (1989) show some results for simulated images using the Lucy method with oversampling. They do not, however, describe in any detail the form the iteration takes when oversampling is included, so it is worthwhile to describe it here.

Assume that the PSF s is known and the object estimate $\hat{o}$ is to be calculated on a grid that is oversampled from the data grid by a factor b. The discrete form of the RL iteration can be rewritten in the following steps:

1. Blur the object estimate with the PSF to obtain $\tilde{i}$, an estimate of the data on the oversampled grid:

$$\tilde{i}_{\mu}^{(k)} = \sum_{\lambda=1}^{bN} \hat{o}_{\nu}^{(k)} s_{\mu\nu}, \qquad \mu = 1, bN. \tag{20}$$

2. Bin the data estimate $\tilde{i}$ in blocks of size b to obtain $\hat{i}$, the estimate of the data on the observed grid:

$$\hat{i}_{m}^{(k)} = \sum_{n=1}^{b} \tilde{i}_{n+bm-1}^{(k)}, \qquad m = 1, N. \tag{21}$$

3. Take the ratio of $i/\hat{\imath}^{(k)}$ on the observed grid and use simple pixel replication to expand it by a factor b, giving ρ, their ratio on the oversampled grid:

$$\rho^{(k)}_{n+bm-1} = i_m/\hat{\imath}^{(k)}_m, \qquad n = 1, b \quad \text{and} \quad m = 1, N. \tag{22}$$

4. Convolve ρ with the transpose of the PSF and multiply by $\hat{o}^{(k)}$ to get $\hat{o}^{(k+1)}$, the next iterate:

$$\hat{o}^{(k+1)}_\mu = \hat{o}^{(k)}_\mu \frac{\displaystyle\sum_{\nu=1}^{bN} \rho_\nu s_{\nu\mu}}{\displaystyle\sum_\nu s_{\nu\mu}}, \qquad \mu = 1, bN. \tag{23}$$

An important feature of restoring images using oversampled grids, especially with undersampled data, is that a significantly better fit to the data is obtained. If the signal-to-noise ratio of the data is high, it is impossible to find an accurate fit of the undersampled estimate of the data because all stars are effectively forced to be in the centers of pixels. A star that is not centered on a pixel must necessarily be restored as a combination of a centered point source and some extended flux that compensates for the centering error; this limits the photometric fidelity of any restoration method. Restoring to a finer grid allows more of the flux in stellar halos to be put back into the center of the star image where it belongs.

G.   SUPPRESSING NOISE AMPLIFICATION

Despite its many advantages and adaptability to the HST image restoration problem, the RL algorithm can yield noise amplification. This is a generic problem for maximum likelihood methods, which attempt to fit the data as closely as possible. If one performs many RL iterations on an image containing an extended object such as a galaxy, the extended emission usually develops a "speckled" appearance. The speckles are not representative of any real structure in the image, but instead are the result of fitting the noise in the data too closely. In order to reproduce a small noise bump in the observed data it is necessary for the object estimate to have a very large noise spike; pixels near the bright spike must then be very black (near zero intensity) in order to conserve flux. The only limit on the amount of noise amplification in the RL method is the requirement that the image not become negative. Thus, once the compensating holes in the image are pushed down to zero flux, nearby spikes cannot grow any further and noise amplification ceases. The positivity constraint alone is sufficient to control noise amplification in images of star fields on a black background; in that case one can perform thousands of RL iterations without generating an unacceptable amount of noise. However, for smooth objects observed at low signal-to-noise ratios, even a modest number (20–30) of RL iterations can produce objectionable noise.

The usual practical approach to limiting noise amplification is simply to stop the iteration when the restored image appears to become too noisy. However, the question of where to stop is a difficult one. The approach suggested by Lucy (1974) was to stop when the reduced χ^2 between the data and the blurred estimate is about 1 per degree of freedom. Unfortunately, one does not really know how many degrees of freedom have been used to fit the data. If one stops after a very few iterations then the estimate is still very smooth and the resulting χ^2 should be comparable to the number of pixels. If one performs many iterations, however, then the object estimate develops a great deal of structure and so the effective number of degrees of freedom used is large; in that case, the fit to the data ought to be considerably better. The maximum entropy method (MEM) and other methods that add regularization to the iteration encourage smoothness in the image and suppress noise amplification, but at the expense of providing poorer quality restorations of objects that are, in fact, not smooth. There is no criterion for the RL method that tells how close the fit ought to be. Note that there is such a criterion built in to the MEMSYS5 maximum entropy package (Gull and Skilling, 1991), and the pixon approach of Piña and Puetter (1992) uses similar ideas.

Another problem is that the answer to the question of how many iterations to perform often is different for different parts of the image. It may require hundreds of iterations to get a good fit to the high signal-to-noise image of a bright star, whereas a smooth, extended object may be fitted well after only a few iterations. In Fig. 3, note how the images of both the central star and the bright star at the top center continue to improve as the number of iterations increases, while the noise amplification in the extended nebulosity is getting much worse. Thus, one would like to be able to slow or stop the iteration automatically in regions where a smooth model fits the data adequately, while continuing to iterate in regions where there are sharp features (edges or point sources).

Another approach to controlling noise amplification is to smooth the final restored image. This method has been developed and mathematically

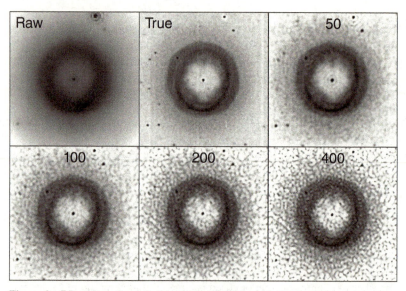

Figure 3 RL restoration of simulated 85 × 85 pixel HST Planetary Camera observation of a planetary nebula. Restored and true images are 256 × 256 (method described in Section VI.F was used to restore the image on a subsampled grid). As the number of iterations (shown at top of each image) increases, the images of bright stars improve, but noise is amplified unacceptably in the nebulosity.

justified by Snyder and Miller (1985) and Snyder *et al.* (1987). Unfortunately, for HST images the amount of smoothing required to reduce the noise amplification is very large. Figure 4 shows the effect of various amounts of smoothing on the restored planetary nebula image. By the time the noise amplification in the nebulosity has been reduced to a visually acceptable level, the images of stars have been grossly blurred. For most purposes, one must pay far too high a price to avoid noise amplification using this method.

An effective approach to such an adaptive stopping criterion is to modify the likelihood function of Eq. (3) so that it becomes flatter in the vicinity of a good fit. The approach we have taken is to use a likelihood function that is identical to Eq. (3) when the difference between the blurred object estimate $\hat{i}$ and the data i is large compared with the noise, but that is essentially constant when the difference is smaller than the

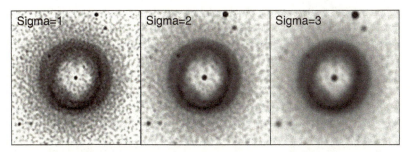

Figure 4 Results of smoothing 400-iteration RL restoration from Fig. 3 with a Gaussian. The width of the Gaussian is $\sigma = 1$, 2, and 3 pixels for the three cases shown. Images of stars are grossly blurred when the RL image is smoothed heavily enough to reduce substantially the noise in the nebulosity.

noise. There are two important advantages to using a modified form of the likelihood function to control noise amplification:

1. The end point of the iteration is well-defined, allowing properties of the final solution to be studied.
2. Acceleration techniques (see Section IV.I) may be used to reach the solution more quickly without changing the final answer.

The damped RL iteration (White, 1994a, c, d) starts from the likelihood function

$$\ln \mathscr{L} = \int f[g(x)]\, dx, \tag{24}$$

where

$$g(x) = -\frac{2}{t^2}\left[i(x)\ln\frac{i(x)}{i(x)} - i(x) + i(x) \right] \tag{25}$$

and

$$f(y) = \begin{cases} \dfrac{\gamma - 1}{\gamma + 1}(1 - y^{\gamma+1}) + y^\gamma, & y < 1, \\ y, & y \geq 1, \end{cases} \tag{26}$$

where $f(y)$ is the "damping function." It is chosen to be a simple function that is linearly proportional to y for $y > 1$, is approximately constant for $y \sim 0$, and has continuous first and second derivatives at $y = 1$ (this allows acceleration techniques to be used to speed convergence of the method.)

The constant γ determines how suddenly the function f becomes flat for $y < 1$. For $\gamma = 0$, $f(y) = y$ and there is no flattening at all. The larger the value of γ, the flatter is the function. The results reported here use $\gamma = 10$, but the value of γ has little effect on the results as long as it is larger than a few.

The quantity $g(x)$ is a slightly modified version of the $\ln \mathcal{L}$ function of Eq. (3). The constants and multiplicative factors are chosen so that the expected value of g in the presence of Poisson noise is unity if the threshold $t = 1$. The threshold t then determines at what level the damping turns on: if $t = 1$ the damping occurs at 1σ, if $t = 2$ at 2σ, etc.

With this new likelihood function, we can follow the steps previously outlined to derive the damped RL iteration:

$$\hat{o}^{(k+1)}(x') = \hat{o}^{(k)}(x') \frac{\int \left\{ 1 + \tilde{g}(x)^{\gamma-1} [\gamma - (\gamma - 1)\tilde{g}(x)] \frac{i(x) - \hat{i}^{(k)}(x)}{\hat{i}^{(k)}(x)} \right\} \times s(x, x') \, dx}{\int s(x, x') \, dx},$$

$$(27)$$

where

$$\tilde{g}(x) = \min[g(x), 1]. \tag{28}$$

Note that in regions where the data and model do *not* agree, $\tilde{g} = 1$ and so this iteration is exactly the same as the standard RL iteration. In regions where the data and model do agree, however, the second term gets multiplied by a factor which is less than 1, and the ratio of the numerator and denominator approaches unity. This has exactly the desired character: it damps changes in the model in regions where the differences are small compared with the noise.

Figure 5 shows the results of applying the damped iteration to the planetary nebula data of Fig. 3. Note that the noise amplification in the nebulosity has been greatly reduced, but that the stars are still very sharp. Figure 6 shows the result of applying both the standard RL method and the damped iteration to HST observations of Saturn; again, the noise in the restored image is greatly reduced without significantly blurring the sharp edges of the planet and its rings. This is perhaps more easily seen in Fig. 7, which shows the intensity of the restored image along a cut through

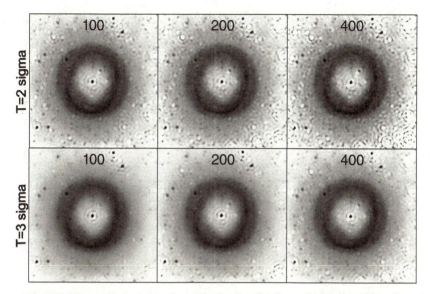

Figure 5 Restoration of data from Fig. 3 using the damped iteration with noise thresholds of 2σ and 3σ. As the number of iterations increases, images of bright stars continue to improve, but noise amplification is much better controlled than for the standard RL iteration.

the planet and rings. Sharp features such as narrow gaps in the rings are essentially identical in the RL and damped images, but the disk of the planet is much smoother in the damped image. Note that the mean intensity in the disk of the planet is the same in the damped and RL images, indicating good photometric linearity.

H. RINGING

Ringing—the appearance of spurious high-frequency structures in a restored image—is not the result of noise amplification, though noise masking can cause ringing. Ringing can occur even in perfectly noiseless data because certain high-frequency structures are completely invisible in the data. (These correspond to frequencies where the Fourier transform of the PSF has zeros.) A graphic example is shown in Fig. 8.

Ringing cannot be controlled by adjusting how the blurred image fits the data, because it is not visible in the blurred image. It can only be eliminated by placing constraints on the object. In the RL method removal

Figure 6   HST Wide-Field Camera observation of Saturn and restorations using the RL iteration ($t = 0$) and the damped iteration with noise thresholds of 1σ and 2σ. In each case, 100 accelerated iterations were used.

of any constant or slowly varying background from the image to be restored is effective in suppressing ringing because the algorithm intrinsically prohibits the restored image intensity to be negative. Other methods of suppressing ringing involve imposition of upper and lower bounds (e.g., Jansson method) and use of bandpass limits.

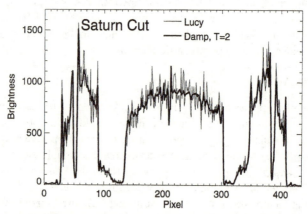

Figure 7   Cut through the rings and disk from upper right to lower left for the restored Saturn images in Fig. 6.

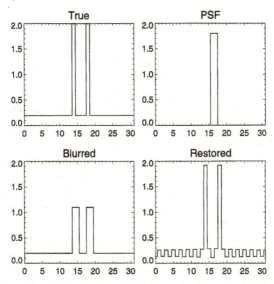

Figure 8 Example of the ringing phenomenon in image restoration. A simple model image composed of two point sources is convolved with a 2-pixel wide PSF. The restoration of the blurred image is contaminated by low-amplitude high-frequency ripples. This structure is "invisible" in the blurred image because the PSF exactly cancels this spatial frequency. The ringing can be suppressed only with the addition of constraints on the restoration.

I. ACCELERATION

A technique for accelerating the RL method has been discovered independently by several researchers (Kaufman, 1987; Holmes and Liu, 1991; Hook and Lucy, 1992; Adorf *et al.*, 1992). The basic idea is that Eq. (10) is replaced by

$$\hat{o}_m^{(k+1)} = \hat{o}_m^{(k)} + \alpha \, \Delta \hat{o}_m^{(k)}, \tag{29}$$

where α is the same for all pixels but varies from one iteration to the next, and $\Delta \hat{o}_m^{(k)}$ is given by

$$\Delta \hat{o}_m^{(k)} = \hat{o}_m^{(k)} \frac{\sum_m (i_m / \hat{\imath}_m^{(k)} - 1) s_{mn}}{\sum_m s_{mn}}. \tag{30}$$

The value of α is chosen to maximize the likelihood $\ln \mathscr{L}$ of Eq. (3). This value is most easily found using Newton's method with the additional constraints that $1 < \alpha < \alpha_{max}$, where α_{max} is determined by the requirement that $\hat{o}_m^{(k+1)} \geq 0$. The search for the best value of α can be performed with little additional computation because the evaluation of $\ln \mathscr{L}$ for different values of α does not require any additional convolutions. The great majority of computing time in the RL iteration goes into computing convolutions, so the time per iteration for the accelerated RL method is only slightly larger than the time per iteration for the standard method.

There is an even more effective acceleration method based on the idea of conjugate directions. The standard RL iteration takes many very small, almost identical steps as it approaches convergence. The line search method described above speeds things up by taking the longest possible step at each iteration, but it usually does not converge very fast either. The problem is similar to that seen in steepest descent minimization approaches: successive steps are rattling back and forth along the bottom of a narrow valley in the function rather than taking long steps along the length of the valley. The solution to this problem is to ensure that steps toward the solution do not ruin the minimization just achieved in the previous step. The mathematical expression of this requirement is that successive steps should be conjugate (Press *et al.*, 1986). Directions Δp_1 and Δp_2 are conjugate if

$$\sum_m \Delta p_1(m) \sum_n \frac{\partial^2 \Phi}{\partial p_m \, \partial p_n} \Delta p_2(n) = 0, \tag{31}$$

where the sum is over all the variables of the function Φ being minimized (or maximized). Suppose two successive steps in the RL iteration are $\Delta \hat{o}^{(k)}$ and $\Delta \hat{o}^{(k+1)}$ [as given by Eq. (30)]. We can force the next step to be conjugate to the previous two by setting

$$\Delta \hat{o} = \Delta \hat{o}^{(k+1)} + \beta \Delta \hat{o}^{(k)}, \tag{32}$$

where

$$\beta = -\frac{\sum\limits_m \left(i_m / \hat{\imath}_m^{(k)^2} \right) \Delta \hat{\imath}_m^{(k)} \, \Delta \hat{\imath}_m^{(k+1)}}{\sum\limits_m \left(i_m / \hat{\imath}_m^{(k)^2} \right) \Delta \hat{\imath}_m^{(k)^2}} \tag{33}$$

and

$$\Delta \hat{\imath}_m = \sum_n \Delta \hat{o}_n s_{mn}. \tag{34}$$

A similar equation can be derived for accelerating the damped method of Section IV.G using conjugate directions.

To accelerate the RL iteration using conjugate directions, save the direction of each step $\Delta \hat{o}$ and adjust the direction of the next step as given in Eq. (32). Then find the best step by doing a line search as described in Eq. (29). This approach typically speeds convergence by a factor of 3 compared to the standard accelerated method (and by a factor of ~ 10 over the standard RL iteration). The rate of convergence for two different images is displayed in Fig. 9.

J. OTHER ADAPTATIONS

Lucy (1992) and Lucy and Hook (1992) extended the basic RL algorithm to a multichannel approach in which smooth, extended structures are restored separately from point sources. This separation helps to avoid the overfitting and noise amplification problems such as those described in Section IV.G. Further development of this approach has led to codes that allow multiple observations of the same field, each with separate PSFs (at possibly very different spatial resolutions), to be optimally co-added (Lucy and Hook, 1992), and for images of the same field observed with subpixel offsets to be combined to achieve higher spatial resolution (Hook and Adorf, 1995).

Starck and Murtagh (1994) have explored modifications of the RL and other methods in which noise amplification is suppressed by applying wavelet-based filters. Their method of image restoration with multiresolution support effects a regularization at each step of the iteration by filtering out statistically insignificant structures at several spatial scales. The method was extended to include Poisson noise distributions as appropriate for HST and other direct optical imaging systems (Murtagh *et al.*, 1995).

V.   Phase Retrieval and Blind Deconvolution

The most difficult practical problem to be solved when restoring HST images is usually not the choice of a restoration algorithm, but rather finding a good point-spread function. Even the most sophisticated algo-

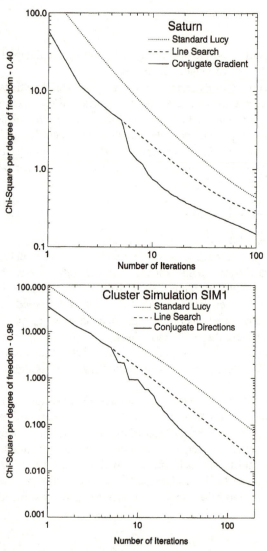

Figure 9 Convergence rates using standard RL iteration, line search acceleration, and conjugate directions acceleration for the HST image of Saturn (Fig. 6) and the simulated HST image of a dense star cluster. Note that a constant has been subtracted from the χ^2 value plotted on the y axis; this allows the χ^2 values among the three solutions to be readily distinguished in a logarithmic plot. The conjugate directions method is faster than the original RL iteration by a factor of 10 and is faster than the line search method by a factor of 3.

rithm cannot produce a decent restored image using a bad PSF, and even the least sophisticated algorithm will be a pretty good job given a good PSF. Obtaining PSFs of sufficient accuracy remains the primary obstacle to restoring the full dynamic range of HST data (both pre- and post-Servicing Mission).

As mentioned in Section III, the HST PSF varies as a function of observing wavelength and filter, intrinsic color of the objects being observed, position of the object in the field of view, and time. There are enough variables that it is not practical to maintain a library of high-quality, observed PSFs covering all possibilities. Computed PSFs (Krist, 1994, 1995; Redding *et al.*, 1994; Hasan and Burrows, 1995) are of good quality, but still frequently fail to match the observed PSFs in all details. Redding *et al.* (1995) are continuing work on refining the optical model of the HST, but the techniques of phase retrieval and blind deconvolution offer alternative approaches to PSF determination given suitable types of images.

The PSF for an optical system is determined by the amplitude $\mathscr{A}(u, v)$ and phase $\phi(u, v)$ of the (approximately) spherical wave front at each point (u, v) on the sphere. The amplitude is usually uniform across the entire pupil except where the pupil is obscured by objects in the light path, such as the secondary mirror and its support structure. The phase measures the deviation of the wave front from the sphere (a perfectly focused wave front has zero phase error). Usually the phase error is measured in units of the wavelength of light being observed.

The PSF s for the given amplitude and phase is

$$s(x, y) = \left| \int \int \mathscr{A}(u, v) \, e^{2\pi j[ux + vy + \phi(u, v)]} \, du \, dv \right|^2$$

$$= \left| \int \int \mathscr{W}(u, v) \, e^{2\pi j(ux + vy)} \, du \, dv \right|^2. \tag{35}$$

Stated simply, the PSF is the square of the amplitude of the Fourier transform of the complex pupil function, $\mathscr{W}(u, v) = \mathscr{A}(u, v) \times \exp[2\pi j\phi(u, v)]$. Note that this equation assumes that the wave front is not too strongly curved over the pupil; if the curvature of the wave front is large one must use a Fresnel transform rather than a Fourier transform.

Phase retrieval is the process of trying to recover the wave front phase $\phi(u, v)$ (and possibly the amplitude $\mathscr{A}$ as well) given a measurement of the PSF. Phase retrieval methods have been used since the discovery of the aberration in the HST primary mirror (Burrows *et al.*, 1991; Fienup *et al.*,

1993) to characterize the HST optical system. Previous phase retrieval efforts have been aimed mainly at an accurate measurement of the spherical aberration. Phase retrieval has much in common with deconvolution, and many of the techniques used for image restoration have counterparts for phase retrieval. However, the equation relating the phase and the observed PSF is nonlinear for the phase retrieval problem, which makes phase retrieval considerably more difficult than image restoration. A particular problem in phase retrieval is that maximum-likelihood approaches to finding the phase tend to get stuck at local maxima of the likelihood rather than finding the globally best solution. Fienup and Wackerman (1986) discuss this and other phase retrieval problems.

A novel phase retrieval iteration (White, 1994b) can be derived from the likelihood equation for Poisson statistics, Eq. (3), here given in discrete form:

$$\ln \mathscr{L} = \sum_x i_x \ln \bar{i}_x - \bar{i}_x - \ln i_x!. \tag{36}$$

The continuous variables (u, v) and (x, y) have been replaced by discrete variables, and the equations are written in one-dimensional form for simplicity of notation. If i_x now represents a scaled version of the PSF, and if the amplitude of the complex pupil function $\mathscr{W}$ is allowed to vary, then the maximum-likelihood solution occurs where all partial derivatives of $\mathscr{L}$ with respect to $\mathscr{W}(u)$ are zero:

$$\frac{\partial \ln \mathscr{L}}{\partial \mathscr{W}(u)} = 0$$

$$= \frac{1}{N} \sum_x (i_x/s_x - 1) e^{-2\pi jux/N} \sum_{u'} \mathscr{W}_{u'} e^{2\pi ju'x/N}. \tag{37}$$

The iteration is simply

$$\mathscr{W}_u^{(k+1)} = \mathscr{W}_u^{(k)} + \frac{\partial \ln \mathscr{L}}{\partial \mathscr{W}_u^{(k)}}. \tag{38}$$

This is a "gradient step" for updating the complex pupil amplitude $\mathscr{W}$. If one wants only the phase ϕ to vary during the iteration, holding the pupil illumination function $\mathscr{A}$ fixed, it is possible to renormalize $\mathscr{W}$ to a constant amplitude after each iteration.

An example of how phase retrieval can improve computed PSFs is shown in Fig. 10. At this wavelength in the middle of the optical band, the computed model PSF is in reasonably good qualitative agreement with the

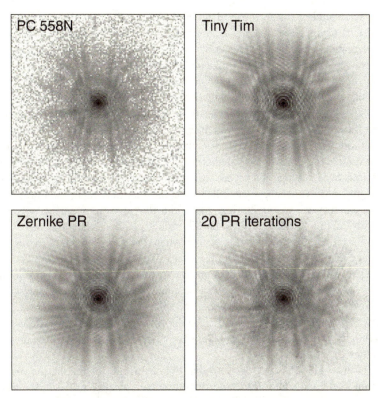

Figure 10 Results of phase retrieval for PC F588N PSF. Top left: observed PSF. Top right: PSF modeled using three wavelengths across filter bandpass. The PSF model was computed using the Tiny Tim software developed by Krist (1994, 1995). Lower left: phase retrieval result using low-order Zernike polynomials to model wave front errors. Lower right: phase retrieval result using 20 iterations of the Poisson maximum-likelihood method. Initial guess was taken from the Zernike phase retrieval model.

observations. However, there are some discrepancies between the observed PSF and the model; note especially that the shape of the PSF core is not well modeled and the position and intensity of the bright ring are not exactly correct. The most likely explanation is that the mirror map and aberration model used in the optical modeling software are not accurate enough. Any discrepancies in the optical model become more obvious at shorter wavelengths.

The simplest phase retrieval method is to model the wave front error $\phi(u, v)$ using a low-order polynomial in (u, v). For HST the appropriate polynomials are the Zernike polynomials given in the *HST Optical Telescope Assembly Instrument Handbook* (Burrows, 1990). The various Zernike polynomial coefficients correspond to wave front tilt (which shifts the position of the object), focus, spherical aberration, astigmatism, and so on.

The lower left panel of Fig. 10 shows the result of adjusting these coefficients to fit the observed PSF. A conjugate gradient optimization method was used to search for the set of Zernike coefficients that maximizes the Poisson log likelihood. The agreement with the observations is improved, though there are still differences. For example, note the bright knot in the observed PSF where the tendril trailing off towards 8 o'clock crosses the bright ring. Also note that the outer dark ring, seen easily in the Zernike PSF, is hardly visible at all in the observed PSF. Other differences are readily visible when the PSFs are blinked on an image display.

The agreement with the data can be improved further by solving for a map of wave front errors across the pupil using the method described above. The results of 20 iterations of such a method are shown in the lower right panel of Fig. 10. Note the improvement in features such as the strength and position of the "tendrils" that extend downward from the center of the PSF.

Holmes (1992) has derived an iterative algorithm for finding the maximum likelihood solution for both the unblurred image and the PSF in the presence of Poisson statistics. This problem is usually referred to as *blind deconvolution*. A maximum likelihood approach to blind deconvolution is certain to fail unless one can place very strong constraints on the properties of the PSF. For example, if one applies the Holmes iteration to HST images with no constraints on the PSF, the solution found is that the unblurred image looks exactly like the data and the PSF is a delta function. (A perfect fit to the data, but ludicrously far from the truth!) If one applies a band-limit constraint on the PSF based on the size of the HST aperture, the result is little better: the PSF solution looks like a diffraction-limited PSF rather than the true spherically aberrated PSF.

If one has either a star within the image or a separate PSF observation, one can do much better by forcing the selected PSF objects to be point sources in the restored image. One simple approach is to start with an initial guess for the image that has a delta function at the position of the star and is zero in the other nearby pixels. Then it is really possible to get estimates of both the unblurred image and the PSF from the Holmes iteration.

This may look to be of only academic interest, but in fact it may be the solution to an important problem for HST images. Most HST deconvolutions have been carried out using observed PSFs because theoretically computed PSFs [using the Tiny Tim software (Krist, 1994, 1995), for example] have not been a very good match to the observations. The noise in observed PSFs presents a problem, however: when using a noisy PSF, the restoration algorithm ought to account for the fact that both the image and the PSF are noisy. The Richardson–Lucy method (and other commonly used image restoration methods) incorrectly assume that the PSF is perfectly known. The blind deconvolution approach allows one to construct an algorithm that takes as input a noisy image and a noisy PSF observation and to construct as output estimates of both the PSF and the unblurred image.

Experiments indicate that this method may be especially useful in crowded stellar fields where it is difficult to find isolated PSF stars. In that case one can start with a very noisy PSF star (either on the edge of the crowded region or from a separate observation) and then can improve the PSF estimate using the overlapping PSFs of stars in the crowded region (Fig. 11). The final PSF is considerably more accurate than the original noisy PSF, which in turn leads to better restorations of the unblurred image. This method has been used only for spatially invariant PSFs thus far, though generalization to space-variant PSFs is feasible (but computationally challenging).

VI. Spectral Deconvolution

Although the majority of image restoration work for the Hubble Space Telescope emphasized true two-dimensional images, substantial effort has also been spent in the area of spectral deconvolution, in particular for the Goddard High-Resolution Spectrograph. The GHRS was studied primarily because comparisons of spectra taken through each of its two apertures—a small science aperture (SSA) that is 0.25 second of arc square, and a large science aperture (LSA) that is 2.0 seconds of arc square—allowed for a rigorous test of the merits of deconvolution techniques. This is because spectral observations made with the SSA of the GHRS are not significantly degraded in resolution from the spherical aberration in the spatial PSF,

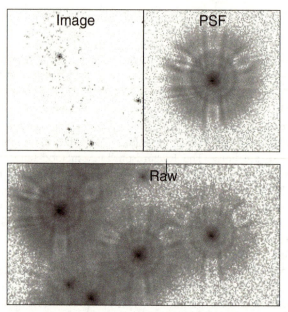

Figure 11   Demonstration of blind deconvolution on a simulated HST Planetary Camera image. The bottom panel shows raw image (wrap-around boundary conditions were used for convenience). The star on the right was used as the PSF star by setting the right half of initial guess to zero except at the pixel where the star is located. Three of the five stars on the left are brighter than the PSF star. The top panels show the image and PSF computed using blind iteration. Note that the computed PSF has a considerably higher signal-to-noise ratio than the PSF star, indicating that the algorithm has managed to use the stellar images of the brighter stars to improve the PSF estimate.

because the SSA samples only the core of the PSF. SSA spectra suffer from a significant loss in sensitivity (factor of 5–6) owing to the large amount of light in the PSF halo. LSA spectra, on the other hand, experience both a factor of 2–3 decrease in resolving power and a loss in sensitivity ($\sim$ 30–40%) that follow directly from the spherical aberration. The immediate implication of this was to ask whether it was better to use the SSA and accept the loss of sensitivity but avoid a loss of resolution, or use the LSA and deconvolve the spectra to restore the spectral resolution. As explained further below, the answer depends on the scientific objectives of the observations.

A. THE LINE SPREAD FUNCTION

The spectral line spread function (LSF) of either the GHRS or FOS is the convolution of the internal line spread function of the spectrograph optics (which can be reasonably represented as a Gaussian function of order 10% the width of a single detector diode: 0.25 second of arc in projection on the sky, or 50 μm in linear dimension) with the spatial PSF light distribution along the direction of spectral dispersion. Sampling by the large diodes is a limiting factor for the resolution, and critical sampling of the spectra is routinely obtained by taking multiple exposures in which the spectra are shifted by one-half or one-fourth diode width along the dispersion. This is shown graphically in Fig. 12 for both the LSA and SSA for the GHRS. As with the cameras, adequate measurement of the spatial PSF at the spectrograph apertures is a difficult task. At the high spectral resolutions of GHRS, and spatial resolution of the HST, few astronomical calibration

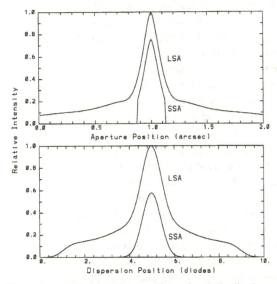

Figure 12   The upper panel shows the one-dimensional distribution of light along the dispersion direction for GHRS apertures. The effective line spread function shown in the lower panel follows from additional smoothing from the spectrograph's internal line spread function and sampling by large diodes. LSA, large science aperture; SSA, small science aperture.

sources exist. For the FOS the gratings introduce variations in the LSF that are dependent on wavelength.

B. ALGORITHMS AND RESULTS

The best restoration results for GHRS spectra have been obtained using Jansson's method (see Chapter 4), in which we iterate

$$\hat{o}^{(k+1)} = \hat{o}^{(k)} + r(\hat{o}^{(k)})(i - s \otimes \hat{o}^{(k)}), \tag{39}$$

with the relaxation function r (due to Frieden, 1975) given by

$$r(\hat{o}) = r_0\left(1 - \frac{2|\hat{o} - (A + B)/2|}{(B - A)}\right). \tag{40}$$

In Eq. (40) r_0 is a constant controlling the step size of each iteration, A is the allowed lower bound (zero enforces positivity), and B is the allowed upper bound (e.g., continuum level for an absorption spectrum). Note that Jansson's original method used a point-successive iteration to speed convergence. Owing to our interest in using deconvolution codes on both images and spectra, each with complex and large PSFs, we have used approaches that enabled FFT implementations rather than requiring direct convolutions in data space.

An example of a deconvolution of a GHRS LSA spectrum is shown in Fig. 13. Although the deconvolution looks very good, the RMS difference of the two spectra is larger than would be expected from photon statistics of two independent spectra taken through the SSA with the same exposure time as used for the LSA spectrum. This suggests that although deconvolution works quite well at restoring resolution, it does so only at the cost of amplifying the noise beyond what would have resulted from a direct SSA observation with the same exposure time.

It is also possible to improve the resolution of spectra observed through the small science aperture in order to compensate for the intrinsic LSF of the instrument and detector sampling. For example, the star χ Lupi was observed with both intermediate resolution ($\lambda/\Delta\lambda \sim 30,000$) and Echelle resolution ($\lambda/\Delta\lambda \sim 90,000$) at the same wavelength through the SSA, each to a signal-to-noise ratio of about 80. This provides an excellent test case, where the deconvolved SSA spectrum can be compared to a higher resolution "truth" spectrum. Figure 14 shows a comparison of these spectra over a 10-Å range; a gaussian function was used to model the LSF of the SSA. Realized resolving powers may be estimated by computation of

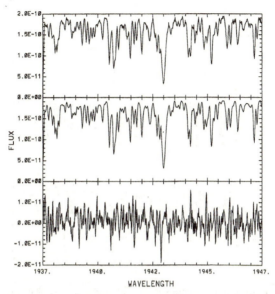

Figure 13 Comparison of GHRS small science aperture (top) and deconvolved large science aperture spectra of χ Lupi taken with grating G160M. The difference spectrum is shown in the bottom panel. (From Gilliland *et al.*, 1992, Astronomical Society of the Pacific; reproduced with permission.)

the full width at half maximum (FWHM) of the autocorrelation functions over the sharp-lined spectral region between 1938.5 and 1939.5 Å. This results in a FWHM of 3.42, 2.18, and 7.45 pixels, for formal resolving powers ($\lambda/\Delta\lambda$) of 33,700, 52,900 (deconvolved), and 51,500 (Echelle). The Echelle spectrum shows some sharper lines suggesting that the above result for resolving power is biased downward by the existence of blended lines over the autocorrelation domain. The deconvolution of SSA data provides remarkably good results, especially when note is taken that the exposure time for the Echelle observation was a factor of 2.5 greater than that for the first-order spectrum.

Deconvolution has been less critical for HST spectroscopy programs than for imaging programs, and in most cases direct observations using the smaller apertures have been superior to deconvolution of data taken through larger apertures. In cases in which throughput was the primary consideration, however, large-aperture observations were successfully deconvolved. This is an interesting contrast to the case of the HST cameras,

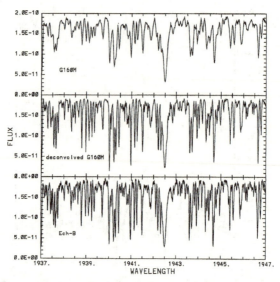

Figure 14 Deconvolution demonstration for SSA spectra. The top panel shows the direct G160M SSA spectrum of χ Lupi, and the lower panel shows the high-resolution "truth" spectrum acquired in Echelle mode (also through the SSA). The central panel shows the deconvolution of the G160M spectrum, which does remarkably well in reproducing features that are obvious in the Echelle data. (From Gilliland *et al.*, 1992, Astronomical Society of the Pacific; reproduced with permission.)

for which, of course, there was no option equivalent to use of the GHRS SSA other than installation of corrective optics.

VII.   Other Algorithms

Although we at the Space Telescope Science Institute have concentrated our image restoration efforts on the Richardson–Lucy algorithm for image deconvolution and the Jansson algorithm for spectral deconvolution, many other methods have been explored by ourselves and our collaborators. It is not an exaggeration to say that the HST spherical aberration problem brought forth an explosion in research on image restoration methods and instigated collaborations between astronomers and experts in signal processing, electrical engineering, computer science, and medical imaging.

Perhaps the major alternatives to RL deconvolution explored for HST images were variations of the maximum entropy method. Weir (1991) obtained good results using a code based on the commercial Maxent package (Maximum Entropy Data Consultants, Ltd.). Wu (1994a, 1995) developed a simpler MEM implementation that is comparable in computational costs to the RL method, and which also provides good photometric integrity. Wu's MEM implementation also provides good spatial resolution (Wu, 1994b). Núñez and Llacer (1991, 1994) developed a maximum *a posteriori* method with a maximum entropy prior.

An alternative approach based on Bayesian statistics was developed by Piña and Puetter (1992) (see also Puetter and Piña, 1993, 1994). Known as the "pixon" method, the essence of this algorithm is to determine the minimum number of degrees of freedom necessary to fit the data. This is done by segmenting the image into "pixons" of varying spatial scale. The segmentation can be done using various criteria (e.g., uniform signal-to-noise ratio). Restoration is performed by the usual maximum-likelihood techniques on the pixon-segmented image.

Katsaggelos and collaborators (Katsaggelos *et al.*, 1994) applied frequency-domain adaptive iterative algorithms to HST images with good results, both in terms of image fidelity and photometric linearity. Katsaggelos's algorithms were extensively evaluated by Busko (1994, 1995) and photometric linearity results are shown in Fig. 15. Lindler (1991) explored block iterative or block Jacobi methods that exploited the cases in which the PSF was much smaller than the total image extent. Lindler *et al.* (1994) further experimented with hybrid approaches (combining linear and non-linear methods) to attain both good photometric and astrometric results on crowded stellar fields.

Coggins *et al.* (1994) developed an iterative/recursive algorithm as an extension to basic iterative approaches. Rather than using the residual image at each step in the iteration to correct the estimate of the true image, they point out that the residual image is itself blurred by the PSF (being the difference between the blurred image and the current estimate of the blurred image). Thus, a better correction to the estimate of the true image is obtained by using a deblurred residual. The deblurred residual is computed recursively at each iteration. The method yields good photometric results on crowded stellar fields, but does not constrain the solution to positive-definite images and thus can have somewhat objectionable residual patterns surrounding bright stars.

Several researchers have investigated blind and semiblind deconvolution methods. In addition to the work described earlier (Section V), Schulz

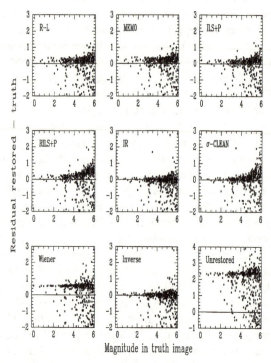

Figure 15 Aperture photometry in restored and unrestored images. A 2.2-pixel-diameter aperture was used to measure the magnitude of each star in an image constructed from a list of 500 stars. In each plot the abscissa is the magnitude in the truth image and the ordinate is the residual (restored magnitude minus true magnitude). A negative residual indicates excess light in the aperture, probably due to crowding. A positive residual means some light is missing from the aperture, either because it was not recovered by the restoration algorithm, or because the restored star profile does not fit inside the aperture. Each plot shows the results from a different restoration method (R–L, Richardson–Lucy; MEM0, maximum entropy; ILS + P, iterative least squares with positivity constraint; RILS + P, ILS + P with regularization by frequency-adaptive filtering; IR, regularized ILS without positivity constraint; σ–CLEAN, CLEAN algorithm adapted for photon-counting imaging; Wiener, Wiener filter; Inverse, iterative least squares without positivity constraint or adaptive filtering; Unrestored, raw data). See Busko (1994, 1995) for details of the restoration methods. Adapted from Busko (1994).

(1993) and Jefferies and Christou (1993) have developed alternative blind deconvolution methods. Schulz's work explicitly models the PSF as resulting from phase errors across the telescope aperture, whereas Jefferies and Christou avoid the trivial solution by applying band-limit and Fourier spectrum constraints.

VIII.   Image Restoration and Photometry

In most astronomical applications of image restoration the preservation of photometric linearity is of the utmost importance. The quantitative analysis of astronomical images, e.g., in measuring stellar magnitudes and colors, in studying the intensity profiles of galaxies, or in measuring the relative intensities of emission and absorption lines in spectra, demands that any restoration algorithm not introduce systematic errors in the photometric scale and that random errors not be increased. Thus, with the development and assessment of each algorithm it has been essential to evaluate the photometric integrity of the method. The importance of this aspect of image restoration is demonstrated by the large number of research efforts in this area (Lindler *et al.*, 1994; Fullton *et al.*, 1994; Linde *et al.*, 1994; Fusi Pecci *et al.*, 1994; Ferrarese and Ford, 1994).

Busko (1994, 1995) has done thorough photometric tests of the photometric properties of a variety of image restoration algorithms applied to simulated HST images. Simulated images are preferable for such studies because only in such cases does one know, absolutely, the true magnitudes or intensity profiles of the objects. Photometric linearity is generally assessed by plotting Δm vs. m—the difference between the measured magnitude of a star after restoration and the true magnitude of the star vs. the true magnitude of the star*. An image restoration algorithm that provides no bias to the photometry yields a plot that is a horizontal line with $\Delta m = 0$. A constant offset in Δm is not a major concern; this can be corrected via an absolute photometric calibration.

Figure 15 shows photometric linearity plots for a variety of restoration algorithms, including the RL method, maximum entropy method, and several of the others previously mentioned. One of the key reasons the RL method has been so broadly used for HST image restoration is because it

*Stellar magnitudes are defined by the relation $m_2 - m_1 = 2.5\log(l_1/l_2)$, where l_1 and l_2 are the luminosities of two stars with magnitudes m_1 and m_2.

is excellent at retaining photometric linearity, as is clearly evidenced by the plots in Fig. 15, while simultaneously controlling noise amplification in the background. Jansson's iteration has proved to be excellent for spectra owing to the added constraint on the continuum level.

IX. Post-Servicing Mission Performance

In December 1993 NASA carried out a stunningly successful Servicing Mission for the HST, during which time a corrective optics package known as COSTAR (Jedrzejewski *et al.*, 1994) and a new camera having its own corrective optics, the Wide-Field/Planetary Camera 2 (Trauger *et al.*, 1994), were installed. COSTAR deployed a set of small corrective mirrors into the light path of each of three other scientific instruments (the Goddard High-Resolution Spectrograph, the Faint Object Spectrograph, and the Faint Object Camera). The corrective mirrors cause a minor degradation in overall throughput, especially in the ultraviolet portion of the spectrum, owing to the additional reflections. However, because the error in the figure of the primary mirror was simple, corrective mirrors could be fabricated to undo the spherical aberration in the primary in a relatively straightforward manner. Within a few weeks after the Servicing Mission was completed, HST imaging had been restored to its nominal specifications for resolution. During the first three years of operations following the Servicing Mission, HST has performed nearly flawlessly.

It is worth noting, however, that the results of restoring pre-Servicing Mission images have been validated in post-Servicing Mission observations. Figure 16 shows a comparison between a pre-Servicing Mission image, its restoration, a post-Servicing Mission image of the same object, and its restoration. Restored pre-Servicing Mission images recovered essentially the full spatial resolution of the telescope, but could not recover sensitivity. Thus, the raw post-Servicing Mission images appear to be very similar to restored pre-Servicing Mission images, but allow the detection of objects ~ 2 magnitudes fainter than was previously possible. Figure 16 also demonstrates the value of restoring even post-Servicing Mission images, especially in cases where high dynamic range is desired, or when dealing with highly crowded fields.

In order to understand the quality of our PSF models and the properties of various image restoration algorithms, we made extensive use of simulated data sets. Some of these have already been shown in this

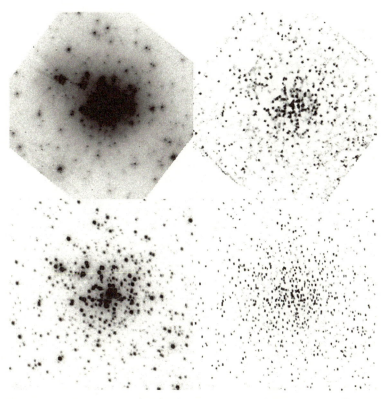

Figure 16 Comparison of images of the star cluster R136 in the Large Magellanic Cloud taken before and after the 1993 HST Servicing Mission (SM). Upper left: pre-SM image, WFPC (image has been rotated to align with post-SM image). Upper right: restoration of pre-SM image. Lower left: post-SM image, WFPC 2. Lower right: restoration of post-SM image. Note that the restored pre-SM image has a spatial resolution equal to that of the post-SM image, but the post-SM image is sensitive to much fainter stars. Restoration of the post-SM image results in further resolution enhancement and some increase in dynamic range in crowded regions.

chapter; however, a simulation of a star cluster similar to that in Fig. 16 shows very graphically the value of working with good simulations (this simulation was done prior to the Servicing Mission). It is very difficult to distinguish between the observed data (Fig. 16) and the simulation (Fig. 17)!

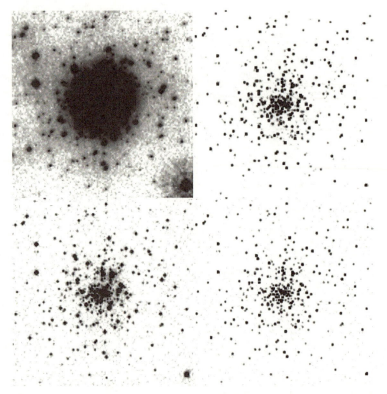

Figure 17 Same as for Fig. 16 but for a simulated star cluster. The simulation, which was computed prior to the Servicing Mission, is virtually indistinguishable from the real data and demonstrates the value of doing detailed modeling and simulations.

X.   Summary and Conclusions

The spherical aberration in the primary mirror of the Hubble Space Telescope resulted in a major change in the kinds of astronomical research programs that could be carried out. Projects that required optimal sensitivity to faint objects or maximum dynamic range were severely compromised, and many were postponed until after the 1993 Servicing Mission and the installation of corrective optics and a new Wide-Field/Planetary Camera.

Nevertheless, an active program of research into image restoration methods (drawing on expertise both within and outside of astronomy) and their adaptation for Hubble imaging and spectroscopy allowed for a maximal scientific return despite the telescope's blurred vision. Image restoration was most successful in morphological studies, but researchers paid close attention to the photometric linearity of the restoration procedures and were able to do both qualitative and quantitative science on deconvolved HST images. Keel (1994) cited some of the research areas where deconvolution played an important role in optimizing HST science: studies of the compact nuclei of galaxies and modeling galaxy brightness profiles, for example. Deconvolution also enhanced HST's images of planets (e.g., Westphal *et al.*, 1991) and improved the measurements of positions and magnitudes of stars in dense star clusters (e.g., Fullton *et al.*, 1995).

In addition, the pursuit of high-quality image restorations for HST required that researchers understand the telescope optical system in great detail. Subtle variations in the point-spread function, with their time and spatial dependencies, needed to be understood in order to be able to compute high-quality PSF models. With more than five years of experience with HST, optical scientists continue to refine their models of the HST optics and are able to predict the PSF more and more accurately.

Good PSF modeling is the key to high-quality image restoration for HST. Both prior to the 1993 Servicing Mission, and now, following that mission, with HST's optical performance restored to design specifications, image deconvolution is limited primarily by the accuracy of the PSFs. We have found a number of restoration algorithms that work well. The algorithms that we believe are the most successful are those most cognizant of the physical constraints on the data, such as minimum and maximum data values and spatial frequency limits, and the noise properties of the data. For example, our success with the Richardson–Lucy algorithm and its extensions results from its nonnegativity constraint and its derivation from a Poisson noise model.

Acknowledgments

The majority of the work described here was done under an augmentation to the Space Telescope Science Institute (STScI) contract with NASA, NAS5–26555, supporting STScI's Image Restoration Project. Project participants, in addition to the authors, included J. Mo, I. Busko, N. Wu, and N. Hamilton. External collaborators included A. Katsaggelos, B. Hunt, J. Núñez, R. Puetter, D. Redding, T. Schulz, and D. Snyder. Significant work on HST image restoration was also carried out at the Space Telescope–European Coordinating Facility by H.-M. Adorf, R. Hook, L. Lucy, and F. Murtagh, and by others in the Space Telescope user community, including T. Cornwell, J. Hollis, I. King, T. Lauer, D. Lindler, W. Keel, and R. Stachnik. The authors also thank S. Stevens–Rayburn for help in preparing the bibliography.

References

Adorf, H.-M., Hook, R. N., Lucy, L. B., and Murtagh, F. D. (1992). *In* "Proceedings of the 4th ESO/ST-ECF Data Analysis Workshop" (P. Grosbøl and R. C. E. de Ruijsscher, eds.), p. 99. European Southern Observatory, Garching, Germany.

Adorf, H.-M., Hook, R. N., and Lucy, L. B. (1993). *Proc. SPIE* **1945**, 75.

Boden, A. F., Redding, D. C., Hanisch, R. J., and Mo, J. (1995). *J. Opt. Soc. Am.*, in press.

Burrows, C. J. (1990). "HST Optical Telescope Assemby Handbook." Space Telescope Science Institute, Baltimore.

Burrows, C. J., Holtzman, J., Faber, S., Bely, P., Hasan, H., Lynds, C., and Schroeder, D. (1991). *Astrophys. J.* **369**, L21.

Busko, I. (1994). *In* "The Restoration of HST Images and Spectra II" (R. J. Hanisch and R. L. White, eds.), p. 279. Space Telescope Science Institute, Baltimore.

Busko, I. (1995). *Publ. Astron. Soc. Pac.* **106**, 1310.

Clark, B. G. (1980). *Astron. Astrophys.* **89**, 377.

Cobb, M. L., Hertz, P. L., Whaley, R. O., and Hoffman, E. A. (1993). *Proc. SPIE* **2029**, 202.

Coggins, J. M., Fullton, L. K., and Carney, B. W. (1994). *In* "The Restoration of HST Images and Spectra II" (R. J. Hanisch and R. L. White, eds.), p. 24. Space Telescope Science Institute, Baltimore.

Cornwell, T. J. (1983). *Astron. Astrophys.* **121**, 281.

Cornwell, T. J., and Evans, K. F. (1985). *Astron. Astrophys.* **143**, 77.

Dempster, A. D., Laird, N. M., and Rubin, D. B. (1977). *J. R. Stat. Soc. B* **39**, 1.

Feller, W. (1968). "An Introduction to Probability Theory and Its Applications." Wiley, New York.

Ferrarese, L., and Ford, H. C. (1994). *In* "The Restoration of HST Images and Spectra II" (R. J. Hanisch and R. L. White, eds.), p. 327. Space Telescope Science Institute, Baltimore.

Fienup, J. R., and Wackerman, C. C. (1986). *J. Opt. Soc. Am. A* **3**, 1897.

Fienup, J. R., Marron, J. C., Schulz, T. J., and Seldin, J. H. (1993). *Appl. Opt.* **32**, 1747.

Frieden, B. R. (1975). *In* "Topics in Applied Physics: Picture Processing and Digital Filtering," (T. S. Huang, ed.), Vol. 6, p. 177. Springer-Verlag, New York.

Fullton, L. K., Carney, B. W., Janes, K. A., Coggins, J. M., and Seitzer, P. (1994). *In* "The Restoration of HST Images and Spectra II" (R. J. Hanisch and R. L. White, eds.), p. 296. Space Telescope Science Institute, Baltimore.

Fullton, L. K., Carney, B. W., Olszewski, E. W., Zinn, R., Demarque, P., Janes, K. A., Da Costa, G. S., and Seitzer, P. (1995). *Astron. J.* **110**, 652.

Fusi Pecci, F., Federici, L., Parmeggiani, G., Ferraro, F. R., Cacciari, C., Iannicola, G., Corsi, C. E., Buonanno, R., Zavatti, F., and Bendinelli, O. (1994). *In* "The Restoration of HST Images and Spectra II" (R. J. Hanisch and R. L. White, eds.), p. 319. Space Telescope Science Institute, Baltimore.

Gilliland, R. (1991). *In* "The Restoration of HST Images and Spectra" (R. L. White and R. J. Allen, eds.), p. 7. Space Telescope Science Institute, Baltimore.

Gilliland, R. L., Morris, S. L., Weymann, R. J., Ebbets, D. C., and Lindler, D. J. (1992). *Astron. J.* **104**, 367.

Gull, S. F., and Skilling, J. (1991). "Quantified Maximum Entropy MemSys5 User's Manual." Maximum Entropy Data Consultants, Royston.

Hanisch, R. J. (1991). *In* "Proceedings of the 3rd ESO/ST-ECF Data Analysis Workshop" (P. Grosbøl and R. H. Warmels, eds.), p. 131. European Southern Observatory, Garching, Germany.

Hanisch, R. J. (1994). *Proc. SPIE* **2198**, 1349.

Hanisch, R. J., and Mo, J. (1993). *In* "Astronomical Data Analysis Software and Systems II" (R. J. Hanisch, R. J. V. Brissenden, and J. Barnes, eds.), *ASP Conf. Ser. Astron. Soc. Pac.* **52**, 524.

Hanisch, R. J., and White, R. L. (1993). *Proc. SPIE* **2029**, 194.

Hanisch, R. J., and White, R. L. (1994). *In* "Proceedings, IAU Symposium No. 158–Very High Angular Resolution Imaging" (J. C. Robertson and W. C. Tango, eds.), p. 61. Kluwer, Dordrecht, The Netherlands.

Hasan, H., and Burrows, C. J. (1995). *Publ. Astron. Soc. Pac.* **107**, 289.

Högbom, J. A. (1974). *Astron. Astrophys. Suppl.* **15**, 417.

Holmes, T. J. (1992). *J. Opt. Soc. Am. A* **9**, 1052.

Holmes, T. J., and Liu, Y.-H. (1991). *J. Opt. Soc. Am. A* **8**, 893.

Hook, R. N., and Lucy, L. B. (1992). *In* "Science with the Hubble Space Telescope" (P. Benvenuti and E. Schreier, eds.), p. 245. European Southern Observatory, Garching, Germany.

Hook, R. N., and Adorf, H.-M. (1995). *In* "Calibrating Hubble Space Telescope: Post Servicing Mission" (A. Koratkar and C. Leitherer, eds.), p. 341. Space Telescope Science Institute, Baltimore.

Jedrzejewski, R. I., Hartig, G., Jakobsen, P., Crocker, J. H., and Ford, H. C. (1994). *Astrophys. J.* **435**, L7.

Jefferies, S. M., and Christou, J. C. (1993). *Astrophys. J.* **415**, 862.

Katsaggelos, A. K., Kang, M. G., and Banham, M. R. (1994). *In* "The Restoration of HST Images and Spectra II" (R. J. Hanisch and R. L. White, eds.), p. 3. Space Telescope Science Institute, Baltimore.

Kaufman, L. (1987). *IEEE Transactions on Medical Imaging* **MI-6**, 37.

Keel, W. C. (1991). *Publ. Astron. Soc. Pac.* **103**, 723.

Keel, W. C. (1994). *In* "The Restoration of HST Images and Spectra II" (R. J. Hanisch and R. L. White, eds.), p. 359. Space Telescope Science Institute, Baltimore.

Krist, J. (1994). "The Tiny Tim User's Manual." Space Telescope Science Institute, Baltimore.

Krist, J. (1995). *In* "Astronomical Data Analysis Software and Systems IV" (R. A. Shaw, H. E. Paynes, and J. J. E. Hayes, eds.), *ASP Conf. Ser. Astron. Soc. Pac.* **77**, 349.

Linde, P., Snel, R., and Spännare, S. (1994). *In* "The Restoration of HST Images and Spectra II" (R. J. Hanisch and R. L. White, eds.), p. 312. Space Telescope Science Institute, Baltimore.

Lindler, D. J. (1991). *In* "The Restoration of HST Images and Spectra" (R. L. White and R. J. Allen, eds.), p. 39. Space Telescope Science Institute, Baltimore.

Lindler, D., Heap, S., Holbrook, J., Malamuth, E., Norman, D., and Vener-Saavedra, P. C. (1994). *In* "The Restoration of HST Images and Spectra II" (R. J. Hanisch and R. L. White, eds.), p. 286. Space Telescope Science Institute, Baltimore.

Lucy, L. B. (1974). *Astron. J.* **79**, 745.

Lucy, L. B. (1992). *In* "Science with the Hubble Space Telescope" (P. Benvenuti and E. Schreier, eds.), p. 207. European Southern Observatory, Garching, Germany.

Lucy, L. B. (1994). *In* "The Restoration of HST Images and Spectra II" (R. J. Hanisch and R. L. White, eds.), p. 79. Space Telescope Science Institute, Baltimore.

Lucy, L. B., and Baade, D. (1989). *In* "Proceedings of the 1st ESO/ST-ECF Data Analysis Workshop" (P. Grosbøl, R. H. Warmels, and F. Murtagh, eds.), p. 219. European Southern Observatory, Garching, Germany.

Lucy, L. B., and Hook, R. L. (1992). *In* "Astronomical Data Analysis Software and Systems I" (D. M. Worrall, C. Biemesderfer, and J. Barnes, eds.), *ASP Conf. Ser. Pac.* **25**, 277.

Murtagh, F. (1993). *In* "Statistical Challenges In Modern Astronomy" (E. D. Feigelson and G. J. Babu, eds.), p. 340. Springer-Verlag, New York.

Murtagh, F., Starck, J.-L., and Bijaoui, A. (1995). *Astron. Astrophys. Suppl.* **112**, 179.

Narayan, R., and Nityananda, R. (1986). *Annu. Rev. Astron. Astrophys.* **24**, 127.

Núñez, J., and Llacer, J. (1991). *In* "Astronomical Data Analysis Software and Systems I" (D. M. Worrall, C. Biemesderfer, and J. Barnes, eds.), *ASP Conf. Ser. Astron. Soc. Pac.* **25**, 210.

Núñez, J., and Llacer, J. (1994). *Proc. SPIE* **2198**, 1357.

Piña, R. K., and Puetter, R. C. (1992). *Publ. Astron. Soc. Pac.* **104**, 1096.

Press, W. H., Flannery, B. P., Teukolsky, S. A., and Vetterling, W. T. (1986). "Numerical Recipes." Cambridge Univ. Press, Cambridge.

Puetter, R. C., and Piña, R. K. (1993). *Proc. SPIE* **1946**, 405.

Puetter, R. C., and Piña, R. K. (1994). *In* "The Restoration of HST Images and Spectra II" (R. J. Hanisch and R. L. White, eds.), p. 14. Space Telescope Science Institute, Baltimore.

Redding, D., Lee, M., and Sirlin, S. (1994). *In* "The Restoration of HST Images and Spectra II" (R. J. Hanisch and R. L. White, eds.), p. 188. Space Telescope Science Institute, Baltimore.

Redding, D., Sirlin, S., Boden, A., Mo, J., Hanisch, R., and Furey, L. (1995). *In* "Calibrating Hubble Space Telescope: Post Servicing Mission" (A. Koratkar and C. Leitherer, eds.), p. 132. Space Telescope Science Institute, Baltimore.

Richardson, B. H. (1972). *J. Opt. Soc. Am.* **62**, 55.

Sault, R. J. (1990). *Astron. J.* **354**, L61.

Schulz, T. J. (1993). *J. Opt. Soc. Am. A* **10**, 1064.

Schwab, F. R. (1984). *Astron. J.* **89**, 1076.

Schwarz, U. J. (1978). *Astron. Astrophys.* **65**, 345.

Shepp, L. A., and Vardi, Y. (1982). *IEEE Transactions on Medical Imaging* **MI-1**, 113.

Snyder, D. (1991). *In* "The Restoration of HST Images and Spectra" (R. L. White and R. J. Allen, eds.), p. 56. Space Telescope Science Institute, Baltimore.

Snyder, D. L., and Miller, M. I. (1985). *IEEE Trans. Nucl. Sci.* **NS-32**, 3864.

Snyder, D. L., Miller, M. I., Thomas, L. J., Jr., and Politte, D. G. (1987). *IEEE Transactions on Medical Imaging* **MI-6**, 228.

Snyder, D. L., Hammoud, A. M., and White, R. L. (1993). *J. Opt. Soc. Am. A* **10**, 1014.

Starck, J.-L., and Murtagh, F. (1994). *Astron. Astrophys.* **288**, 342.

Trauger, J. T. *et al.* (1994). *Astrophys. J.* **435**, L3.

Trussell, H. J., and Hunt, B. R. (1978a). *IEEE Trans. Acoust. Speech Signal Process.* **26**, 157.

Trussell, H. J., and Hunt, B. R. (1978b). *IEEE Trans. Acoust. Speech Signal Process.* **26**, 608.

Weir, N. (1991). *In* "Proceedings of the 3rd ESO/ST-ECF Data Analysis Workshop" (P. Grosbøl and R. H. Warmels, eds.), p. 115. European Southern Observatory, Garching, Germany.

Westphal, J. A. *et al.* (1991). *Astrophys. J.* **369**, L51.

White, R. L. (1994a). *In* "The Restoration of HST Images and Spectra II" (R. J. Hanisch and R. L. White, eds.), p. 104. Space Telescope Science Institute, Baltimore.

White, R. L. (1994b). *In* "The Restoration of HST Images and Spectra II" (R. J. Hanisch and R. L. White, eds.), p. 198. Space Telescope Science Institute, Baltimore.

White, R. L. (1994c). *In* "Astronomical Data Analysis Software and Systems III" (D. R. Crabtree, R. J. Hanisch, and J. Barnes, eds.), *ASP Conf. Ser. Astron. Soc. Pac.* **61**, 292.

White, R. L. (1994d). *Proc. SPIE* **2198**, 1342.

White, R. L. and Hanisch, R. J. (1991). *Proc. SPIE* **1567**, 308.

Wu, C. F. J. (1983). *Annals of Statistics* **11**, 95.

Wu, N. (1994a). *In* "The Restoration of HST Images and Spectra II" (R. J. Hanisch and R. L. White, eds.), p. 58. Space Telescope Science Institute, Baltimore.

Wu, N. (1994b). *In* "Astronomical Data Analysis Software and Systems III" (D. R. Crabtree, R. J. Hanisch, and J. Barnes, eds.), *ASP Conf. Ser. Astron. Soc. Pac.* **61**, 300.

Wu, N. (1995). *In* "Astronomical Data Analysis Software and Systems IV" (R. A. Shaw, H. E. Payne, and J. J. E. Hayes, eds.), *ASP Conf. Ser. Astron. Soc. Pac.* **77**, 305.

Chapter 11 | Maximum Probable Estimates of Spectra

B. Roy Frieden

Optical Sciences Center, The University of Arizona
Tucson, Arizona

List of Symbols

A	slit area
c	speed of light
C	normalization constant
df	degrees of freedom
e_m	standard deviation of noise at mth point in image data
$\{\varepsilon_m\}$	noise in image data
h	Planck's constant

H	entropy, Jaynes form
H_1	entropy, Burg form
$\{i_m\}$	image data
k	Boltzmann's constant
L	total number of photons in empirical object
m	general resolution cell in object
$\{m_m\}$	empirical object
M	number of resolution cells constituting object
MP	maximum probable
n_m	number of object photons in mth resolution cell
$\hat{n}_m$	estimated version of n_m
$\bar{n}_m$	average number of photons in mth resolution cell
n_m/z_m	occupancy ratio of mth resolution cell
N	total number of photons in object
o_m	object or true spectrum, mth value
$\hat{o}_m$	deconvolved, restored, or estimated (all equivalent) version of o_m
$p(q_1,\ldots,q_M)$	object-class probability law
P_N	probability of noise
$P_1(\{\varepsilon_m\}|\{n_m\})$	probability of noise conditional upon object
$P(n_1,\ldots,n_M)$	total probability of n_1 photons in cell $m = 1$ of object, and $\ldots$, and n_M photons in cell $m = M$ of object
$P_0(n_1,\ldots,n_M)$	probability of $\{n_m\}$ photons incident upon input slit
$P_0(n_1,\ldots,n_M|q_1,\ldots,q_M)$	$P_0(n_1,\ldots,n_M)$ conditional upon definite numbers $\{q_m\}$
$\mathscr{P}(n_1,\ldots,n_M,\varepsilon_1,\ldots,\varepsilon_M)$	joint probability of object and noise
q_m	probability that a randomly selected object photon will reside in cell m; relative intensity or "signal" intensity of cell m
$\{q_m\}$	prior object; prior spectrum
Q_m	prior object; prior spectrum; biases
R	distance between spectral source and entrance slit
rect	rectangular function
s_{mn}	sampled version $s(x_m - x_n)$
$s(x)$	point-spread function of spectroscope
t	exposure time
T	temperature of a blackbody
w^2	area of spectral source
W	quantum-mechanical degeneracy factor
$W(n_1,\ldots,n_M)$	degeneracy as a function of $\{n_m\}$
x_m	generalized wavelength or wave number, mth value
z_m	number of quantum degrees of freedom in mth resolution cell
α_m	parameter used in forming $\{\hat{n}_m\}$ when empirical object given
δ	Dirac δ function
Δx	resolution "cell" length, set by user

$\Delta \lambda_m$	spectral wavelength range
$\Delta \nu_m$	spectral frequency range
$\Delta \tau_m$	coherence time for mth resolution cell
λ_m	wavelength, mth value
$\{\lambda_m\}$	Lagrange multipliers on image constraints
μ	Lagrange multiplier on normalization constraint
ν	frequency (temporal)
$\bar{\nu}_m$	wave number, mth value
σ	standard deviation
σ_m	coherence area for mth resolution cell

I. Introduction

Every spectrum is obtained in a unique physical setting, defined by a unique set of *physical effects*. Source temperature, state of coherence, entrance-slit size, and so forth, all affect spectral shape. Therefore, the knowledge that the spectrum was formed under given circumstances should be used to form an *estimate* of the spectrum. In effect, the user hopes to "work backwards" from the given image data, through the physical effects, to the true spectrum. It follows that each estimation algorithm must be tailored to specific physical knowledge about the given spectrum. There should be no single best algorithm for use on all spectra.

Physical knowledge is part of a larger body of knowledge about the spectrum, called prior knowledge. (This is in the sense of prior to, or aside from, knowledge of the data.) A second component of prior knowledge is *statistical* in nature. The unknown spectrum may be imagined to belong to a "class" of spectra, all having the same statistical properties. For example, they may all have a small, but unknown, number of lines. Or the spectrum in question may have a small, but unknown, number of lines. Or the spectrum in question may have previously been estimated, but from observation of only a small number of photons. This would give us a random approximation to the true spectrum. Or the user may want to admit "maximum prior ignorance" about the probable spectrum present. This weakest form of prior knowledge will be discussed in detail. Finally, the reader may simply have some other form of prior knowledge at hand.

To date, the prior knowledge built into deconvolution algorithms has mainly been deterministic in nature, such as by use of positivity or boundedness. But a unified approach to estimating spectra should accommodate all possible physical and statistical prior knowledge about formation of the spectra. In particular, the approach should be physical, based on the Bose–Einstein nature of photons. Such an approach will be presented here. One general restoring principle will be derived, from which particular estimators follow, depending on what kind of prior knowledge is at hand. A maximum probable* (MP) criterion will be the overall estimation criterion in use. This gives the estimates the following desirable property: if all the known deterministic and statistical conditions for forming the given spectrum were repeated over and over, the one spectrum that *actually occurred* the largest number of times would coincide with the MP estimate. That is, the estimate will be "right" more often than any other estimate. We suggest that this is about the most that one could ask of an estimate.

Depending on prior knowledge, the MP estimator will be either Jaynes maximum entropy, Kikuchi–Soffer maximum entropy, maximum weighted-Burg entropy, minimum "photon-site" entropy, minimum "empirical" entropy, some combination of these, or something altogether different. It behooves the user to specify most accurately and completely *his* state of prior knowledge, so as to tailor an MP estimator to the given problem. We shall show in detail how to do this.

II. Orientation

This work will be restricted to the consideration of emission spectra. Absorption spectra and other kinds of spectra could be treated analogously but really need a separate physical development.

We will assume that the spectrometer basically consists of an entrance slit of area A focused on an exit slit at $1:1$ magnification (Fig. 1). The recorded image is blurred because of diffraction and aberrations in the optics between the two slits and because of electronic blur in the readout circuit. The ultimate aim is to deconvolve or restore the spectral image so as to retrieve the spectral pattern at the entrance slit; the latter pattern is called the object.

The object consists of intensities o_m, $m = 1, \ldots, M$, or $\{o_m\}$ for short. These intensities are at wavelengths $\{\lambda_m\}$ or wave numbers $\{\bar{\nu}_m\}$; the choice

---

*A Bayesian approach will be used, where all data and prior information are included. The approach is philosophically like maximum-*a posteriori* (MAP) estimation. Hence, our MP (or MAP) estimate is *maximum probable given the data and prior information.*

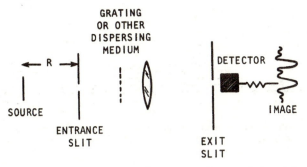

Figure 1 Basic spectrometer layout. The source is extended and incoherent. Entrance and exit slits are at equal conjugates. The image is formed electronically after detection at the exit slit.

is arbitrary. To keep the analysis general, we will call these abscissas $\{x_m\}$. As is conventional, the deconvolved or restored version of the object is denoted by $\{\hat{o}_m\}$.

The subdivision of the $\{x_m\}$ abscissas is uniform,

$$x_m = m \, \Delta x, \tag{1}$$

so that each x_m is centered on a resolution "cell" of length Δx (in either wavelength or wave-number units). The observer is assumed satisfied to resolve the spectrum at this ultimate level of resolution.

At a given x_m, the energy per photon is of course $h\nu_m$, where h is Planck's constant and ν_m is the known frequency. Hence, if n_m denotes the number of object photons that radiate from the mth cell during a fixed exposure time, o_m obeys

$$o_m = n_m h \nu_m. \tag{2}$$

Obviously, if n_m is estimated, this relation allows o_m to be estimated as well (see Fig. 2).

Going one step further, suppose that the object consists of N radiated photons in all, that is,

$$\sum_{m=1}^{M} n_m = N. \tag{3}$$

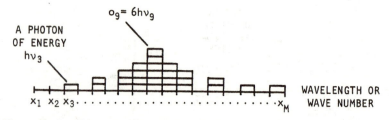

Figure 2 An object spectrum to be estimated. Note that here $n_1 = n_2 = 0$, $n_3 = 1$, $n_4 = 0$, $n_5 = 2, \ldots$.

Quantity N is assumed to be (approximately) known by conservation of energy from object to image.

In summary, then, we will seek estimates $\{\hat{n}_m\}$ obeying normalization condition (3) and recorded image data. Once the $\{\hat{n}_m\}$ are known, Eq. (2) may be used to arrive at the object estimate $\{\hat{o}_m\}$.

III. Physical Model for the Object

Imagine photons to be streaming from the entrance slit of area A toward the exit slit. These of course include all wavelengths of the source. Picture next just photons of one wavelength (or wave number) x_m as they flow from the entrance slit. Because these are quantum-mechanical entities, *they cannot occupy continuously* all positions in space during their flow. Instead, they may occupy only finite positions in space, called degrees of freedom (df) or modes. These are shown schematically as cubes in Fig. 3. Note that there are but a finite number z_m of such cubes and that we must subscript z because the number of modes will vary from one wavelength to another.

Each photon that flows from the entrance slit may exist only in such a mode. However, more than one may crowd in. This is a property of Bose–Einstein particles, to which photons belong as a class. Photons are also known to be indistinguishable microscopically, so that a given configuration of them within the modes cannot be distinguished from the same configuration where some of the photons have interchanged mode positions. As we shall see, the MP solution that we seek will depend vitally on this Bose–Einstein aspect of photon statistics.

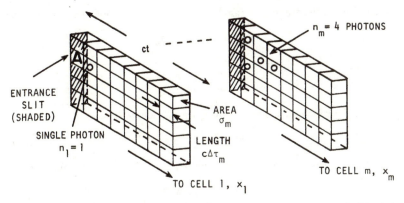

Figure 3 Degrees of freedom (small cube) for light from the entrance slit (shaded area A) at wavelengths x_1 and x_m.

A. COUNTING THE DEGREES OF FREEDOM z*

Consider one resolution cell x_m of the spectrum. The number of df, z_m, is defined as the number of "independent" volumes that exist over the volume swept out by photons leaving the entrance slit during one exposure time t (see Fig. 3). An "independent" volume is the three-dimensional space over which a photon is coherent. Hence, if it has a coherence length $c \, \Delta\tau_m$, with c the speed of light and $\Delta\tau_m$ the coherence time, and if it has a coherence area σ_m, the coherence volume is simply their product, or $c \, \Delta\tau_m \sigma_m$. On the other hand, the total volume swept out by a wave front of photons all leaving the aperture of area A during exposure time t is ctA. Hence the number of coherence volumes is

$$\frac{ctA}{c \, \Delta\tau_m \sigma_m} \equiv z_m = \frac{t}{\Delta\tau_m} \frac{A}{\sigma_m}. \tag{4}$$

Of these factors, A and t are of course directly measurable and need no further consideration.

The coherence time $\Delta\tau_m$ relates to abscissa x_m and resolution interval Δx as follows. By the Heisenberg uncertainty principle, $\Delta\tau_m$ goes inversely with spectral purity, or

$$\Delta\tau_m \approx 1/\Delta\nu_m, \tag{5}$$

*Kikuchi and Soffer, 1977.

where $\Delta \nu_m$ is the frequency range consistent with x_m and Δx. By $\lambda \nu = c$,

$$\Delta \nu_m = (c/\lambda_m^2) \, \Delta \lambda_m,$$

so that by Eq. (5) we have

$$\Delta \tau_m \approx \lambda_m^2/c \, \Delta \lambda_m. \tag{6}$$

Then, if x_m represents wavelength λ_m, we conclude that

$$\Delta \tau_m \approx x_m^2/c \, \Delta x. \tag{7a}$$

But if x_m instead represents wave number $\bar{\nu}_m \equiv 1/\lambda_m$, we obtain

$$\Delta \tau_m \approx 1/c \, \Delta x. \tag{7b}$$

In either case [Eq. (7a) or (7b)], $\Delta \tau_m$ is related to observables.

The coherence area σ_m is measured at the entrance slit, because this is where the object is imagined to exist. From Fig. 1, this slit is distance R from the spectral source, which is incoherent and of area w^2. Then the Van Cittert–Zernike theorem (Born and Wolf, 1959) gives

$$\sigma_m \approx R^2 \lambda_m^2/w^2. \tag{8}$$

Then, if x_m represents a wavelength, we obtain

$$\sigma_m \approx R^2 x_m^2/w^2, \tag{9a}$$

whereas if it represents a wave number,

$$\sigma_m \approx R^2/x_m^2 w^2 \tag{9b}$$

results.

The preceding is meant to be merely a representative calculation of σ_m. If instead a lens is interposed between source and entrance slit, σ_m will obey a different expression. The essential point is that σ_m is a computable quantity, whatever the setup.

The number of degrees of freedom, z_m, for a given resolution cell x_m may then be computed by combining Eq. (4) with Eq. (7a) or (7b) and Eq. (9a) or (9b).

It is sometimes not necessary to know the absolute value of z_m but rather the order of magnitude of *ratio* n_m/z_m, which might be called the occupancy ratio of the degrees of freedom by photons. This may be estimated as follows (Goodman, 1985). Suppose that we knew that the spectrum was like that of a blackbody at temperature T. We could then use the well-known relation for the *average number* of photons per mode (degree of freedom),

$$\bar{n}_m/z_m = [\exp(h\nu_m/kT) - 1]^{-1}. \tag{10}$$

Parameters h and k are Planck's and Boltzmann's constants, respectively. This mean ratio can often be regarded as a useful approximation to the ratio n_m/z_m. In particular, it shows that for viewing the spectrum of the sun ($T = 6000$ K) in the visible region the ratio is small,

$$n_m/z_m \ll 1, \tag{11a}$$

whereas at wavelengths exceeding 3 μm the ratio is large,

$$n_m/z_m \gg 1. \tag{11b}$$

Other kinds of sources are considered in Kikuchi and Soffer (1977).

B. DEGENERACY FACTOR W

Figure 3 shows $n_1 = 1$ photon for cell 1 located at a particular df position. It also shows $n_m = 4$ photons located at four particular positions for cell m. In how many ways may one photon locate in one particular df position of cell 1 and four photons locate in four particular positions of cell m? Because we cannot label the photons in any way (they are indistinguishable), there is only one way in which these placements may be made.

By contrast, suppose the $n_1 = 1$ photon may be *at any* df position of cell 1. Now, in how many ways can this photon be placed within the cell? Because it may occupy any of z_1 positions, there are z_1 such ways. More generally, the number of ways n_m photons may locate in cell m is given by the Bose–Einstein factor (Kikuchi and Soffer, 1977)

$$(n_m + z_m - 1)!/n_m!(z_m - 1)!.$$

This is the number of ways n_m entities may be distributed among z_m positions, where any number may occupy a given position but where interchanges of the entities do not count as new ways.

Finally, we ask how many ways W can n_1 photons locate in cell 1, *and...*, and n_M photons locate in cell M. Because these photon placements are independent from cell to cell, the answer is a product of the single-cell answers:

$$W(n_1, \ldots, n_M) = \prod_{m=1}^{M} \frac{(n_m + z_m - 1)!}{n_m!(z_m - 1)!}. \tag{12}$$

This is the net Bose–Einstein factor. It will be used in the following to form Bayesian estimators (see, e.g., Frieden, 1983) of the object $\{n_m\}$.

IV. Most Probable Object: Definition

The deconvolved or restored object that we seek is *the most probable number-count set* $\{n_m\}$. This is called the maximum probable (MP) estimate of the object. It obeys simply

$$P(n_1, \ldots, n_M) = \text{maximum} \tag{13}$$

through choice of the $\{n_m\}$. What is probability P?

We shall first find a component $P_0(n_1, \ldots, n_M)$ of P. Quantity P_0 describes the *a priori* probability of the $\{n_m\}$ photons *as they are incident upon the input slit*. This is independent of the df. Hence, what is the probability P_0 that n_1 slit photons will enter cell 1, and..., and n_M photons will enter cell M? One possibility is that P_0 is a constant, independent of $\{n_m\}$. But that actually assumes the unknown spectrum to be equal-energy white (see Section VII), quite a restrictive assumption. We find next an expression for P_0 that allows for a more general state of prior knowledge about the spectrum, in fact the most general.

Let us regard the "fate" of each photon incident upon the entrance slit. Let q_m be the probability that such a photon will occupy general cell x_m of the object (its fate). The probabilities $\{q_m\}$ define an object that (temporarily) will be assumed to be known. It is the ideal, underlying, signal* spectrum that ultimately gives rise to the unknown number-count spectrum $\{n_m\}$. This ideal spectrum is not necessarily flat (the conventional assumption). It could, for example, be the Balmer line series.

The $\{q_m\}$ may be called the prior object or prior spectrum. We temporarily assume it to be known aside from the image data. Indeed, in the

*By the law of large numbers (see, e.g., Frieden, 1983) $q_m \to n_m/N$ as $N \to \infty$, q_m describes the relative *intensity* of resolution cell x_m in the limit of an infinite number of photons. This intensity is called the signal.

MP principle given later, the $\{q_m\}$ supplement the data and are inserted independently of them. We shall show in later sections how the $\{q_m\}$, or at least a probability law for $\{q_m\}$, may be estimated.

With definite values for $\{q_m\}$, probability P_0 obeys a simple chain rule

$$q_1^{n_1} q_2^{n_2} \cdots q_M^{n_M}.$$

This assumes that each photon placement is independent of all others, as would be true for an incoherently radiating source. Stated as an equality,

$$P_0(n_1, \ldots, n_M | q_1, \ldots, q_M) = q_1^{n_1} q_2^{n_2} \cdots q_M^{n_M}, \tag{14}$$

because this is the probability of the $\{n_m\}$ *conditional upon* definite numbers $\{q_m\}$. (In standard probabilistic notation, the vertical bar denotes "conditional upon.")

However, more generally the user has uncertainty about *what* prior spectrum $\{q_m\}$ may be present. That is, the user can prescribe only a probability law

$$p(q_1, \ldots, q_M) \tag{15}$$

on possible spectra.[†] This law defines the body of statistical prior knowledge at hand defining the "object class." It includes all biases and prejudgments that the user may have about the unknown object. We shall assume different forms for this object-class law later. For now, we shall see how it fits into the overall MP approach.

The existence of a probability law for $\{q_m\}$ affects expression (14) for P_0. By the total probability law of statistics, Eqs. (14) and (15) may be combined to yield a net

$$P_0(n_1, \ldots, n_M) = \int_{\Sigma q_m = 1} \cdots \int q_1^{n_1} \cdots q_M^{n_M} p(q_1, \ldots, q_M) \, dq_1 \cdots dq_M. \tag{16}$$

Again, this is the *a priori* probability of a number-count object $\{n_m\}$ if the unknown spectrum $\{q_m\}$ is chosen randomly from a class of spectra defined by $p(q_1, \ldots, q_M)$.

[†]Readers who wish thoroughly to confuse themselves can dwell on the fact that expression (15) is also a "probability of probabilities." This concept is in fact philosophically sound, and the basis of all Bayesian estimation methods. However, it will be more fruitful to regard expression (15) as simply a probability of signal *intensities*.

According to statistical mechanics the total probability P of a number-count object $\{n_m\}$ is the simple product of Eq. (16) with the degeneracy law $W(n_1, \ldots, n_M)$ of Eq. (12). Thus

$$P(n_1, \ldots, n_M) = \prod_{m=1}^{M} \frac{(n_m + z_m - 1)!}{n_m!(z_m - 1)!}$$

$$\times \int_{\Sigma q_m = 1} \cdots \int q_1^{n_1} \cdots q_M^{n_M} p(q_1, \ldots, q_M) \, dq_1 \cdots dq_M$$

$$= \text{maximum} \tag{17}$$

is our MP principle for the unknown spectrum $\{n_m\}$. The maximum is sought through choice of $\{n_m\}$. This defines the MP answer in the absence of image data, but with knowledge of object class (15).*

V. Most Probable Object in the Presence of Data

The data $\{i_m\}$ are the measured output spectrum at the exit slit. This is simply the convolution of the number-count object at the entrance slit with the point-spread function $s(x)$ of the intervening optics and follow-on electronics,

$$i_m = \sum_{n=1}^{M} o_n s_{mn} = h \sum_n n_n \nu_n s_{mn}, \quad m = 1, \ldots, M,$$

$$s_{mn} \equiv s(x_m - x_n). \tag{18}$$

The presence of noise in $\{i_m\}$ has been temporarily ignored. Note that the convolution has a discrete form because of the discrete size Δx of the resolution cells in object and image space. Also, it is assumed for simplicity that the form of $s(x)$ does not vary with abscissa x_m.

Hence, the object $\{o_m\}$ or $\{n_m\}$ that satisfies the MP principle (17) must also obey the M constraint equations (18). In addition, there is the normalization constraint (3). These may be tagged onto Eq. (17) by the conventional use of Lagrange multipliers. In addition, it turns out to be

---

*Bayesian estimation in general permits estimation in the absence of data. However, we are not advocating the abandonment of spectrometers; see Section V.

more convenient to maximize $\ln P$ instead of P directly. Then the MP principle becomes the following. Vary the $\{n_m\}$ to attain

$$\ln P(n_1, \ldots, n_M) + \mu\left(\sum n_m - N\right) + \sum_{m=1}^{M} \lambda_m\left(h \sum_n n_n \nu_n s_{mn} - i_m\right)$$

$$= \text{maximum}, \tag{19}$$

with P given by Eq. (17). This states that the estimated object is the most probable member of object class $p(q_1, \ldots, q_M)$ that also obeys the image data. *The estimate now combines the known physics of photons, prior knowledge $\{z_m\}$ of degrees of freedom, prior knowledge $p(q_1, \ldots, q_M)$ of object class, and posterior knowledge of image data.* The remaining task is to establish probability law $p(q_1, \ldots, q_M)$.

VI. Object-Class Law

Probability law (15) describes the user's judgment as to what signal object is present. More specifically, it displays the user's judgment as to the probable values of the prior signal $\{q_m\}$ *and* the user's assessment of the strength of this judgment. This is clarified as follows.

Suppose the user thinks that the prior $\{q_m\}$ ought to have values near a definite set of number $\{Q_m\}$, called biases. Suppose also that the user wants to express this suspicion with the *highest possible conviction*. Then the user asserts that there is *no spread* possible about the $\{Q_m\}$, or

$$p(q_1, \ldots, q_M) = \delta(q_1 - Q_1) \cdots \delta(q_M - Q_M). \tag{20}$$

The user's opinion is exceedingly strong: no other signal object can be present. Note that there is still an estimation problem, however. Even with one spectrum $\{Q_m\}$ present *any number-count object* $\{n_m\}$ may result. If, for example, the signal object $\{Q_m\}$ is a true rect function, the *number-count* object $\{n_m\}$ may be a rectangle with one or more grooves and bumps on it. This is the object that emerges from the entrance slit and is later imaged; hence it is the one that is sought.

As to which object $\{n_m\}$ is present, this is where the image data in principle (19) enter in. Their effect is to force the estimate toward the true $\{n_m\}$ present. The effect of the $\{Q_m\}$ is to exert merely a beneficial[†] biasing effect on the estimate, as shown later.

[†]Conversely, if the user has the highest conviction *but is wrong*, the user's $\{Q_m\}$ will exert a deleterious effect on the estimate (see Section XI.B).

An analogy is the corresponding problem of two-dimensional object restoration. Suppose that the object really *is* a gray transparency with a dark letter A imprinted on it. The $\{Q_{mn}\}$ now describe the ideal signal A and its background. But because the number N of photons passing through the transparency is finite, the actual number-count object $\{n_m\}$ will randomly depart from the ideal A.

Alternatively, the user may have *less conviction* than is expressed by Eq. (20) about the unknown $\{q_m\}$. If so, the user should permit q_1 (for example) to have a finite range of values, perhaps with Q_1 as now the most likely value (formerly it was the *only* value permitted). An example is

$$p(q_1) = (1/\sqrt{2\pi})\sigma^{-1} \exp\left[-(q_1 - Q_1)^2/2\sigma^2\right],$$

the normal law. (The same goes for the other $\{q_m\}$.) The size of σ defines the user's conviction about Q_1 as the dominant value for q_1. The smaller it is made, the more conviction is expressed. In the limit $\sigma \to 0$, the state (20) of utmost conviction is achieved once again. Of course, forms for probability law $p(q_1)$ other than the normal can also express these tendencies. These are discussed in the next two sections.

VII.   Case of a "White" Object; Maximum Entropy

Suppose that prior knowledge is of the maximum-conviction form (20), where in addition all

$$Q_m = 1/M.$$

This corresponds to prior knowledge of an equal-energy white spectrum. Combining Eqs. (16) and (20) then yields

$$P_0(n_1, \ldots, n_M) = (1/M)^{\Sigma n_m}.$$

But Σn_m is the constant N, by Eq. (3). Hence, P_0 is now a constant, independent of the $\{n_m\}$. This has an important ramification.

Principle (17) now consists of just the Bose–Einstein degeneracy factor, exactly Kikuchi and Soffer's form (1977). Also, as these authors showed (see also Section IX.B), in the case (11a) of sparsely occupied df it becomes Jaynes's maximum-entropy form [see Eq. (39) in Section IX.B and Jaynes (1968)]. Hence, both the Kikuchi–Soffer and Jaynes estimators are special cases of the MP approach, corresponding to the prior knowledge that the unknown spectrum is equal-energy white with the highest conviction.

This is a satisfying result. It has long been known that maximum entropy exerts a smoothing tendency on its estimate (Frieden, 1972). Now we see where this originates: in the high-conviction assertion of an equal-energy white object. If the underlying object is "known" to be flat, the most probable estimate will also tend to be flat.

It has also been observed in past empirical work with maximum entropy that the outputs are quite accurate estimates of the true objects, often exhibiting resolution beyond (it would seem) the realm of possibility; see Frieden (1972) or Wernecke and D'Addario (1977). Now we can see why this should be true. These high-resolution estimates occurred only when the object consisted of isolated impulses against basically a flat background. That is, the object was essentially "equal-energy white." In this case, as just discussed, the MP estimate *is* Kikuchi–Soffer or maximum entropy. These high-resolution outputs were therefore maximum probable as well!

The reader should be cautioned at this point, however, that not all MP estimates (19) will turn out to obey maximum entropy (see Section IX).

VIII.   Scenarios for Knowing $p(q_1, \ldots, q_M)$

We shall next investigate a limited but interesting number of cases for which specific object-class laws $p(q_1, \ldots, q_M)$ are formed. Remarkably, in some cases the estimation principle (19) may itself be used to form p. In these cases we formally set all $\lambda_m = 0$, because object class is defined independent of knowing the data.

An interesting case that we shall *not* consider is where the user knows *power spectra* for the object and the noise and from these wants to infer $p(q_1, \ldots, q_M)$.* Indeed, the problem has not been solved yet, to our knowledge.

The problem of forming $p(q_1, \ldots, q_M)$ is central to our whole approach. It is also the most difficult step. Once known and substituted into Eq. (19), a solution $\{\hat{n}_m\}$ could always be at least *numerically* found—but not so with forming $p(q_1, \ldots, q_M)$. It often takes some ingenuity to frame the given prior knowledge in the form of a probability law. Some of the easier

---

*Problem posed by P. A. Jansson.

cases are taken up next. Luckily, the reader will also find them applicable in many cases.

A. KNOWLEDGE OF EMPIRICAL DATA

One scenario on which to build a $p(q_1, \ldots, q_M)$ is *empirical* evidence, that is, *actual observation of a number-count object* $\{n_m\}$ prior to estimating the object. The proviso is that object $\{n_m\}$ belongs to the "same class" as the unknown object. We shall assume this. There are two different assertions that the user may make, given the information $\{n_m\}$. These are that the data represent a MP estimate of the object or that the data only imply a fixed probability law (to be found) on possible objects. We consider these alternative assertions in Sections VIII.A.1 and VIII.A.2.

1. Maximum Conviction That Empirical Data Represent the Object

First consider the assertion that, with strongest conviction, the $\{n_m\}$ represent the MP estimate of the unknown object in the absence of image data. We found that a state of strongest conviction is represented by form (20) with biases $\{Q_m\}$. Then what are the $\{Q_m\}$ in terms of observables $\{n_m\}$?

The MP principle (19) gives the MP solution $\{n_m\}$ either in the presence or in the absence of image data. In the latter case, the $\{\lambda_m\}$ are merely all set equal to zero. Although Eq. (19) is usually used to predict values $\{n_m\}$ for a known $p(q_1, \ldots, q_M)$, we shall instead use it in reverse, finding probabilities $\{Q_m\}$ from observed $\{n_m\}$. This use follows the assumption that the *observed* $\{n_m\}$ were *maximum probable* as well.

The principle (19) with $P(n_1, \ldots, n_M)$ given by Eq. (17) and the use of maximum conviction (20) results in a statement

$$\ln \prod_{m=1}^{M} \frac{(n_m + z_m - 1)!}{n_m!(z_m - 1)!} Q_m^{n_m} + \mu\left(\sum n_m - N\right) = \text{maximum}, \quad (21)$$

through choice of the $\{n_m\}$. The $\{\lambda_m\}$ were set equal to zero because the situation is prior to observing the image.

The solution may be found by the use of Stirling's approximation to the factorial,

$$\ln n! \approx n \ln n. \quad (22)$$

Expanding the logarithm of the product and using identity (22) yields

$$\sum (n_m + z_m - 1)\ln(n_m + z_m - 1) - \sum n_m \ln n_m$$
$$+ \sum n_m \ln Q_m + \mu\left(\sum n_m - N\right) = \text{maximum}. \tag{23}$$

This estimator, without the first sum, was first derived by Hershel (1971). We have ignored a term in the $\{z_m\}$ because these do not affect the maximization. The solution is obtained merely be setting the derivative $\partial/\partial n_m$ of Eq. (23) equal to zero,

$$Q_m = Cn_m/(n_m + z_m - 1). \tag{24}$$

Parameter C is a normalization constant incorporating μ. [That Eq. (24) attains a maximum and not a minimum for Eq. (23) may be verified by taking $\partial^2/\partial n_m \, \partial n_k$ of Eq. (23) and noticing that this gives a diagonal matrix of numbers, all of which are negative.]

As mentioned before, normally Eq. (24) would be used to express n_m in terms of a known Q_m. Here, we use it the other way around because it is assumed instead that n_m is known (observed in the prior object) and Q_m is to be estimated. Equation (24) thereby provides a means of knowing the prior spectrum $\{Q_m\}$. We see that each Q_m depends not only on the observed prior object value n_m but also on the number z_m of degrees of freedom. In particular, each Q_m is not simply linear in the corresponding prior object value n_m, which intuition might otherwise suggest. The essential reason is that n_m represents the number of photons in all possible arrangements over the z_m df, whereas Q_m represents simply the probability of occupation of cell m *prior to* the df (image) phenomenon. Note, in particular, that $Q_m \propto n_m$ when the df are sparsely occupied ($z_m \gg n_m$), as intuition suggests.

In summary, with the $\{n_m\}$ known as empirical observables, Eq. (24) permits the $\{Q_m\}$ to be known. Finally, by maximum conviction, $p(q_1, \ldots, q_M)$ is of the form (20).

2. Impartial Use of Empirical Data

Having considered maximum conviction, we now consider the opposite situation, called fair or impartial use of the data. Here, our conviction will only be as strong as the data permit. The motivation is as follows.

If a coin is flipped 100 times and 80 of the outcomes are heads, we cannot say that the probability q_1 of a head is 0.80 for sure. For example, even if $q_1 = 0.001$, there would still be a finite chance of obtaining the 80 heads. Hence, all q_1 between 0 and 1 are possible, and the evidence of $n_1 = 80$ heads *implies no more than a probability law* $p(q_1)$ describing the chance that any one value of q_1 is the true one. This law will have its peak at $q_1 = 0.80$, of course, so that value $q_1 = 0.80$ is maximum probable. However, it is not the only possibility.

So it is with knowledge of an empirical prior object $\{n_m\}$. If $N = 100$ photons come from the object, and $n_1 = 80$ are from cell 1, this does not imply that for sure $q_1 = 80/100$. All values of q_1 between 0 and 1 are still possible, according to a probability law. Hence, considering all the cells simultaneously, we seek the probability law

$$p(q_1,\ldots,q_M|n_1,\ldots,n_M), \tag{25}$$

that is, $p(q_1,\ldots,q_M)$ contingent upon the evidence $\{n_m\}$. As a matter of nomenclature, because we are allowing *all* q values to exist, we call this fair or impartial conviction about the $\{q_m\}$.

This problem may be solved using elementary probability theory. Bayes' rule states that the answer may be formed as

$$p(q_1,\ldots,q_M|n_1,\ldots,n_M) = \frac{P(n_1,\ldots,n_M|q_1,\ldots,q_M)p_0(q_1,\ldots,q_M)}{P(n_1,\ldots,n_M)} \tag{26}$$

(see, e.g., Frieden, 1983). The right-hand quantities have to be found.

Probability $P(n_1,\ldots,n_M|q_1,\ldots,q_M)$ describes Eq. (17) when one particular set $\{q_m\}$ is present,

$$P(n_1,\ldots,n_M|q_1,\ldots,q_M) = \prod_{m=1}^{M} \frac{(n_m + z_m - 1)!}{n_m!(z_m - 1)!} q_m^{n_m}. \tag{27}$$

Probability $p_0(q_1,\ldots,q_M)$ represents what the user supposes the probability of the $\{q_m\}$ would be prior to observation of the empirical $\{n_m\}$ (hence it is a "preprior" probability law). Prior to these observations, it is reasonable to assign all possible values of q_m the *same* probability of occurrence. This is the MacQueen–Marschak (1975) definition of maximum prior ignorance,

$$p_0(q_1, \ldots, q_M) = \prod_{m=1}^{M} \text{rect}(q_m - \tfrac{1}{2}),$$

$$\text{rect}(x) \equiv \begin{cases} 1 & \text{for} \quad |x| < \tfrac{1}{2} \\ \tfrac{1}{2} & \text{for} \quad |x| = \tfrac{1}{2} \\ 0 & \text{for} \quad |x| > \tfrac{1}{2} \end{cases}. \tag{28}$$

This says that the prior spectrum is *anything* with equal probability. Note that this assumption is consistent with the overall stance of fair conviction in this section.

The denominator $P(n_1, \ldots, n_M)$ in Eq. (26) is the integral of the numerator over all $\{q_m\}$. By Eqs. (27) and (28) the integral is Dirichlet's (see Gradshteyn and Ryzhik, 1965), with the result

$$P(n_1, \ldots, n_M) = \frac{(M-1)!}{(L+M-1)!} \prod_{m=1}^{M} \frac{(n_m + z_m - 1)!}{(z_m - 1)!}, \quad L \equiv \sum n_m. \tag{29}$$

Note the distinction between L and N: the empirical object prior to data $\{i_m\}$ has L photons, whereas the unknown object that forms $\{i_m\}$ has N photons.

With all the right-hand members of Eq. (26) now known, it becomes

$$p(q_1, \ldots, q_M) = \frac{(L+M-1)!}{(M-1)!} \frac{q_1^{n_1} \cdots q_M^{n_M}}{n_1! \cdots n_M!}. \tag{30}$$

(The contingency upon $\{n_m\}$ is, for brevity, suppressed in the left-hand argument.) This is the prior-knowledge probability law that we sought. As a check, notice that in the limit of no prior data $\{n_m\} = 0$, $L = 0$, it becomes a constant. This is consistent with the maximum-prior-ignorance condition (28) previously adopted, which supposed no prior data present.

B. MAXIMUM CONVICTION WITHOUT EMPIRICAL DATA

Perhaps the commonest circumstance is where *no empirical data* $\{n_m\}$ are at hand, but the user has an *opinion* as to what the object is. A way of phrasing this opinion is, "Before seeing the image data, I highly suspect the object to be of such and such a shape $\{n_m\}$." If, in fact, the user thinks it *most probable* that the object has this shape, then the derivation of the preceding section again applies. [Note that the $\{n_m\}$ in principle (21) are mere assertions. They may be real data or, as in this case, imagined or anticipated data.] Hence the result (24) may again be used to find the prior spectrum $\{Q_m\}$.

A word of assurance is perhaps necessary here. The reader might suppose that because the particular values $\{Q_m\}$ are input into the principle (19), the answer $\{\hat{n}_m\}$ that results will necessarily be dominated by the $\{Q_m\}$. This is not true. The image data in Eq. (19) exert their own, independent influence in biasing the answer toward peaks and valleys that are consistent with *it*. Thus, the $\{Q_m\}$ exert a biasing tendency but *only* a tendency. The main influence on the solution $\{\hat{n}_m\}$ is exerted by the image data.

The reader may surmise that perhaps the image data may be used as the prior spectrum $\{Q_m\}$. Of course the image is a blurred version of the object, but nevertheless it does bear a resemblance to it. In this case the estimated object $\{\hat{n}_m\}$ will be biased toward the image values, which actually is a helpful tendency because the image is relatively smooth. Empirically, this helps to keep down noise and artifact oscillations in the $\{\hat{n}_m\}$, as we find when testing out this idea in Section XI.

C. ZERO PRIOR KNOWLEDGE; MAXIMUM IGNORANCE

Probably the most common state of prior knowledge is a state of zero prior knowledge (sometimes also called maximum ignorance). In this state the $\{z_m\}$ are known, defining the coherence of the source and the geometry of the spectrometer (as previously found), but "nothing else" is known about the object. In this "nothing else" lies a mystery. What does it mean? Jaynes (1968) and Kikuchi and Soffer (1977) implicitly assumed it to mean an equal-energy white object

$$p(q_1,\ldots,q_M) = \delta(q_1 - 1/M) \cdots \delta(q_M - 1/M), \qquad (31)$$

as previously discussed. We shall see that this is consistent with the assertion that

$$n_m \propto z_m - 1 \qquad (32)$$

is the MP object in the absence of image data [see Eq. (38)]. This looks reasonable because this prior object is then simply proportional to the number of degrees of freedom.

However, consider the opposing argument. Because the $\{z_m\}$ are known, by Eq. (32) a *unique object* is MP in the absence of data. Can this represent a state of *maximum ignorance*?

In particular, suppose that all the $\{z_m\}$ are equal. Then by Eq. (32) the prior object is imagined to be uniform. Why should such a choice of object be preferred over any other shape? Would it not make more sense to allow for many possible MP objects to be present *a priori*, each with the *same* probability? (MP solutions do not have to be unique, as discussed later.)

Perhaps the greatest fault with the choice (31) for representing zero prior knowledge is its assertion, *with maximum conviction*, that the prior object is equal-energy white. *This is a definite statement about shape.* How can this represent a state of *zero* prior knowledge?

For these reasons, following MacQueen and Marschak (1975), we use as an alternative definition of maximum ignorance

$$p(q_1, \ldots, q_M) = \prod_{m=1}^{M} \text{rect}(q_m - \tfrac{1}{2}). \tag{33}$$

Now *any* distribution $\{q_m\}$ of prior objects can be present (so long as they obey normalization), and *each distribution* has the same probability of occurring. Further, the use of this definition in principle (19) (with all $\lambda_m = 0$ because there are no image data) yields an MP prior object obeying

$$\sum (n_m + z_m - 1)\ln(n_m + z_m - 1) + \lambda\left(\sum n_m - N\right) = \text{maximum} \tag{34}$$

[see Eq. (47b).] This equation has an interesting property. Suppose, in particular, that all the $\{z_m\}$ are equal. Then Eq. (34) is a *global criterion*; that is, it is blind to interchanges of n_i with n_j, for any i, j. For example, if the solution to Eq. (34) were $\{n_m\} = (4, 2, \ldots)$, then the different object $(2, 4, \ldots)$ would also be a solution. They only way such interchanges would not result in a new object is if all the $\{n_m\}$ were equal. But such an object is actually the solution to

$$\sum n_m \ln n_m + \lambda\left(\sum n_m - N\right) = \text{minimum} \tag{35}$$

and so cannot also satisfy Eq. (34) (compare right-hand sides).

The result is that the MP prior object that is consistent with definition (33) is actually an *ensemble* of objects, each occurring with the *same* probability. *This* is more in the spirit of maximum ignorance than was the preceding definition (31).

But if this is true, how could its use in the MP principle (19) (including data) result in a *unique* object estimate $\{n_m\}$? The answer is that the image data will constrain the estimate toward the one object that is consistent with *its* shape. This will rule out the other members of the object ensemble.

In summary, it makes more sense for "maximum ignorance" about a spectrum to mean an infinity of equally probable possibilities (33) than to imply the unique, equal-energy white spectrum (31).

IX. Estimators

The aim of this chapter has been to show how MP estimators may be formed for spectroscope outputs. Now we are in a position to do this. Having established a general MP principle (19) and specific forms of prior knowledge (20), (24), (30), (31), and (33), we now can combine results. What does estimator (19) become under different forms of prior knowledge? And what is its solution, in terms of data $\{i_m\}$?

A. HIGH-CONVICTION CASES

Here the prior probability law is of the general form (20). To review, this includes the case where the $\{Q_m\}$ are simply guessed at, based on the user's expectations, or where the image data $\{i_m\}$ are used to represent $\{Q_m\}$, or the case (24) of empirical data and high conviction, or the case (31) of an equal-energy white signal spectrum.

Substitution of Eq. (20) into principle (19) yields

$$\ln \prod_{m=1}^{M} \frac{(n_m + z_m - 1)!}{n_m!(z_m - 1)!} Q_m^{n_m} + \mu\left(\sum n_m - N\right)$$
$$+ \sum \lambda_m \left(h \sum_n n_n \nu_n s_{mn} - i_m\right) = \text{maximum.} \tag{36}$$

Expanding the logarithm of the product, using Stirling's approximation (22) and ignoring terms that are constant in n_m, yields

$$\sum (n_m + z_m - 1)\ln(n_m + z_m - 1) + \sum n_m \ln(Q_m/n_m)$$
$$+ \mu \sum n_m + \sum \lambda_m h \sum_n n_n \nu_n s_{mn} = \text{maximum.} \tag{37}$$

This is the restoring principle for this case.

Its solution is obtained by setting $\partial/\partial n_m$ of Eq. (37) equal to zero, resulting in

$$\hat{n}_m = \frac{(z_m - 1)Q_m}{\exp(-\mu - h\nu_m \Sigma_n \lambda_n s_{nm}) - Q_m}. \tag{38}$$

This resembles the Bose factor $[\exp(h\nu/kT) - 1]^{-1}$ of quantum optics. Of course, the resemblance is no coincidence, both deriving from similar physics.

In the case (24) of empirical data, substitution into Eq. (38) gives $\hat{n}_m$ in terms of the observed *prior* object $\{n_m\}$.

The unknowns in Eq. (38) are μ and $\{\lambda_n\}$. These are found by demanding $\hat{n}_m$ to obey the image constraint equations (18) and normalization (3). Because the unknowns enter in a nonlinear way, the resulting $M + 1$ equations were solved by an iterative technique—Newton–Raphson relaxation (see, e.g., Hildebrand, 1956). Empirical cases are studied in Section XI.

B. ENTROPY-LIKE ESTIMATORS

The entropy H of an object $\{n_m\}$ is defined as

$$H = -\sum_{m=1}^{M} n_m \ln n_m. \tag{39}$$

A principle of estimation $H = $ maximum has long been advocated by Jaynes (1968). We note that the estimator (37) is *not* of this simple form, although one sum (the second) is close to it. The conclusion is that Jaynes's principle is not generally MP.

In this vein, however, consider limiting cases (11a) of sparsely occupied df sites. With the approximation

$$\ln(n_m + z_m - 1) = \ln\left[(z_m - 1)\left(1 + \frac{n_m}{z_m - 1}\right)\right] \approx \ln(z_m - 1) + \frac{n_m}{z_m - 1},$$

it follows that

$$(n_m + z_m - 1)\ln(n_m + z_m - 1) \approx (n_m + z_m - 1)\ln(z_m - 1) + n_m \tag{40}$$

after ignoring the small term proportional to $n_m/(z_m - 1)$.

Substitution of Eq. (40) into principle (37) yields an estimator

$$\sum n_m \ln(Q_m/n_m) + \text{(constraint terms)} = \text{maximum}, \tag{41}$$

the constraint terms all linear in the $\{n_m\}$. We call this a principle of maximum cross entropy.* In the limit of all $Q_m = 1/M$ it becomes simply maximum entropy.

We can summarize this situation by the statement that the MP object obeys Jaynes's maximum-entropy principle when the "white" object definition (31) of maximum ignorance is used and when the object is of such low intensity that the df sites are mostly unoccupied. The latter situation is obeyed by weak astronomical objects such as planets in the visible and IR regions and the sun in the visible region (see Kikuchi and Soffer, 1977).

The "Burg entropy" is of the form

$$H_1 \equiv + \sum_{m=1}^{N} \ln n_m.$$

A principle H_1 = maximum was first proposed by J. P. Burg (1967) in the estimation of power spectra from autocorrelation data. There is one condition under which estimator (37) goes into the Burg form, described next.

Consider limiting cases (11b) of highly occupied df sites. With the approximation

$$\ln(n_m + z_m - 1) = \ln\left[n_m\left(1 + \frac{z_m + 1}{n_m}\right)\right] \approx \ln n_m + \frac{z_m - 1}{n_m},$$

it follows that

$$(n_m + z_m - 1)\ln(n_m + z_m - 1) \approx (n_m + z_m - 1)\ln n_m + z_m - 1 \quad (42)$$

after ignoring the small terms proportional to $(z_m - 1)/n_m$.

Substitution of Eq. (42) into principle (37) yields an estimator

$$\sum(z_m - 1)\ln n_m + \sum n_m \ln Q_m + \text{(constraint terms)} = \text{maximum}, \quad (43)$$

the constraint terms all linear in the $\{n_m\}$. Terms in purely $z_m - 1$ have been dropped because these do not affect the maximization.

In the limit of all $Q_m = 1/M$, because $\sum n_m$ obeys normalization, the net principle is maximization of the first term in Eq. (43). This is a weighted form of Burg's entropy, with weights $z_m - 1$.

We can summarize this section as follows. The MP object obeys maximum entropy of type H (Jaynes) or type H_1 (Burg) when the "white"

*As a matter of nomenclature, Shore (1981) calls it *minimum* cross-entropy.

object definition (31) of maximum ignorance is used and when the df sites are sparsely or highly populated, respectively.

C. *ESTIMATORS FOR EMPIRICAL DATA; IMPARTIAL CONVICTION*

In a previous section we found the object class (30) that is implied by observation of an empirical object $\{m_m\}$,

$$p(q_1, \ldots, q_M) = \frac{(L + M - 1)!}{(M - 1)!} \frac{q_1^{m_1} \cdots q_m^{m_M}}{m_1! \cdots m_M!}, \quad L = \sum m_n. \quad (44)$$

We now want the estimator implied by this form of prior knowledge. This is obtained by substituting Eq. (44) into principle (19). After expanding the logarithm of the product, evaluating the integral by the use of Dirichlet's equality, and using Stirling's approximation (22), we obtain

$$\sum (n_m + z_m - 1)\ln(n_m + z_m - 1)$$
$$+ \sum (n_m + m_m)\ln(n_m + m_m) - \sum n_m \ln n_m$$
$$+ \mu\left(\sum n_m - N\right) + \sum \lambda_m\left(h \sum n_n v_n s_{mn} - i_m\right) = \text{maximum}. \quad (45a)$$

Quantities constant in the $\{n_m\}$ have been ignored. This is the general estimator for this state of prior knowledge.

We note that the first sum in Eq. (45a) has the form of the negative of an entropy [see Eq. (39)]. This entropy will be called photon-site entropy because it involves the df sites $\{z_m\}$. Likewise the second sum in Eq. (45a) is the negative of "empirical" entropy, in that it involves the empirical data $\{m_m\}$. Hence, overall, estimator (45a) is one of *minimum* photon-site entropy plus minimum empirical entropy plus maximum Jaynes entropy (third sum).

It is interesting that the solution to Eq. (45a) may sometimes be obtained by differentiation and sometimes not. Taking the second derivative of Eq. (45a) shows that the first derivative yields a maximum if and only if

$$n_m/(z_m - 1) < m_m/n_m. \quad (45b)$$

This is an interesting requirement relating the relative sparsity of df sites (left side) to the "strength" of the prior data $\{m_m\}$ (right side). When differentiation does work, it leads to a quadratic equation for the $\{\hat{n}_m\}$,

$$\hat{n}_m^2 + \hat{n}_m(m_m + z_m - 1 - \alpha_m) + m_m(z_m - 1) = 0. \quad (46a)$$

Parameter α_m lumps together all the unknowns μ and $\{\lambda_m\}$, through

$$\alpha_m \equiv \exp\left(-1 - \mu - h\nu_m \sum \lambda_n s_{mn}\right). \qquad (46b)$$

Of course, the roots of Eq. (46a) may easily be found. The unknown quantities μ and $\{\lambda_m\}$ may be found by substituting $\{\hat{n}_m\}$ from Eq. (46a) into the imaging equations (18) and normalization equation (3). Because of the nonlinear nature of these equations, a Newton–Raphson (or other iterative) method of solution would be necessary.

Limiting forms of principle (45a) are easily established, as before. If, for example, the exposure time is so short that there is no photon degeneracy, i.e., $z_m = 1$ for all m, it becomes

$$\sum (n_n + m_m)\ln(n_m + m_m) + \text{(constraint terms)} = \text{maximum.} \quad (47a)$$

This is a principle of *minimum* empirical entropy. The solution $\{\hat{n}_m\}$ to Eq. (47a) cannot be obtained by setting its derivative equal to zero. That would instead *minimize* Eq. (47a). In fact, the solution $\{\hat{n}_m\}$ tends to be extreme and bumpy, the opposite tendency of maximizing entropy (39) (as might be expected).

In the limiting case of no prior data, all $m_m = 0$. This describes zero prior knowledge by MacQueen and Marschak's criterion. Then Eq. (45a) becomes a principle of minimum photon-site entropy,

$$\sum (n_m + z_m - 1)\ln(n_m + z_m - 1) + \text{(constraint terms)} = \text{maximum.}$$
$$(47b)$$

Tendencies of this solution were discussed following Eq. (34) (see also Section XI.B).

If condition (11a) of sparsely populated df holds true, as shown in Eq. (40) the first sum in Eq. (45a) becomes absorbed into the linear constraints so that the estimator is

$$\sum (n_m + m_n)\ln(n_m + m_m) - \sum n_m \ln n_m + \text{(constraint terms)}$$
$$= \text{maximum.} \qquad (48)$$

The solution may again be found by differentiation, because the second derivative of Eq. (48) is always negative, as is easily verified.

If there is high df occupancy (11b), estimator (45a) becomes

$$\sum (n_m + m_m)\ln(n_m + m_m) + \sum (z_m - 1)\ln n_m + \text{(constraint terms)}$$
$$= \text{maximum.} \qquad (49a)$$

This is a principle of minimum empirical entropy (first sum) plus maximum weighted-Burg entropy (second sum). Once again, differentiation does not always yield the solution. Taking the second derivative of Eq. (49a) shows that the first derivative attains a maximum if and only if

$$n_m/(z_m - 1) < m_m/n_m + 1. \tag{49b}$$

X. How to Handle Noise

The preceding MP estimators have assumed no noise to be present in the image data $\{i_m\}$. However, the incorporation of noise presents little problem. The estimators will simply be revised by having something added to them in all cases. They therefore will keep the forms previously derived. This is shown next.

Let the image data suffer from noise $\{\varepsilon_m\}$. Then the imaging equations are [compare Eq. (18)]

$$i_m = h \sum n_n \nu_n s_{mn} + \varepsilon_m, \tag{50}$$

where ε_m is the noise.

Now we have two sets of unknowns, object $\{n_m\}$ and noise $\{\varepsilon_m\}$. The proper MP principle to use is therefore that *the* $\{n_m\}$ *and* $\{\varepsilon_m\}$ *that occurred are both (jointly) most probable,*

$$\mathscr{P}(n_1, \ldots, n_M, \varepsilon_1, \ldots, \varepsilon_M) \equiv \mathscr{P}(\{n_m\}, \{\varepsilon_m\}) = \text{maximum}. \tag{51}$$

The definition of conditional probability gives

$$\mathscr{P}(\{n_m\}, \{\varepsilon_m\}) = P(\{n_m\})P_1(\{\varepsilon_m\}|\{n_m\}), \tag{52}$$

where P is the probability of an object, as considered in previous sections, and P_1 is the probability of the noise conditional upon that object (thereby allowing for signal-dependent noise). Now $\mathscr{P}$ is a maximum if and only if $\ln \mathscr{P}$ is. Then Eqs. (51) and (52) yield an MP principle

$$\ln P(\{n_m\}) + \ln P_1(\{\varepsilon_m\}|\{n_m\}) = \text{maximum} \tag{53}$$

through choice of $\{n_m\}$ and $\{\varepsilon_m\}$.

Of these two contributors to the maximum, the first is already given by Eq. (17). It takes on the particular forms previously given under different states of object class $p(q_1, \ldots, q_M)$. Hence, this part of the MP principle is already known.

The second contributor to Eq. (53) has to do with the type of noise assumed to be present. To keep things simple, we shall assume it to be additive and independent gaussian but with generally position-dependent variance $\{e_m^2\}$. This describes, for example, Johnson noise and other noise types that enter into the output of the spectrometer. Thus, we have as the probability P_N of noise

$$P_N(\varepsilon_1, \ldots, \varepsilon_M) = \left[\prod_{m=1}^{M} (\sqrt{2\pi} e_m)^{-1} \right] \exp\left(- \sum_{m=1}^{M} \frac{\varepsilon_m^2}{2 e_m^2} \right). \tag{54}$$

The tie-in between this P_N and P_1 is found next.

Because the noise is given to be additive, it is independent of the underlying object $\{n_m\}$:

$$P_1(\{\varepsilon_m\}|\{n_m\}) = P_N(\{\varepsilon_m\}). \tag{55}$$

Accordingly, the overall MP principle (53) becomes

$$\ln P(\{n_m\}) - \sum_{m=1}^{M} \frac{\varepsilon_m^2}{2 e_m^2} = \text{maximum}, \tag{56}$$

after using Eq. (54) and ignoring a sum in $\{e_m\}$ that does not affect the solution. As in preceding sections, the image data are added on as Lagrange constraints

$$\sum \lambda_m \left(i_m - h \sum n_n \nu_n s_{mn} - \varepsilon_m \right) + \mu \left(N - \sum n_m \right) \tag{57}$$

to the left side of Eq. (56).

Equations (56) and (57) constitute the overall MP principle that we sought. This principle embodies all the physical and prior knowledge that we have about the object and noise. The specific $\ln P(\{n_m\})$ to be used will depend on the particular prior knowledge $p(q_1, \ldots, q_M)$ at hand. Specific $\ln P(\{n_m\})$ were derived in preceding sections. We shall carry through some particular cases next.

XI. Test Cases

We will find estimates $\{\hat{n}_m\}$ in numerical cases consisting of experimental data (Section XI.A) and simulated data (Section XI.B). In all cases, high-conviction form (20) will be assumed because the solutions $\{\hat{n}_m\}$ are relatively easy to form and because of the ease with which prior knowledge

$\{Q_m\}$ may be varied. In Section XI.A low photon occupancy n_m/z_m will be assumed, whereas in Section XI.B both low and high occupancies will be allowed. Note that low occupancy is consistent with maximum entropy according to Section IX.B, that is, particle-like behavior for the photons, whereas high occupancy is consistent with wavelike behavior for the photons. How, then, does the wave or particle nature of photons affect the estimates?

A. APPLICATION TO EXPERIMENTAL DATA

Under the prior-knowledge conditions mentioned earlier, the principle given by Eqs. (56) and (57) is augmented by Eq. (41) to become

$$\sum_m n_m \ln(Q_m/n_m) - \sum_m \frac{\varepsilon_m^2}{2e_m^2} + \sum_m \lambda_m \left(i_m - h \sum_n n_n \nu_n s_{nm} - \varepsilon_m \right)$$

$$+ \mu \left(\sum_m n_m - N \right) = \text{maximum.} \qquad (58)$$

By the form of the first sum, the object is being estimated by a principle of maximum cross entropy. By the second sum, the noise is being estimated by a principle of least squares (notice the minus sign). The df have dropped out because of the sparsity constraint (11a). This is convenient, because it avoids the need for their detailed calculation.

The solution $\{\hat{n}_m\}$, $\{\hat{\varepsilon}_m\}$ to Eq. (58) is obtained merely by setting $\partial/\partial n_m$ (all $\{\varepsilon_m\}$ fixed) of Eq. (58) equal to zero and by setting $\partial/\partial \varepsilon_m$ (all n_m fixed) equal to zero. These result, respectively, in solutions

$$\hat{n}_m = Q_m \exp\left(-1 - \mu - h\nu_m \sum_n \lambda_n s_{nm} \right) \qquad (59a)$$

and

$$\hat{\varepsilon}_m = -\lambda_m e_m^2. \qquad (59b)$$

Note from the form of Eq. (59a) that $\hat{n}_m$ can only be positive, regardless of parameters μ and $\{\lambda_n\}$. Hence, a positivity constraint is a *natural consequence* of the overall approach. For other cases of prior knowledge, such as the case of emission spectra, *boundedness* would result. By comparison, all other restoration methods that enforce positivity or boundedness do it in an ad hoc manner.

It only remains to find the unknown parameters μ and $\{\lambda_n\}$. These are found by substituting the forms (59a) and (59b) for $\{n_m\}$ and $\{\varepsilon_m\}$ into the constraint equations, resulting in conditions

$$i_m = h \sum_n \nu_n s_{mn} Q_n \exp\left(-1 - \mu - h\nu_n \sum_k \lambda_k s_{kn}\right) - \lambda_m e_m^2,$$

$$N = \sum_n Q_n \exp\left(-1 - \mu - h\nu_n \sum_k \lambda_k s_{kn}\right). \tag{60}$$

This is a system of $M + 1$ equations in the $M + 1$ unknowns μ and $\{\lambda_m\}$. It may be solved by regarding the known left-hand sides as target values in a relaxation program that iteratively homes in on the solution. The Newton–Raphson method works fine for this job. With an initial trial solution of all $\lambda_m = 0$ and $\mu = N$, the final solution is attained in usually about 10 iterations.

To proceed further, we have to make assumptions about the forms for $\{Q_n\}$ and $\{e_m\}$. These describe prior knowledge about the object and the noise. For simplicity, let all $\{Q_m\}$ be equal. This means that we have maximum conviction that the unknown spectrum is a flat one. Such a spectrum may be called "quasi-featureless." This prior knowledge should exert a smoothing influence on the solution. Sure enough, by Eq. (58) the object is now being restored by a principle of maximum entropy H, known to foster smoothness in its outputs (Frieden, 1972).

Let us assume the noise to be strict-sense stationary of order 1. This means that all the $\{e_m\}$ values are equal. Such a situation is very common.

Under these prior conditions, the restoring algorithm given by Eqs. (59) and (60) becomes very close to one previously used by Frieden (1972). That one was also a maximum-entropy restoring algorithm in the presence of additive noise. We present some results of the latter method applied to spectral data.

Using spectral image data provided by Jansson et al. (1970), we restored a portion of the Q branch of the ν_3 band of CH_4. Results are shown in Fig. 4. Note the smoothness of the output, no doubt enforced by the prior conviction of a flat object. But, despite this smoothness, the right-hand line has been split into two components. These same data were also restored by Jansson, using his bound-constrained algorithm (see Chapters 4, 6, and 7 in this text). The restoration by Jansson is very similar to ours, with relative differences in the principal peaks amounting to a few percent.

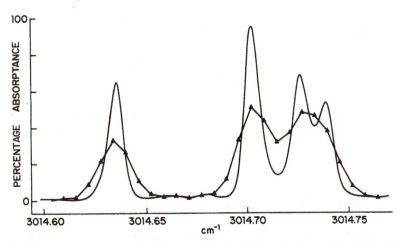

Figure 4 Restoration by maximum entropy (solid curve) of a portion of the Q branch of the ν_3 band of the infrared spectrum of CH_4 near 3014.7 cm^{-1}. Triangles indicate the experimental inputs $\{i_m\}$ used in the restoring scheme, some of which lie outside the plotted region and are not shown.

It is important to realize, however, that this similarity is a coincidence of the case. Jansson's boundedness constraint *always* tends toward smoothness near the bounds, regardless of *a priori* knowledge of object class, whereas our MP estimator [Eqs. (17) and (19)] can be constrained toward bumpy solutions, the bumps consistent with the prior knowledge at hand. In the high-conviction case that we are treating here, the bumps would be manifest in the inputs $\{Q_m\}$. We next study the effect of varying the $\{Q_m\}$ in some computer simulations.

B. EFFECT OF CHOICE OF PRIOR OBJECT $\{Q_m\}$ AND OF z

How do the form $\{Q_m\}$ of a prior object and the number z of degrees of freedom affect the estimate $\{\hat{n}_m\}$? This was the problem (37) augmented by data constraints (50). Suppose in addition that the image data $\{i_m\}$ suffer from Gaussian additive noise, so that Eq. (54) is true. Then the total estimator given by Eqs. (56) and (57) becomes

$$-\sum n_m \ln(n_m/Q_m) + \sum (n_m + z - 1)\ln(n_m + z - 1) - \frac{1}{2e^2}\sum \varepsilon_m^2$$
$$+ \mu\left(\sum n_m - N\right) + \sum \lambda_m\left(\sum n_n s_{nm} + \varepsilon_m - i_m\right) = \text{maximum}. \quad (61)$$

(For simplicity we have taken all $h\nu_m = 1$, $z_m = z$, and $e_m = e$.) This has a solution

$$\hat{n}_m = (z - 1)Q_m / \left[\exp\left(-\mu - \sum \lambda_n s_{nm}\right) - Q_m\right], \qquad (62)$$

$$\hat{\varepsilon}_m = \lambda_m e^2 \qquad (63)$$

after setting the derivatives $\partial/\partial n_m$ and $\partial/\partial \varepsilon_m$ of Eq. (61) equal to zero. The unknown parameters μ and $\{\lambda_m\}$ number $M + 1$ and must satisfy the $M + 1$ constraint equations

$$\sum \hat{n}_n s_{nm} + \hat{\varepsilon}_m = i_m, \quad m = 1, \ldots, M, \qquad (64)$$

and Eq. (3). Solutions to these equations were obtained using the Newton–Raphson method.

The algorithm was applied to computer-simulated image data, represented by the dashed curve in Fig. 5. The corresponding object is represented by the shaded area. The latter was convolved with a sinc-squared kernel so as to form a diffraction-blurred signal image. Diffraction is particularly difficult to overcome in a restoration, because all spatial frequencies beyond that imposed by the finite lens pupil are *missing* from

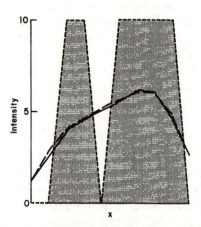

Figure 5 Object (shaded) used in computer simulations. Its diffraction image is the solid curve. The data image (dashed curve) is the diffraction image plus 4% amplitude random noise. All plotted points are spaced by one-half the Nyquist interval. Hence to resolve the central dip in the object would require super-resolution.

the signal image. Recovery of these frequencies is called superresolution. Until the pioneer work of Biraud (1969), Jansson (1968), and Jansson *et al.* (1970) practical superresolution was thought to be impossible.

The image data were formed by adding 4% (of peak image value) Gaussian noise to the signal image. Sampling was at one-half the Nyquist interval so as to permit superresolution of the central dip in the object.

The outputs (62) are shown in Fig. 6, representing four assumed forms for $\{Q_m\}$, each form accompanied by either a low z ($z = 2$) or a high z ($z = 100$). Intermediate z values did not cause significantly different results.

In Fig. 6a, $\{Q_m\}$ was made constant, corresponding to prior knowledge of a flat spectrum. This is a rather conservative guess at the object.* Both restorations exhibit resolution of the central object dip, indicating super-resolution.

The restoration for low z suffers a bit more overshoot at the gradient edges. This is consistent with a "photon-bunching" phenomenon that occurs as the solution to estimator (34). Notice that our estimator (61) includes Eq. (34). If Eq. (34) per se is solved, the answer is that all N photons jam into the resolution cell x_m that has the largest degeneracy z_m over $m = 1, \ldots, M$. The other cells are empty, and the estimated spectrum is a spike. The presence of image data, and of noise, will compromise this extreme behavior. However, there is still a tendency to bunch photons in some cells and deplete them from others, as seen in Fig. 6a–c.

By contrast, the solution for high z ($z = 100$) corresponds to maximum entropy [see the derivation of Eq. (41)], and hence is smoother in the figure.

In summary, then, MP estimates for particle-like photons are smoother than for wavelike photons. This perhaps goes against intuition, but it was true in other test cases as well (Fig. 6b and c).

With nothing else known about object class, one tactic is to make $\{Q_m\}$ proportional to the image data, on the basis that the object ought to look something like the image. Certainly the image should provide more "ground truth" about the object than did the complete greyness used previously. This was tried in Fig. 6b. Indeed, the central dip is resolved better than in Fig. 6a, and the plateau tops are better approximated as well, indicating

---

*But only as far as resolution goes. By contrast, we showed in Section VIII.C that the "flat spectrum" hypothesis is actually a strong assumption about *shape*.

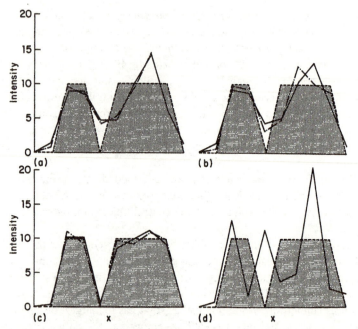

Figure 6 Maximum-probable restorations (62) (solid and dot-dashed curves) of the data image in Fig. 5. Each solid curve is for $z = 100$, each dot-dashed curve for $z = 2$, indicating particle-like and highly Bose photons, respectively. The root-mean-square error (RMSE) of each restoration from the true object (shaded) is as follows. Each part (a–d) results from a different assumed form for biases $\{Q_m\}$, as indicated.

		RMSE	
Part	$\{Q_m\}$	$z = 100$	$z = 2$
(a)	constant	2.9	2.8
(b)	image	2.6	2.2
(c)	true object	0.9	0.6
(d)	true object, displaced	6.5	—

that the tactic has some merit. The root-mean-square errors from the true object are lower than in Fig. 6a as well.

The *best* $\{Q_m\}$ to use ought to be the true object, of course, and we tried this out in Fig. 6c. As expected, results are much better than in Fig. 6a or 6b.

Finally, in Fig. 6d we asked what the effect would be of inputting for $\{Q_m\}$ the true object but displaced to the left by one point. This would test the sensitivity of the tactic to registration error. Results show that the output is very sensitive to such error. If a high-gradient $\{Q_m\}$ is used, it had better be in the correct registration. (The restoration for $z = 2$ is not shown because the algorithm would not converge under these contradictory conditions.)

The benchmark comparison with these results would be the ordinary least-squares restoration of the given image. This was carried through. However, the results cannot be plotted because, owing to the noise level in the data, the output was highly oscillatory, varying in size by $\pm 10^6$, and showing no resemblance to the true object.

Table I **Estimators for Different Prior-Knowledge Scenarios**

Prior knowledge	*Maximum-probability estimator (equation)*
High conviction of object $\{Q_m\}$, general occupancy ratio n_m/z_m	Minimum photon-site entropy plus maximum cross entropy [(37)]
High conviction of object $\{Q_m\}$, low occupancy ratio n_m/z_m	Maximum cross entropy [(41)]
High conviction of white object $Q_m = 1/M$ for all m, low occupancy ratio n_m/z_m	Maximum Jaynes entropy [(39)]
High conviction of white object $Q_m = 1/M$ for all m, high occupancy ratio n_m/z_m	Maximum weighted Burg entropy [(43)]
Maximum ignorance (MacQueen–Marschak), general occupancy ratio n_m/z_m	Minimum photon-site entropy [(47b)]
Impartial conviction, empirical data, general occupancy ratio n_m/z_m	Minimum photon-site entropy plus minimum empirical entropy plus maximum Jaynes entropy [(45a)]
Impartial conviction, empirical data, low occupancy ratio n_m/z_m	Minimum empirical entropy plus maximum Jaynes entropy [(48)]
Impartial conviction, empirical data, high occupancy ratio n_m/z_m	Minimum empirical entropy plus maximum weighted Burg entropy [(49a)]

XII. Conclusion

We have shown that it is possible to combine the physics of the "taking" conditions with prior knowledge of "object class" to come up with a maximum probable estimate of a spectrum. The resulting estimation principle depends strongly on what these quantities are in a given problem. No one algorithm can apply to all problems and still be called maximum probable. We have summarized some of the more important cases in Table I. It is up to the user, ultimately, to define his or her particular case of prior knowledge. The more descriptive of the unknown it is, the better will be the resulting MP estimate.

References

Biraud, Y. (1969). *J. Astron. Astrophys.* **1**, 124–127.

Born, M., and Wolf, E. (1959). "Principles of Optics." Macmillan, New York.

Burg, J. P. (1967). "Maximum Entropy Spectral Analysis." Paper presented at the 37th Annual Society of Exploration Geophysicists Meeting, Oklahoma City.

Frieden, B. R. (1972). *J. Opt. Soc. Am.* **62**, 511–517.

Frieden, B. R. (1991). "Probability, Statistical Optics and Data Testing," 2nd ed. Springer-Verlag, New York.

Goodman, J. W. (1985). "Statistical Optics." John Wiley, New York.

Gradshteyn, I. S., and Ryzhik, I. M. (1965). "Tables of Integrals, Series and Products." Academic Press, New York.

Hershel, R. S. (1971). Ph.D. Dissertation, University of Arizona, Tucson.

Hildebrand, F. B. (1956). "Introduction to Numerical Analysis." McGraw-Hill, New York.

Jansson, P. A. (1968). Ph.D. Dissertation, Florida State University, Tallahassee.

Jansson, P. A., Hunt, R. H., and Plyler, E. K. (1970). *J. Opt. Soc. Am.* **60**, 596–599.

Jaynes, E. T. (1968). *IEEE Trans. Syst. Sci. Cybernetics* **SSC-4**, 227–241.

Kikuchi, R., and Soffer, B. H. (1977). *J. Opt. Soc. Am.* **67**, 1656–1665.

MacQueen, J., and Marschak, J. (1975). *Proc. Natl. Acad. Sci. U.S.A.* **72**, 3819–3824.

Shore, J. E. (1981). *IEEE Trans. Acoust. Speech, Signal Process.* **ASSP-29**, 230–237.

Wernecke, S. J., and D'Addario, L. R. (1977). *IEEE Trans. Comput.* **C-26**, 351–364.

Chapter 12 | Fourier Spectrum Continuation

Samuel J. Howard

Physics Department, The Florida State University
Tallahassee, Florida

List of Symbols

a_n	coefficients of cosine terms of Fourier series
A_n	coefficients of cosine terms of discrete Fourier series
b	number of unique coefficients in complex Fourier series expression for $u(k)$
b_n	coefficients of sine terms of Fourier series
B_n	coefficients of sine terms of discrete Fourier series
c	number of unique coefficients in complex Fourier series expression for $v(k)$
$C_n, F(n)$	discrete Fourier spectral components as given by discrete Fourier transform, which are the coefficients of a complex Fourier series
$\mathcal{F}$	forward Fourier transform operator

$\mathscr{F}^{-1}$	inverse Fourier transform operator
$f(k)$	general function of integer variable k
$f(x)$	general function
$F(w)$	Fourier transform of $f(x)$
$i(x), i$	"image" data recorded after degradation by convolution and noise
$I(w), O(w)$ $\tau(w), N(w)$	Fourier transforms of $i(x)$, $o(x)$, $s(x)$, $n(x)$
j	imaginary number $\sqrt{-1}$
k	integer variable for spatial function, the index of sampled values
n	integer Fourier spectral variable
$n(x), n$	noise, which together with the convolution of the object with the impulse response makes up the image
N	number of sample points, which is the same as the number of complex coefficients
$\hat{o}$	inverse-filtered estimate of object
$o(x), o$	"object" or function sought by deconvolution
$\text{rect}(x)$	rectangular function with height and width of unity
$s(x), s$	impulse response function
$u(k)$	discrete Fourier series composed of only low-frequency and dc terms
$U(n)$	discrete Fourier transform of $u(k)$
$v(k)$	discrete Fourier series composed of high-frequency terms such that when added to $u(k)$ a series of a complete Fourier spectral band is formed
$V(n)$	discrete Fourier transform of $v(k)$
w	Fourier spectral variable
x	generalized independent variable, in this work the wave number for spectroscopy data, but for other research areas time, mass number, wavelength, angle, etc.
x'	variable of integration
Δw	standard deviation of Fourier transform
Δx	standard deviation of spatial function
Ω	highest frequency present in band-limited data
Ω_p	cutoff frequency defining Frieden's optimum processing bandwidth
$\text{III}(x)$	Dirac "comb" $\sum_{n=-\infty}^{\infty} \delta(x-n)$

I. Introduction

Experimental data are not always displayed in a form that reveals the maximum available information. A modified form may make more of the inherent information manifest and accessible. A different form may also provide easier comparison or more convenient mathematical manipulation and a simpler theoretical treatment. For these reasons, further processing of experimental data may be desired.

Other than the operations that emphasize certain aspects of the data, there are destructive influences that reduce our knowledge of the features of the data under investigation. These influences introduce deviations from the true values of the quantity being measured called experimental error, or simply error. Many of these degrading operations are selective, in that their effects are more damaging to certain features of the data. For example, the errors that occur in most experimental data are most damaging to the fine detail. This selective effect is most clearly displayed under the Fourier transform, where it is seen that the lower frequencies of the Fourier spectrum of the data are little affected but that the higher frequencies are nearly all in error. Also, these errors render ineffective many possible mathematical operations or transformations by introducing much larger errors into the output.

The deviations due to some of these destructive influences are reversible. These are usually described as systematic errors. Many of the degradation processes that affect images and most recorded data are classified as systematic errors. For many of these cases the error may be expressed as a function known as the impulse response function. Much mathematical theory has been devoted to its description and correction of the degradation due to its influence. This has been discussed in some detail by Jansson in Chapter 1 of this volume. In that correction of this type of error usually involves increasing the higher frequencies of the Fourier spectrum relative to the lower frequencies, this operation (deconvolution) may also be classified as an example of "form alteration."

The remaining errors in the data are usually described as random, their properties ultimately attributable to the nature of our physical world. Random errors do not lend themselves easily to quantitative correction. However, certain aspects of random error exhibit a consistency of behavior in repeated trials under the same experimental conditions, which allows more probable values of the data elements to be obtained by "averaging" processes. The behavior of random phenomena is common to all experimental data and has given rise to the well-known branch of mathematical analysis known as statistics. Statistical quantities, unfortunately, cannot be assigned definite values. They can only be discussed in terms of probabilities. Because (random) uncertainties exist in all experimentally measured quantities, a restoration with all the possible constraints applied cannot yield an exact solution. The best that may be obtained in practice is the solution that is "most probable." Actually, whether an error is classified as systematic or random depends on the extent of our knowledge of the data and the influences on them. All unaccounted errors are generally classified as part of the "random" component. Further knowledge determines many errors to be systematic that were previously classified as random.

More generally, any additional knowledge of the data and influences on them counters the effects of the destructive influences on the data, both random and systematic, to increase our knowledge of the important parameters in the data. Prior knowledge of the physical limits to the values that the data may assume, for example, may be employed as constraints to recover additional information not given in the original data.

II. Advantages of Fourier Analysis

One advantage of Fourier analysis is that it provides a way of exhibiting the data in another form, so that different aspects of the data become evident. An important example of this concerns the interferogram taken with a Michelson interferometer, for which only limited interpretation is allowed. One cannot locate or measure the electromagnetic spectral components until the Fourier transform is computed to separate the various components into distinct "lines." It is also noted that many of the processes that degrade experimental data are manifested largely in the higher-frequency portion of the Fourier spectrum. That much of the error may be localized in a narrow band of high frequencies, along with the fact that this band may be adequately represented by a small number of discrete Fourier spectral components, allows considerable reduction of the computational burden of restoration problems. Here we are making the not unreasonable assumption for most experimental data that the informational content of the very highest frequencies (as recorded) is nominal. An additional advantage of Fourier analysis is that many of the physical processes that affect optical images and many other types of experimental data have a simple mathematical form under Fourier transformation. Thus they lend themselves to convenient mathematical manipulation. This has given rise to a large body of mathematical theory relating to restoration and other enhancement of degraded data.

III. Restoration and Enhancement of Experimental Data

A. INTRODUCTORY REMARKS

Given degraded experimental data, most researchers desire a restoration of the original appearance of the quantities being measured. Many, however, are satisfied with the results of almost any ad hoc method that improves the resolution. Another important consideration for many researchers is the unambiguous display of the quantities of interest in the data. Each of these items will be addressed in turn.

B. DECONVOLUTION

For many data of interest, the degrading effects can be expressed as a convolution of the original appearance of the data with the appropriate impulse response function plus additive noise,

$$i(x) = \int_{-\infty}^{\infty} o(x')s(x - x')\, dx' + n(x), \tag{1}$$

where o is the original appearance of the data, s the functional form of the impulse response function, n the noise term, and i the final appearance of the data recorded. This model for the degradation has been thoroughly discussed earlier by Jansson, and only a brief summary of the important results will be given here. There is an advantage to expressing Eq. (1) in terms of the Fourier transform because many of the restoration operations discussed in this chapter are most efficiently performed in the Fourier domain. This yields

$$I(w) = O(w)\tau(w) + N(w), \tag{2}$$

where capital letters generally denote the transformed quantities. One exception is the transform of the impulse response, which is denoted by τ. The convolution becomes a simple multiplication in the Fourier domain. This suggests the equally simple inverse operation for recovering the original appearance of the data o by dividing this equation through by τ and solving for O:

$$O(w) = I(w)/\tau(w) - N(w)/\tau(w). \tag{3}$$

In principle, inverse transformation would produce the restoration. The noise term in this equation, however, is predominantly random and increases with frequency for the deconvolution of nearly all experimental data. Random error is almost impossible to treat quantitatively and is the primary obstacle to an accurate restoration. The determination of optimum solutions would require the inclusion of statistical criteria (see Chapter 11). Nevertheless, for many data of interest, simply dividing the transform of the data by the transform of the impulse response function, $I(w)/\tau(w)$, and inverse transforming (often called inverse filtering) yields a good approximation to o when the noise level is low. For higher noise levels, the increasing noise term rapidly obscures all information in the deconvolved results, making truncating or filtering of the resulting spectrum necessary to achieve meaningful results.

There are a number of methods, nearly all iterative, that treat the inverse problem of recovery and successfully (some quite imaginatively) deal with the noise problem. The straightforward inverse-filtered estimate was adhered to in this research because of the possibility of saving computational time in the overall restoration. In practice, only discrete data are taken.

The discrete Fourier transform (DFT) of the data is evaluated to take advantage of the considerable speed and accuracy of the fast Fourier-transform algorithm as calculated by modern digital computers. For most data (and most restorations) encountered in practice, it is found that only a relatively small number of Fourier spectral components are required to represent the data adequately, the higher frequencies being mostly noise. Because of the small number of Fourier spectral components involved, as well as the fast speed of the FFT for determining the DFT, inverse filtering is a computationally economical restoration procedure. However, the increasing noise prevents the most desirable restoration when the noise level is appreciable. In the restorations treated here, the most noise-free low-frequency components of the inverse-filtered spectrum will be used as the initial stage in the restoration because of the considerable saving of computer time. The Fourier spectrum will then be continued from the truncated inverse-filtered estimate to yield the complete spectrum. Very fast algorithms have been developed for recovering a band of frequencies beyond the given low-frequency band (Howard, 1981a, b) to yield an overall restoration procedure that is very economical computationally.

Function continuation procedures are applied to many other problems besides inverse-filtered Fourier spectral continuation and will be discussed in a separate section.

C. APODIZATION

To display unambiguously the quantities of interest in the data implies that some operation, or series of operations, must be performed to alter the form of the data. In most spectroscopy and chromatography data, the quantities of interest are discrete and are usually displayed as sharply peaked functions or "lines." However, all recording instruments broaden and alter the functional form to a certain extent, so that the data recorded for a monochromatic source would not be an infinitely sharp line, or Dirac δ function. The lines recorded would be broadened and generally have artifacts around them (such as sidelobes, or "ringing"). It is important to display each discrete component in the data unambiguously so that important parameters (such as wave number, intensity, or chemical compound) may be determined from them. For sharply peaked functions, this would generally involve improving the resolution and removing the artifacts, so that overlapping with adjacent peaked functions would be minimized.

It is found that multiplication of the Fourier transform of the data by a carefully chosen window function is very effective in removing the artifacts around peaked functions. This process is called apodization. Apodization with the triangular window function is often applied to Fourier transform spectroscopy interferograms to remove the ringing around the infrared spectral lines obtained from the transformation. However, other, more appropriate window functions for these data are currently enjoying more widespread use (Bell, 1972). Apodization is often done at the expense of resolution because the peaked functions are usually considerably broadened in the process. However, apodization usually produces a less ambiguous result on the whole, for artifacts such as ringing (characteristic of convolution with the sinc function) extend a considerable distance from each peak. This results in overlapping with adjacent peaks. However, we shall find later that both removal of artifacts *and* resolution improvement are possible within the context of the Fourier spectrum continuation techniques.

A general discussion of resolution is provided in the following section.

D. RESOLUTION AND THE UNCERTAINTY PRINCIPLE

Deconvolution, the inverse operation of recovering the original function o from the convolution model as given in Eq. (1), employs procedures that almost always result in an increase in resolution of the various components of interest in the data. However, there are many broadening and degrading effects that cannot be explicitly expressed as a convolution integral. To consider resolution improvement alone, it is instructive to consider other viewpoints. The *uncertainty principle* of Fourier analysis provides an interesting perspective on this question.

Letting Δx denote the standard deviation of the spatial function and Δw the standard deviation of the spectral function, we must necessarily have

$$\Delta x \, \Delta w \geq 1/4\pi. \qquad (4)$$

A proof of this relation may be found in Bracewell (1978). Note that the spectral variable used in this and the next chapter is the same as that defined in Eqs. (7) and (8). Now consider a spatial distribution $f(x)$ and its Fourier spectrum $F(w)$ that come close to satisfying the equality in Eq. (4). We may take Δx and Δw as measures of the width, and hence the resolution, of the respective functions. To see how this relates to more realistic data, such as infrared spectral lines, consider shifting the peak function $f(x)$ by various amounts and then superimposing all these shifted functions. This will give a reasonable approximation to a set of infrared lines. To discuss quantitatively what is occurring in the frequency domain, note that the Fourier spectrum of each shifted function by the shift theorem is given simply by the spectrum of the unshifted function multiplied by a constant phase factor. The superimposed spectrum would then be

$$[1 + \exp(j2\pi wx_1) + \exp(j2\pi wx_2) + \cdots]F(w),$$

where $x_1, x_2, \ldots$ are the amounts that the spatial functions are shifted. As Δx becomes smaller, each peak becomes narrower, and merged peaks become resolved. The Δw of each superimposed peak would necessarily have to increase all by the same amount to satisfy the uncertainty principle. With the aid of the foregoing relation, it is not too difficult to show that the superimposed Fourier spectrum is generally broadened also.

These arguments may be generalized to show that any overall increase in resolution is accompanied generally by a broadening of the Fourier spectrum. This seems to be borne out in practice because almost any operation that alters the magnitude of the high frequencies relative to the low frequencies affects the resolution. Broadening the Fourier spectrum or boosting the higher frequencies relative to the lower frequencies generally improves resolution and increases the detail, and narrowing the Fourier spectrum or reducing the highest frequencies almost always degrades resolution and causes smoothing. This is apparently the reason for the success of such ad hoc methods as zeroing the dc component and lowest frequencies of the Fourier spectrum, using high-pass filters, and the unconstrained Van Cittert method for the improvement of resolution. However, it needs to be pointed out that any boosting of the higher frequencies relative to the lower frequencies in an effort to improve resolution must be done in a meaningful way and not with random magnitudes and phases such as one would obtain by the addition of random noise. Actually, multiplication by the simple window functions discussed earlier is one example of a meaningful way. These window functions affect the magnitudes only, which are altered in some smooth way. This seems to have the effect of producing smooth results without violent oscillations, and without the introduction of extraneous detail (which random phase changes would introduce).

Incidentally, the uncertainty principle associated with the name of Heisenberg, well known in quantum mechanics, follows from the expression given here when de Broglie's relationship connecting the momentum of a particle with its wavelength is included.

E. CONTINUATION OF THE FOURIER SPECTRUM

This subject has been discussed by Jansson in Chapter 4, so the present discussion will be kept brief.

Because no unique extension of the Fourier spectrum exists in the absence of additional information, the impetus of the research described in this chapter has been to discover as much prior information as possible about the data and the influences on them, and to find ways to formulate this information as "constraints" to produce increasingly more probable values for the Fourier spectrum. Ways of successfully applying six constraints to experimental data for spectral restoration have been developed. All of these constraints cannot generally be applied to any given set of data. The constraints must be appropriate to the data under consideration.

The first constraint developed by the author was based on prior knowledge of the correctly restored function. The first step in the restoration was the calculation of the inverse-filtered estimate with only the most noise-free low-frequency spectral components retained. Prior knowledge of the correctly restored function was used to construct an artificial function that, it was hoped, would closely resemble the correctly restored function. Specifically, it was hoped that this function's spectrum beyond the truncation frequency Ω_p would resemble the original spectrum of the undistorted function, so that it might be substituted into the spectrum of the inverse-filtered estimate from Ω_p on to yield a complete set of spectral components that were reasonably correct. This procedure was successful in significantly improving gas chromatographic data (Howard, 1978; Howard and Rayborn, 1980). The artificial function was created by measuring the heights, widths, and locations of the inverse-filtered peaks (which represented the various chemical components of the sample) and constructing Gaussian functions with these parameters. Superimposing these Gaussian functions, which of course had no sidelobes or other artifacts, yielded the artificial function, which at least crudely resembled the correctly restored function for most cases. The disadvantage of this procedure is that it is very difficult to create an artificial function for the general case. If the peaks in the data are not fairly well separated, it is very difficult to obtain the required parameters. This method thus has limited application.

The second constraint restricted the Fourier spectrum. This was an ad hoc filter that was applied to the entire inverse-filtered spectrum to bring the magnitudes of the high-frequency values of the spectrum (which were mostly noise) to values much closer to the correctly restored ones. This procedure resulted in observable improvement over the inverse-filtered estimate for infrared lines obtained from grating spectroscopy (Howard, 1982).

Many of the most effective constraints set well-defined limits to the data function (or its spectrum) beyond which the correct function is not allowed to go. An important example of this type of constraint is "nonnegativity," whereby the correctly restored function is not allowed to extend below the zero base line and thereby take on nonphysical negative values. This is an appropriate constraint for spectroscopy and optical images. A further example of the constraints of fixed limits is that of an "upper bound" to the values of the restoration. Another important constraint of this type is that of "finite extent," for which no deviations from zero are allowed for the spatial function over those intervals on the spatial axis that lie outside the known extent of the original object. This constraint would be appropriate for spectroscopy in which the electromagnetic radiation in the spectral bands outside the band of interest is filtered out.

It is found that recovering only a band of spectral components beyond the truncation frequency of the inverse-filtered spectrum, along with the use of discrete components (as given by the DFT), considerably reduces the computational burden. Incidentally, recovering only a finite band of frequencies implies the additional constraint of restricting the highest frequencies to zero. This is not a serious restriction, because nearly all of the detail of interest resides in the relatively narrow band restored. It is usually very difficult to restore the highest frequency spectral components accurately. Actually, recovery of only a narrow band of spectral components is often necessary for stability in the equations resulting from the procedures that apply many of these constraints. This is especially true for the constraint of finite extent. Also, recovering only a narrow band always results in a smoother restored (spatial) function.

Very efficient ways of recovering only a band of discrete spectral components with the application of constraints of well-defined limits were developed in this research. These constraints are generally applied in the spatial domain and begin with the function (in the spatial domain) formed from only the low-frequency band of Fourier spectral components. This function is held fixed, and the function formed from a band of high-frequency spectral components immediately adjacent to the low-frequency band is added to this (fixed spatial) function and allowed to vary until the total function is out of the nonallowed region, or as far out of it as the noise and other error will allow it to go. Within this very loose general framework, a variety of approaches may be developed. Mathematically, this scheme may be formulated as a minimization procedure in which the expression to be minimized is given as some function of the deviations into the nonallowed region. Such functions could be the familiar sum of squares, the sum of absolute values, or other expressions. (For the constraint of finite extent, one may wish to minimize the maximum deviation of the sum of the two functions corresponding to the two spectral bands.) Various techniques, both numerical and analytical, may then be applied to find the continued spectrum that minimizes one of these "measures" of the error. A very direct numerical technique may be used to minimize the largest deviation of the error. In this research all further calculations were done with the sum-of-squares expression because many of these calculations could be done analytically *once and for all*.

The explicit sum-of-squares expression for each constraint of well-defined limits will be more fully discussed in succeeding sections.

IV.   Discrete Fourier Transform and Discrete Function Continuation

A.   *DISCRETE FOURIER TRANSFORM*

As with the continuous Fourier transform, we could treat the equations of the discrete Fourier transform (DFT) completely independently, derive all the required theorems for them, and work entirely within this "closed system." However, because the data from which the discrete samples are taken are usually continuous, some discussion of sampling error is warranted. Further, the DFT is inherently periodic, and the limitations and possible error associated with a periodic function should be discussed.

The DFT is defined as

$$C_n \equiv \frac{1}{N} \sum_{k=0}^{N-1} f(k) \exp\left(-j\frac{2\pi}{N}nk\right) = F(n). \tag{5}$$

Its companion, the inverse DFT, is defined as

$$f(k) \equiv \sum_{n=0}^{N-1} C_n \exp\left(j\frac{2\pi}{N}nk\right). \tag{6}$$

In these equations, N is the number of sample points, n and k are integers, and j is the imaginary number $\sqrt{-1}$.

To illustrate more appropriately the relationship between the (continuous) Fourier transform and the DFT, the alternative form given in Chapter 1 will be employed. Accordingly, we define this transform as

$$\mathscr{F}\{f(x)\} \equiv \int_{-\infty}^{\infty} f(x) \exp(-j2\pi xw)\, dx = F(w). \tag{7}$$

Then to recover the original function $f(x)$ we must define the inverse transform as

$$\mathscr{F}^{-1}\{F(w)\} \equiv \int_{-\infty}^{\infty} F(w) \exp(j2\pi xw)\, dw. \tag{8}$$

The forward and inverse Fourier transforms are denoted by $\mathscr{F}$ and $\mathscr{F}^{-1}$, respectively. In the foregoing equations, we identify w with ω used in Chapter 1 via the relation $w = \omega/2\pi$.

The Fourier series, which has a discrete spectrum but periodic spatial function, is actually a special case of the Fourier transform. (Note that an equally spaced discrete spectrum necessarily implies a periodic function having a finite period given by the wavelength of the lowest frequency.) See Bracewell (1978) to see how the explicit form of the Fourier series may be obtained from the Fourier transform. Taking discrete, equally spaced samples of the periodic function (or samples over one period because the series repeats indefinitely), we arrive at the final form of the DFT (see, e.g., Howard, 1978). We may even skip the intermediate step of obtaining the Fourier series and directly consider discrete values of *both* the function and its spectrum as special cases of the Fourier transform. With the DFT, both domains must obviously be periodic as well as discrete. See also Fig. 3 of Chapter 1 to see how the DFT may be acquired from the Fourier transform by sampling and/or periodicity.

Addressing first the limitations of a periodic representation, such as with the DFT or Fourier series, we see that it is evident that these forms are adequate only to represent either periodic functions or data over a finite interval. Because data can be taken only over a finite interval, this is not in itself a serious drawback. However, under convolution, because the function represented over the interval repeats indefinitely, serious overlapping with the adjacent periods could occur. This is generally true for deconvolution also, because it is simply convolution with the inverse filter $\mathscr{F}^{-1}\{1/\tau(w)\}$. If the data go to zero at the end points, one way of minimizing this type of error is simply to pad more zeros beyond one or both end points to minimize overlapping. Making the separation across the end points between the respective functions equal to the effective width of the impulse response function is usually sufficient for most practical purposes. See Stockham (1966) for further discussion of end-point extension of the data in cyclic convolution.

Another problem with a discrete periodic representation is that it is very difficult to represent functions that are not equal at their end points. An abrupt discontinuity across the end points of a periodic representation causes considerable amplification of the high frequencies, which, in general, are very difficult to work with. Also, unless the sampling interval is taken very fine, this function cannot be adequately represented by the DFT. However, this problem can be dealt with by extending each end point in a smooth curve that joins smoothly with the next period so as to yield an overall periodic function that is reasonably smooth. This is a valid operation because it is only necessary that the function be adequately represented over the interval of interest. The advantage of a smooth curve is that the high frequencies are now much smaller, and the overall numerical problem much more tractable.

With convolution and deconvolution, one must be careful to avoid end-point error with this type of function. Convolution with the function beyond the end point of the data will extend inside the interval containing the data about half the length of the impulse response function, so the error will extend about half the length of the impulse response function also (assuming the impulse response function is approximately symmetrical). To minimize this error, the function extending beyond the end points should approximate the true function if anything is known about it. However, simply extending a very smooth curve across the end points is usually sufficient for most practical purposes. A good discussion of the end-point error involved in the deconvolution of molecular scattering data is provided by Sheen and Skofronick (1974). Incidentally, a discrete function for extending the data in a very satisfactory manner is provided in a versatile spline-fitting computer program developed by De Boor (1978). It minimizes the curvature by minimizing the second derivative of the discrete data function. Weights can be assigned to particular points, such as the end points, to assure small deviations there, and small weights can be assigned to the points beyond the end points so as to have sufficient flexibility to allow the discrete function there to form a very smooth curve. This program is also useful for smoothing and base-line fitting. This technique for end-point extension that we have just discussed provides an alternative to data windowing.

Next, we shall discuss sampling error. Fortunately, Fourier analysis is one of the few disciplines that provide a quantitative treatment of this. As discussed earlier, sharp corners, discontinuities, and fine detail require the highest frequencies to represent them adequately. Smooth data may usually be represented adequately by a lower frequency band. On sampling with a coarse interval, it is evident that all fine detail is missed. This error enters into the Fourier expression of the data as an impersonation, or "aliasing," of the lower frequencies by the higher frequencies; that is, for every low frequency in the Fourier expression, there is a set of high-frequency components that exhibit exactly the same behavior under the Fourier transform. From the sampled data alone it would be impossible to determine how much of the lower frequency is present or how much of any of the other frequencies in the higher frequency set is present. All one would know is their sum. However, if all of the higher frequencies were absent, that is, if the data were "band limited," then a complete recovery of the continuous function would be allowed by the discrete samples.

One procedure for recovering the continuous (band-limited) function exactly is provided by the *Whittaker–Shannon sampling theorem*, which is expressed by the equation

$$f(x) = \sum_{n=-\infty}^{\infty} f(n/2\Omega)\mathrm{sinc}(2\Omega x - n). \tag{9}$$

This formula tells us that when the data are band limited with $w = \Omega$, the highest frequency present, the spacing of the sampling interval may be as large as $\Delta x = 1/2\Omega$ and yet allow complete recovery of the original continuous function. A proof of the sampling theorem may be found in Hamming (1962) and Papoulis (1962). An interesting discussion of sampling and aliasing error in terms of the comb (III) function is provided by Bracewell (1978). The DFT representation, however, allows a more convenient and faster interpolation of periodic data. To see how this can be, we shall write a few terms of the inverse DFT, which is a discrete Fourier series:

$$
\begin{aligned}
f(k) &= \sum_{n=0}^{N-1} C_n \exp\left(j\frac{2\pi}{N}nk \right) = C_0 + C_1 \exp\left(j\frac{2\pi}{N}k \right) + \cdots \\
&= A_0 + A_1 \cos\left(\frac{2\pi}{N}k \right) + B_1 \sin\left(\frac{2\pi}{N}k \right) + A_2 \cos\left(\frac{2\pi}{N}2k \right) + \cdots .
\end{aligned}
\tag{10}
$$

Note that we could let k be a continuous rather than a discrete variable, and we would then obtain a continuous function

$$
f(x) = A_0 + A_1 \cos\left(\frac{2\pi}{N}x \right) + B_1 \sin\left(\frac{2\pi}{N}x \right) + A_2 \cos\left(\frac{2\pi}{N}2x \right) + \cdots ,
\tag{11}
$$

where the continuous variable x has replaced k. This is the same functional form as the continuous Fourier series with the period $L = N$:

$$
f(x) = a_0 + a_1 \cos\left(\frac{2\pi}{L}x \right) + b_1 \sin\left(\frac{2\pi}{L}x \right) + a_2 \cos\left(\frac{2\pi}{L}2x \right) + \cdots .
\tag{12}
$$

The continuous function consisting of only N terms can at best only approximate the most general (continuous) function, for, as we know, the Fourier series of the most general periodic function requires an infinite number of terms. However, consider the periodic function under examination made up of only N or fewer terms of its Fourier series, that is, a band-limited function. Then, because of the linear independence of the sinusoidal terms, the coefficients of the DFT are exactly the same as those of the Fourier series of the continuous function. That is, we find

$$
A_0 = a_0, \qquad A_1 = a_1, \qquad B_1 = b_1, \qquad A_2 = a_2, \ldots ,
\tag{13}
$$

and the continuous function with the DFT coefficients produces exactly the same continuous function as the Fourier series. No information is lost on taking the Fourier transform. [For an interesting proof of this and other important theorems for the DFT, see Bracewell (1978).] Therefore the N sample points obtained by taking the inverse DFT of these coefficients completely represent the original DFT coefficients, and hence the continuous function formed from them. We may state this another way. If the continuous periodic function is band limited, having N or fewer discrete spectral components, then N samples equally spaced over one period are sufficient for perfect interpolation between the points. This may be done by constructing a continuous series from the DFT coefficients. Note that only N samples are necessary because one cycle is fully characteristic of the period function. For an interesting derivation of the relationship between the coefficients of the Fourier series and the DFT, see Hamming (1962), whose explicit expression shows exactly which set of coefficients aliases a particular given coefficient.

Closely associated with sampling error is another question: to what extent is the information given in the DFT, a finite number of discrete spectral components, capable of representing the continuous function over an infinite interval? A periodic function is certainly capable of being represented by a finite number of discrete samples. In the preceding paragraphs we have discussed how this is possible. For the nonperiodic function, though, even if it is band limited and the sampling interval is sufficiently small, a finite number of samples is insufficient to represent the continuous function correctly. The sampling theorem tells us that an infinite number of samples is required. There is an aspect of a finite band, though, that we have not yet considered in these problems. In particular, the property of finite extent of the band may be used as a constraint, as discussed earlier, to extrapolate the sampled function. We would hope that an infinite number of sample points may be obtained in this way, and that they will be unique. Unfortunately, this is not the case. Fiddy and Hall (1981) and Schafer et al. (1981) have pointed out that, given a finite number of samples of a function, the constraints of square integrability and band limitation are not sufficient to determine that function uniquely. A family of functions is the best that may be determined.

With experimentally obtained data, which always contain random noise, questions such as uniqueness are academic. With noise, an exact solution is impossible to obtain, and the best that one may do, in a statistical sense, is to find the solution that is most probable. Even a solution that is most probable (overall) is not always what is desired. A unique solution is forced by the criterion or method of solution adopted, such as minimizing the sum of squares or maximum deviation. Each of these criteria produces a solution that is optimum in some sense. Various aspects of the error will be minimized in each of these slightly differing solutions. We may want to choose the criterion that minimizes some aspect of this error (or emphasizes some other aspect of the restoration). However, we may want to choose some particular criterion simply because it leads to more convenient mathematical manipulation or faster numerical calculation.

Except for special cases, the differences among the restorations produced by the various criteria are usually small and inconsequential. In a gross sense, the restorations produced by the methods mentioned in this section that apply many of the important constraints are usually much closer to the original undistorted function, even in the presence of moderate to high noise levels. The primary objective in initiating these restoration procedures is thus largely accomplished.

For the representation of almost all experimental data over an interval, the DFT has been found to be very adequate and useful. The errors peculiar to the DFT representation, such as those due to periodicity and a finite number of samples, are usually small compared with the errors due to noise. Restoration operations, such as deconvolution, are usually adequately performed within the DFT formulation.

As discussed earlier, the information-containing components of the Fourier spectrum of experimental data (even after restoration) become smaller with increasing frequency and merge into the noise in the high-frequency portion of the spectrum. After taking the DFT of experimental data, it is found that the useful information is represented in most cases by a quite small number of discrete Fourier components. The information-bearing components that are aliased would necessarily be very small, and would contribute negligible error. However, the noise spectrum decays very slowly (if at all) and is still significantly large at the higher frequencies, and the highest of the noise frequencies would be aliased (for any sampling interval). Taking a finer sampling interval minimizes this aliasing. With a finer sampling interval, more of the high-frequency noise spectral components that were previously aliased now show up in the Fourier spectrum. (That they are now present in the Fourier spectrum implies that they are certainly not aliased.) Taking increasingly finer sampling intervals therefore further minimizes the aliasing of the low-frequency informational components by the high-frequency noise components. The reader should be aware that this has little effect on the low-frequency noise, however. Note that this reduction of the aliasing with a finer sampling interval closely corresponds to minimization of the error in the data in a statistical sense in that simply "getting more points" on the data allows a reduction of the error by least-squares calculations.

Straightforward inverse filtering with the spectral components given by the DFT would thus involve very few spectral components with the type of experimental data discussed here.

When using the fast-Fourier-transform algorithm to calculate the DFT, inverse filtering can be very fast indeed. By keeping the most noise-free inverse-filtered spectral components, and adding to these an additional band of restored spectral components, it is usually found that only a small number of components are needed to produce a result that closely approximates the original function. This is an additional reason for the efficiency of the method developed in this research.

The preceding discussion also suggests a way of "tightly packing" the data, that is, a way of representing the significant information in the data by a much smaller number of points. Truncating the Fourier spectrum at some point before the noise assumes a significant fraction of the magnitude and saving this much smaller number of discrete components is a procedure that usually preserves all the important information in the data.

Extrapolation of discrete Fourier spectra is accomplished by applying the constraints to the discrete Fourier series given by the inverse DFT.

B. EXTRAPOLATION OF DISCRETE FOURIER SPECTRA

The discrete Fourier spectrum as given by the DFT is represented by the coefficients of a complex Fourier series. The inverse DFT is, of course, this series:

$$f(k) = \sum_{n=0}^{N-1} C_n \exp\left(j\frac{2\pi}{N}nk\right) = C_0 + C_1 \exp\left(j\frac{2\pi}{N}k\right) + \cdots$$

$$= A_0 + A_1 \cos\left(\frac{2\pi}{N}k\right) + B_1 \sin\left(\frac{2\pi}{N}k\right) + A_2 \cos\left(\frac{2\pi}{N}2k\right)$$

$$+ B_2 \sin\left(\frac{2\pi}{N}2k\right) + \cdots + A_{N-1} \cos\left(\frac{2\pi}{N}\right)(N/2)k. \qquad (14)$$

The coefficients of the sines and cosines will be real for real data. Restoring a high-frequency band of c (unique complex) discrete spectral components to a low-frequency band of b (unique complex) spectral components will be the same (when transformed) as forming the discrete Fourier series from the high-frequency band and adding this function to the series formed from the low-frequency band. When applying the constraints in the spatial domain, the Fourier series representation will be used.

In all restorations treated in this research, the series formed from the low-frequency band is held fixed and the series formed from the high-frequency band altered by allowing the coefficients to vary until the particular constraint or constraints are satisfied. Let $u(k)$ denote the series for the low-frequency band and $v(k)$ the series for the high-frequency band. The explicit Fourier series expression for each would be

$$u(k) = A_0 + A_1 \cos\left(\frac{2\pi}{N}k\right) + B_1 \sin\left(\frac{2\pi}{N}k\right) + A_2 \cos\left(\frac{2\pi}{N}2k\right)$$

$$+ \cdots + B_{b-1} \sin\left(\frac{2\pi}{N}\right)(b-1)k, \tag{15}$$

$$v(k) = A_b \cos\left(\frac{2\pi}{N}bk\right) + B_b \sin\left(\frac{2\pi}{N}bk\right) + A_{b+1} \cos\left(\frac{2\pi}{N}(b+1)k\right)$$

$$+ \cdots + B_{b+c-1} \sin\left[\frac{2\pi}{N}(b+c-1)k\right]. \tag{16}$$

The sum $u(k) + v(k)$ yields the restored function. The coefficients in $v(k)$ are varied to satisfy the constraints. Because $u(k)$ is constant for all function continuations discussed here, no useful purpose is served by writing out its Fourier series expression, and so the series representation will always be suppressed.

Let the discrete spectrum, which consists of the coefficients of $u(k)$ and $v(k)$, be denoted by $U(n)$ and $V(n)$, respectively. The low-frequency spectral components $U(n)$ are most often given by the most noise-free Fourier spectral components that have undergone inverse filtering. For these cases $V(n)$ would then be the restored spectrum. However, for Fourier transform spectroscopy data, $U(n)$ would be the finite number of samples that make up the interferogram. For these cases $V(n)$ would then represent the interferogram extension.

As mentioned earlier, the sum of the squared error is found to be the most convenient measure of the error because much of the calculation may be done analytically. In the following sections the sum of the squared error will be formulated for each of the constraints, and the form of the equations in the unknown Fourier coefficients for each constraint will be determined. Values of both artificial and experimental data will then be substituted in these equations to determine these unknown Fourier spectral components of the extended spectrum. From these, the completely restored function may be determined.

V.   Constraint of Finite Extent

A.   THEORY

The constraint of finite extent applies to data that exist only over a finite interval and have zero values elsewhere. Let N1 and N2 denote the nonzero extent of the original undistorted data. The restored function $u(k) + v(k)$, then, should have no deviations from zero outside the known extent of the data. To find the coefficients in $v(k)$ that best satisfy this constraint, we should minimize the sum of the squared points outside the known extent of the object. Actually, recovering only a band of frequencies in $v(k)$ implies the additional constraint of holding all higher frequencies above this band equal to zero. This is necessary for stability and is an example of one of the smoothing constraints discussed earlier. We minimize the expression

$$
\sum_{\substack{k<N1,\\k>N2}} [u(k) + v(k)]^2 = \sum_{\substack{k<N1,\\k>N2}} \left\{ u(k) + A_b \cos\left(\frac{2\pi}{N}bk\right) \right.
$$

$$
+ B_b \sin\left(\frac{2\pi}{N}bk\right) + A_{b+1}\cos\left[\frac{2\pi}{N}(b+1)k\right]
$$

$$
\left. + \cdots + B_{b+c-1}\sin\left[\frac{2\pi}{N}(b+c-1)k\right] \right\}^2 .
$$

The standard procedure for finding the minimum of a function of several variables is to take the partial derivative with respect to each variable and set the result equal to zero. Taking the derivative with respect to each unknown coefficient and setting the result equal to zero gives (summation interval suppressed)

$$\frac{\partial}{\partial A_b} \sum [u(k) + v(k)]^2 = 0,$$

$$\frac{\partial}{\partial B_b} \sum [u(k) + v(k)]^2 = 0,$$

$$\frac{\partial}{\partial A_{b+1}} \sum [u(k) + v(k)]^2 = 0, \tag{17}$$

$$\vdots$$

$$\frac{\partial}{\partial B_{b+c-1}} \sum [u(k) + v(k)]^2 = 0.$$

Considering in detail the derivative with respect to A_b, we have

$$\sum \frac{\partial}{\partial A_b} \left[u(k) + A_b \cos\left(\frac{2\pi}{N} bk\right) + \cdots \right]^2$$

$$= 2\sum [u(k) + v(k)] \cos\left(\frac{2\pi}{N} bk\right) = 0. \tag{18}$$

The other derivatives are calculated in a similar manner, and we find for the complete set of equations

$$\sum [u(k) + v(k)] \cos\left(\frac{2\pi}{N} bk\right) = 0,$$

$$\sum [u(k) + v(k)] \sin\left(\frac{2\pi}{N} bk\right) = 0,$$

$$\sum [u(k) + v(k)] \cos\left[\frac{2\pi}{N} (b+1)k\right] = 0, \tag{19}$$

$$\vdots$$

$$\sum [u(k) + v(k)] \sin\left[\frac{2\pi}{N} (b+c-1)k\right] = 0.$$

Because of aliasing, the total number of coefficients obtained should not be greater than N. We have a set of $2c$ linear equations for the $2c$

unknown coefficients. A number of standard methods are available for solving a set of linear equations. We used the Gauss–Jordan matrix reduction method. However, there are important advantages to iterative methods when the number of equations to be solved is large. Once the coefficients have been obtained, they may be converted to complex form and added to the original spectrum. Taking the inverse DFT would then yield the restored function. However, if the number of solved coefficients is small, it may be quicker simply to substitute the coefficients into the series representation for $v(k)$ and add this series to $u(k)$.

B. APPLICATION

1. Artificial Data

To illustrate the effectiveness of the finite-extent constraint, we show the restoration of two merged, almost completely noise-free Gaussian peaks in Fig. 1. These data, like all data discussed in this chapter, make up a 256-point data field. The original peaks are shown in Fig. 1a. The Fourier spectrum truncated after the seventh (complex) coefficient is shown in Fig. 1b. The restoration of Fig. 1c was accomplished by restoring 16

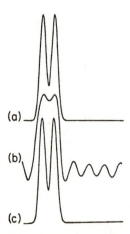

Figure 1   Restoration of a Fourier spectrum of almost completely noise-free data. (a) Original two Gaussian peaks. (b) Same peaks with the spectrum truncated after the seventh (complex) coefficient. (c) Peaks in b with 16 (complex) coefficients restored.

complex coefficients to the Fourier spectrum with the summation region over the last 128 points of the data field of Fig. 1b. Note that an almost perfect restoration was accomplished. The Fourier spectrum of a Gaussian function is also a Gaussian and dies out very quickly, so restoring only 32 (16 unique complex) coefficients produced a very good approximation to the original data. It was found that when the minimization procedure involved only the last quarter of the data field (64 data points), a good restoration was nevertheless obtained. The solved coefficients were almost the same as in the previous case. Even when the summation was computed over only one-eighth of the data field (32 points), a crude approximation to the first few coefficients after the truncation frequency was obtained. However, the higher-frequency components were much larger than their true size.

It is important to note that these data are not completely error-free, and that some approximations are involved. There are several sources of small error, computational roundoff probably being dominant. Also, a Gaussian function never dies out exactly to zero. Its values, however, are very small even two or three standard deviations away from its peak. If the noise and other error are very much larger than this, however, severe computational difficulties ensue. The following figures illustrate the effect of moderate amounts of noise on the restoration.

The effects of a measured amount of Gaussian noise on the deconvolution of a single Gaussian peak are shown in Fig. 2 to enable a more quantitative discussion of the results. The original Gaussian function is shown in Fig. 2a. The deconvolution of this function (with another Gaussian) is shown in Fig. 2b. This shows that the correctly restored result should be Gaussian in form and have a width about half of the original function. The restoration from the noisy data will be compared with this result. See Chapter 1 for further discussion of the relationship among the widths of the various functions under convolution.

Note that inverse filtering is sufficient for the restoration if the data are largely noise-free. We shall soon see, however, that even relatively small amounts of noise necessitate truncation of all but a few coefficients of the Fourier spectrum. Figure 2c shows the same function shown in Fig. 2a, with Gaussian noise of root-mean-square (RMS) amplitude $\frac{1}{20}$ of the amplitude of the peak superimposed, a rather high level. On deconvolution of this noisy function, it was found that meaningful results could not be obtained unless the spectrum of the inverse-filtered result was truncated after the sixth (complex) coefficient. The reason for this is quite apparent

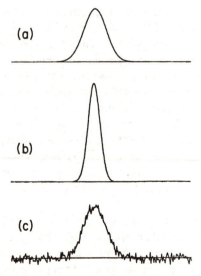

(a)

(b)

(c)

Figure 2 Deconvolution of a single Gaussian peak. (a) Noise-free Gaussian peak. (b) Inverse filtering of the peak in a with another Gaussian function. The resulting peak is also Gaussian in form. (c) Peak in a with Gaussian noise of RMS amplitude $\frac{1}{20}$ of the amplitude of the Gaussian peak superimposed.

when the spectrum of the noise in the inverse-filtered result is examined. Figure 3 illustrates this problem. Figure 3a shows the magnitude of the first nine Fourier spectral components of the noise in the original data (Fig. 2c). Figure 3b shows the magnitude of these same spectral components after inverse filtering. From examination of Fig. 3b it is evident that keeping more than six (complex) coefficients in the inverse-filtered result would contribute unacceptably large noise error to the restoration. Even when only six or fewer coefficients are kept, the high noise level contributes considerable instability to the solution. This is especially apparent when the summation interval is decreased. In Figs. 4 and 5, the extent of the summation interval, the region to which the constraint is applied, is indicated by the tick marks. Figure 4a shows the inverse-filtered function with only six coefficients kept in its Fourier spectrum. Recall that peak broadening and ringing around the peak are characteristics of a function with a truncated spectrum. They come from the theoretical interpretation of convolution of the original function with the sinc function.

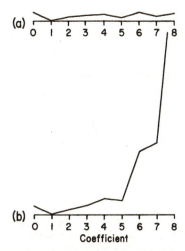

Figure 3 Fourier spectrum of the noise that was superimposed on the Gaussian peak in Fig. 2c. (a) Spectrum of the noise in the original peak. (b) Spectrum of the noise after inverse filtering. It is evident that the noise error increases considerably after the sixth complex coefficient.

Letting the summation interval include the first negative sidelobe on each side of this function and solving for 32 (16 complex) coefficients produced the result shown in Fig. 4b, a reasonably good result. However, taking the summation interval farther away from the peak produced a highly erroneous result, as shown in Fig. 5. For the summation interval defined by the tick marks, restoring 16 (complex) coefficients produced the result shown in Fig. 5a, a very unsatisfactory result. To achieve improvement for this summation interval, no more than six (three complex) coefficients were sought. The result is shown in Fig 5b. As stated earlier, recovering only a small band of frequencies is an additional constraint that is necessary in many cases of heavily error-laden data.

2. Experimental Data

The experimental data chosen for improvement were two merged infrared spectral lines, shown in Fig. 6a. These were methane spectral lines taken with a two-pass Littrow-type diffraction grating spectrometer. Like all the grating spectrometer data discussed here, these data show isolated sets of

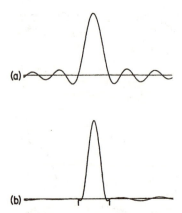

Figure 4 Restoration by inverse filtering of the low-frequency band followed by spectral restoration of the high-frequency band with the constraint of finite extent. (a) Function produced by inverse filtering of the peak in Fig. 2c with six (complex) coefficients retained in the Fourier spectrum. (b) Improved function resulting from the restoration of 16 (complex) coefficients to the spectrum with the constraint of finite extent applied to the region indicated by the tick marks.

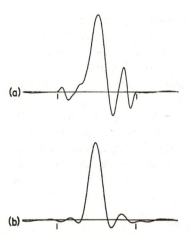

Figure 5 Restoration of the inverse-filtered result shown in Fig. 4a with the required summation for the constraint of finite extent taken over the interval indicated by the tick marks. (a) Restoration of 16 (complex) coefficients to the inverse-filtered estimate. (b) Restored function produced by restoring only three (complex) coefficients to the inverse-filtered estimate.

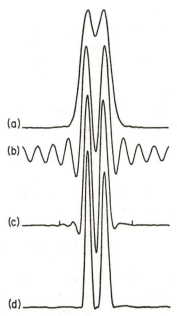

Figure 6 Restoration of a Fourier spectrum of inverse-filtered noisy infrared peaks with the constraint of finite extent. (a) Two merged infrared peaks. (b) Inverse-filtered infrared peaks with the spectrum truncated after the 10th (complex) coefficient. (c) Spectrum restored by applying the constraint outside the marked region. Five (complex) coefficients were restored. (d) Spectrum restored with the constrained region including the first negative sidelobes and the dip between the peaks as well as all other regions outside the peaks. Sixteen (complex) coefficients were recovered.

infrared lines excerpted from data recorded and described by Hunt *et al.* (1978). To improve the resolution, inverse filtering was first performed. A Gaussian function was chosen as the impulse response function, which reasonably approximates the true impulse response for these lines (Jansson, 1968). The Fourier spectrum of a Gaussian is usually an adequate approximation to most impulse response functions at their lower frequencies. However, because this spectrum dies out quickly, the approximation is usually poor at higher frequencies. If inverse filtering is not followed by spectral restoration and more high frequencies are retained, a better approximation to the correct impulse response will probably be needed. To minimize the error due to the noise in these methane lines, the deconvolution was truncated after the 10th (complex) coefficient, as shown in Fig. 6b. To restore the spectrum after the 10th coefficient, the summation interval was chosen to cover the first 65 and the last 64 data points. This interval was chosen because it is evident from an examination of the original data that no information exists in these regions. Attempting to restore 32 coefficients produced highly erroneous results. A much narrower band of frequencies must be attempted if a reasonable result is to be obtained. Solving for 10 (five unique complex) coefficients produced the restored peaks shown in Fig. 6c. A considerable improvement in resolution was obtained. The constraint assured that the base line was straight in the summation region, as it should be. However, the first negative sidelobes around the peaks were still rather large and the function dipped strongly negative between the peaks. More of the function will have to be constrained for additional improvement. Extending the summation region to include the first negative sidelobe on each side of the peaks and most of the negative dip between the peaks produced the restoration shown in Fig. 6d, which has a much-improved appearance. However, this operation is of questionable validity, because we suspect that information may exist in these included regions. For further improvement, without assumptions, additional constraints are needed. We shall see in the next chapter that the constraint of minimum negativity yields a better result for the restoration of these infrared lines.

In data-point units, the original infrared peaks were about 34 units wide (full width at half maximum). This corresponds to an actual width of approximately 0.024 cm^{-1}. The impulse response function was about 25 units wide. After inverse filtering and restoration of the Fourier spectrum, the resolved peaks were 11 and 14 units wide, respectively. This is close to the Doppler width of these lines.

VI. Concluding Remarks

Most of the original work of this chapter involves extrapolation of discrete Fourier spectra from some "region of support." This region of support may be the interferogram from Fourier transform spectroscopy. It may also be the inverse-filtered Fourier spectrum determined according to the convolution model, with the high-frequency spectrum truncated because of the increasing noise. A somewhat detailed discussion of the convolution model was included, because both the inverse filtering and the subsequent extension of the Fourier spectrum for restoration may be quantitatively discussed in terms of this model.

Even though restoration in two distinct spectral bands leads to very fast algorithms, it is still not optimum because of the residual error in the low-frequency band of spectral components used as a region of support. Perhaps the requirement that the inverse-filtered low-frequency spectrum (or, equivalently, its corresponding spatial function) be held constant for the restoration could be relaxed to allow small variations in the low-frequency spectral components closest to the truncation point. This could effect an improved solution that better satisfies the relevant constraints. Ways of weighting some spectral components more than others are currently being explored.

We shall end this chapter with a few practical remarks concerning the calculation of the inverse-filtered spectrum. In this research the Fourier transform of the data is divided by the Fourier transform of the impulse response function for the low frequencies. Letting $\hat{o}$ denote the inverse-filtered estimate and n the discrete integral spectral variable, we would have for the inverse-filtered Fourier spectrum

$$\hat{O}(n) = I(n)/\tau(n). \tag{20}$$

For the cases where τ is very small or zero at certain frequencies, the indeterminate discrete Fourier spectral components corresponding to these values of n may be determined along with the high-frequency band when the Fourier spectrum is continued. It is necessary to make sure that these coefficients, along with their associated sine and cosine terms, are included in the resulting equations that enforce the constraints.

Data are often normalized so that the area under the curve is preserved. This area is given by the dc spectral term, that is, for $n = 0$. To preserve the area in the discrete inverse-filtered result, every term should be multiplied by the dc spectral components of the impulse response function (if the impulse response function has not been normalized earlier). We would then have for $\hat{o}$

$$\hat{O}(n) = [I(n)/\tau(n)]\tau(0). \tag{21}$$

If unity is to be preserved for the area throughout, then each term of the inverse-filtered result should be multiplied by $\tau(0)/I(0)$. Note that spectrum continuation, because it does not involve the dc term, does not change the area under the data curve.

References

Bell, R. J. (1972). "Introductory Fourier Transform Spectroscopy." Academic Press, New York.

Bracewell, R. N. (1978). "The Fourier Transform and Its Applications." McGraw-Hill, New York.

De Boor, C. (1978). "A Practical Guide to Splines." Springer-Verlag, New York.

Fiddy, M. A., and Hall, T. J. (1981). *J. Opt. Soc. Am.* **71**, 1406.

Hamming, R. W. (1962). "Numerical Methods for Scientists and Engineers." McGraw-Hill, New York.

Howard, S. J. (1978). M.S. Thesis, University of Southern Mississippi, Hattiesburg.

Howard, S. J. (1981a). *J. Opt. Soc. Am.* **71**, 95.

Howard, S. J. (1981b). *J. Opt. Soc. Am.* **71**, 819.

Howard, S. J. (1982). Ph.D. Dissertation, Florida State University, Tallahassee.

Howard, S., and Rayborn, G. H. (1980). Deconvolution of Gas Chromatographic Data. NASA Contractor Report CR-3229.

Hunt, R. B., Brown, L. R., and Toth, R. A. (1978). *J. Mol. Spectrosc.* **69**, 482.

Jansson, P. A. (1968). Ph.D. Dissertation, Florida State University, Tallahassee.

Papoulis, A. (1962). "The Fourier Integral and Its Applications." McGraw-Hill, New York.

Schafer, R. W., Mersereau, R. M., and Richards, M. A. (1981). *Proc. IEEE* **69**, 432.

Sheen, S. H., and Skofronick, J. G. (1974). *J. Chem. Phys.* **61**, 1430.

Stockham, T. G., Jr. (1966). *1966 Spring Joint Computer Conference Proceedings* **28**, 229.

Chapter 13 | Minimum-Negativity-Constrained Fourier Spectrum Continuation

Physics Department, The Florida State University
Tallahassee, Florida

List of Symbols

A_n — coefficients of cosine terms of discrete Fourier series
b — number of unique coefficients in complex Fourier series expression for $u(k)$
B_n — coefficients of sine terms of discrete Fourier series
c — number of unique coefficients in complex Fourier series expression for $v(k)$
$g(k)$ — function of $u(k) + v(k)$ defined as $g(k) = 1/(1 + \exp\{K[u(k) + (k)]\})$
H — Heaviside step function: $H(x) = 1$ when $x > 0$, 0 when $x \leq 0$
k — integer variable for spatial function, the index of sampled values
K — variable parameter in expression for $w(k)$
n — integer Fourier spectral variable
N — number of sample points, which is the same as the number of complex coefficients
T_m — measured transmittance for absorption spectroscopy
$u(k)$ — discrete Fourier series composed of only low-frequency and dc terms
U — observed flux after absorption
U_0 — incident flux on sample gas
$v(k)$ — discrete Fourier series composed of high-frequency terms such that when added to $u(k)$ a series of a complete Fourier spectral band is formed
x_c — denotes the original extent of the interferogram

I. Theory

This chapter continues the development of restoration with the application of constraints that began in the preceding chapter. See Chapter 12 for an explanation of terms.

We discovered in Chapter 12 that the spatial function as given by the discrete Fourier transform (DFT) is a discrete Fourier series. Letting $u(k)$ denote the (known) series consisting of only low-frequency terms and $v(k)$ the series consisting of only high-frequency terms, we want to determine the unknown coefficients in $v(k)$ that best satisfy the constraints. Expressing deviations of the total function forbidden by the constraints as some function of $u(k) + v(k)$, we shall try to determine the coefficients of $v(k)$ that minimize these deviations. Sum-of-squares expressions for these measures of the error have been found to result in the most efficient computational schemes.

Minimizing the sum-of-squares expression for the constraint of finite extent resulted in a set of linear equations in the unknown coefficients. Chapter 12 closes with applications of the procedure that enforce this constraint. In this chapter we deal primarily with the constraint of minimum negativity, although we shall see later how other constraints may be included in the resulting formulation. The resulting much more general procedure has proved far more effective in correctly restoring the Fourier spectrum than the earlier methods mentioned. This observation is the motivation for reserving a separate chapter to discuss this method and its applications.

The constraint of minimum negativity (Howard, 1981) applies to data for which it is known that the correctly restored function should be all positive. For our formulation, we want to find the coefficients of $v(k)$ that best satisfy this constraint. These coefficients will be those that minimize the negative deviations in the total function $u(k) + v(k)$. The sum of the squared values of the negative deviations is given by

$$\sum_{k=0}^{N-1} \{H[-u(k) - v(k)][u(k) + v(k)]\}^2,$$

where H is the unit step function such that $H(x) = 1$ when $x > 0$ and $H(x) = 0$ when $x \le 0$. The set of coefficients in $v(k)$ that minimizes this expression is the desired solution. However, we cannot carry out the minimization procedure with this expression. We cannot take the partial derivatives with respect to this function because it is not continuous. However, consider the alternative expression

$$\sum_{k=0}^{N-1} \left(\frac{1}{1 + \exp\{K[u(k) + v(k)]\}} [u(k) + v(k)] \right)^2,$$

where K is a variable parameter. Note that $1/[1 + \exp(Ky)]$ approaches $H(-y)$ as $K \to \infty$ for arbitrary y. Thus, as K in this summation approaches infinity, the entire expression approaches arbitrarily close to the original expression with the step function. Now we have an expression in closed form that consists entirely of analytic functions. We may carry out the minimization procedure by taking the derivatives with respect to the unknown coefficients and setting the results equal to zero. For simplification let

$$g(k) = 1/(1 + \exp\{K[u(k) + v(k)]\}).$$

Then we may express the resulting set of equations as follows:

$$\frac{\partial}{\partial A_b} \sum_{k=0}^{N-1} \{g(k)[u(k) + v(k)]\}^2 = 0,$$

$$\frac{\partial}{\partial B_b} \sum_{k=0}^{N-1} \{g(k)[u(k) + v(k)]\}^2 = 0,$$

$$\frac{\partial}{\partial A_{b+1}} \sum_{k=0}^{N-1} \{g(k)[u(k) + v(k)]\}^2 = 0, \tag{1}$$

$$\vdots$$

$$\frac{\partial}{\partial B_{b+c-1}} \sum_{k=0}^{N-1} \{g(k)[u(k) + v(k)]\}^2 = 0.$$

Considering in detail the derivative with respect to A_b, we have

$$\sum_{k=0}^{N-1} \frac{\partial}{\partial A_b} \left(\frac{1}{1 + \exp\{K[u(k) + v(k)]\}} [u(k) + v(k)] \right)^2$$

$$= 2 \sum_{k=0}^{N-1} \cos\left(\frac{2\pi}{N} bk \right) \frac{u(k) + v(k)}{1 + \exp\{K[u(k) + v(k)]\}}$$

$$\times \left(\frac{1}{1 + \exp\{K[u(k) + v(k)]\}} - K \exp\{K[u(k) + v(k)]\} \right)$$

$$\times \frac{u(k) + v(k)}{1 + \exp\{K[u(k) + v(k)]\}} \right) = 0. \tag{2}$$

The second term in the rightmost set of large parentheses approaches zero as $K \to \infty$, so that for K large we have a good approximation

$$\sum_{k=0}^{N-1} \frac{u(k) + v(k)}{(1 + \exp\{K[u(k) + v(k)]\})^2} \cos\left(\frac{2\pi}{N} bk\right) = 0. \qquad (3)$$

The other derivatives are calculated in a similar manner, and we have for the complete set of equations

$$\sum_{k=0}^{N-1} g^2(k)[u(k) + v(k)]\cos\left(\frac{2\pi}{N} bk\right) = 0,$$

$$\sum_{k=0}^{N-1} g^2(k)[u(k) + v(k)]\sin\left(\frac{2\pi}{N} bk\right) = 0,$$

$$\sum_{k=0}^{N-1} g^2(k)[u(k) + v(k)]\cos\left[\frac{2\pi}{N} (b+1)k\right] = 0, \qquad (4)$$

$$\vdots$$

$$\sum_{k=0}^{N-1} g^2(k)[u(k) + v(k)]\sin\left[\frac{2\pi}{N} (b+c-1)k\right] = 0,$$

where $b + c < N/2$. Because the unknown coefficients are also in the exponents, these equations are nonlinear in the coefficients and would ordinarily be difficult to solve. However, certain iterative techniques may be applied that not only yield a converging solution but also bring about considerable simplification in the equations. A form of the method of successive substitutions was developed for this research. The Newton–Raphson method also shows promise. However, because a matrix square in the number of equations, $2c$, must be solved for each iteration, the Newton–Raphson method may not be appropriate when c is very large. It does, however, converge faster than the method of successive substitutions. (See Hildebrand, 1965, for a discussion of the method of successive substitutions and the Newton–Raphson method.)

We now discuss in detail the particular iterative method chosen for this research. First, each of the equations is solved for one of the unknown coefficients within the first set of brackets. There are a number of ways of doing this. However, a reduction in the number of calculations results from solving for the coefficient associated with the multiplicative sinusoidal factor in each equation. Only the first equation of the resulting set will be written, because it illustrates all the salient features:

$$
A_b = -\sum_{k=0}^{N-1} \frac{u(k) + v(k) - A_b \cos[(2\pi/N)bk]}{(1 + \exp\{K[u(k) + v(k)]\})^2}
$$

$$
\div \sum_{k=0}^{N-1} \frac{\cos^2[(2\pi/N)bk]}{(1 + \exp\{K[u(k) + v(k)]\})^2} .
$$

(5)

Next, the initial values of the coefficients are substituted in the right-hand sides of the equations. Zeros are substituted if no better initial values are known. The substantial advantage of this approach now becomes apparent. For K sufficiently large, if the function $u(k) + v(k)$ is negative by even the smallest amount, the exponential will be extremely small and the factor $1/(1 + \exp\{K[u(k) + v(k)]\})$ will be approximately unity. If $u(k) + v(k)$ is positive by even the smallest amount, the factor $1/(1 + \exp\{K[u(k) + v(k)]\})$ essentially vanishes. These approximations approach exactness as $K \to \infty$. Thus the equations may be expressed essentially as follows:

$$
A_b = -\sum_{k=0}^{N-1} \left[u(k) + v(k) - A_b \cos\left(\frac{2\pi}{N}bk\right) \right]
$$

$$
\times \cos\left(\frac{2\pi}{N}bk\right) \left[\sum_{k=0}^{N-1} \cos^2\left(\frac{2\pi}{N}bk\right) \right]^{-1},
$$

$$
B_b = -\sum_{k=0}^{N-1} \left[u(k) + v(k) - B_b \sin\left(\frac{2\pi}{N}bk\right) \right]
$$

$$
\times \sin\left(\frac{2\pi}{N}bk\right) \left[\sum_{k=0}^{N-1} \sin^2\left(\frac{2\pi}{N}bk\right) \right]^{-1},
$$

$$
A_{b+1} = -\sum_{k=0}^{N-1} \left\{ u(k) + v(k) - A_{b+1} \cos\left[\frac{2\pi}{N}(b+1)\right] \right\}
$$

$$
\times \cos\left[\frac{2\pi}{N}(b+1)k\right] \left\{ \sum_{k=0}^{N-1} \cos^2\left[\frac{2\pi}{N}(b+1)k\right] \right\}^{-1},
$$

(6)

$$
\vdots
$$

$$
B_{b+c-1} = -\sum_{k=0}^{N-1} \left\{ u(k) + v(k) - B_{b+c-1} \sin\left[\frac{2\pi}{N}(b+c-1)k\right] \right\}
$$

$$
\times \sin\left[\frac{2\pi}{N}(b+c-1)k\right] \left\{ \sum_{k=0}^{N-1} \sin^2\left[\frac{2\pi}{N}(b+c-1)k\right] \right\}^{-1}
$$

for $u(k) + v(k) < 0$. That is, take the summation only over those data points for which the function $u(k) + v(k)$ is negative. It is evident that considerable computation is saved by avoiding calculation of the exponentials. Note that the negative values of $u(k) + v(k)$ change in each iteration as the new values of the coefficients are determined, so that a different summation is made in each iteration. For this reason it is necessary to test $u(k) + v(k)$ for its negative values before summing over them.

Although this procedure was developed from the constraint of minimum negativity, it will easily accommodate other constraints also, with only slight modification. Note that if the summation is not over a different set of data points for each iteration, but over a fixed set of points, the summation need be computed only once, because $u(k)$ and the sinusoids are constant for each value of the variable k. This is true for the finite-extent constraint, in which the summation is taken over the fixed interval outside the known extent of the data. The form of Eqs. (6) would be the same as that taken by the equations of finite extent when the method of successive substitutions is used. The constant summation interval of finite extent yields the constant coefficients (not to be confused with the sought coefficients) of the set of linear equations in the unknowns. The form of the method of successive substitutions for a set of linear equations is known as the Gauss–Seidel or the Jacobi method. In the former method each new unknown determined is used in improving the unknown in the succeeding equation. In the latter method every new unknown is determined from all the preceding ones before another iteration is begun. That both constraints result in very similar equations under the method of successive substitutions strongly suggests that both constraints could probably be included in one formulation, so as to yield the optimum resolution for a given set of data. It is found in practice that all that is necessary to implement successfully both constraints on the same set of data is to take the summation over all data points outside the known extent of the data (for the constraint of finite extent) while summing the negative values only over all other intervals (for the constraint of minimum negativity).

Further, if the data have an upper bound as well as a lower bound, this additional constraint may also be applied by summing over the data points above the upper bound [after subtracting the upper-bound value from $u(k) + v(k)$] as well as those below the lower bound. Although this additional constraint has been successful in minimizing values above the upper bound, it has resulted in little overall improvement in the restoration for the small number of data sets attempted here.

The modified procedure discussed here can successfully accommodate any of the constraints mentioned or any combination of them in a relatively uncomplicated way. Applying all the constraints simultaneously produces a result much more likely to be correct than would be the result of applying each constraint separately. The general procedure developed permits great versatility.

There are two different procedures by which the iteration may be carried out. One may substitute each improved coefficient, along with all the others, into the succeeding equation. This is known as the *point successive* procedure. Alternatively, one may determine all the unknown coefficients in all the equations using only the set of coefficients determined from the previous iteration. This is known as the *point simultaneous* procedure. Undesirable convergence properties of the point simultaneous procedure motivated its abandonment early in the research, however. The former procedure converges almost always when the tolerance is not chosen exceedingly small. The tolerance determines when the iteration is stopped. The iterations continue until the maximum change in any coefficient over its value in the previous iteration is less than the tolerance. With this procedure, convergence is quick to a large tolerance and much slower to a smaller tolerance. In this research the tolerance was usually taken to be about $\frac{1}{100}$ of the values close to the truncation point. With this tolerance, most of the restorations shown in this work usually required 10–50 iterations. However, convergence for some data was very slow. Occasionally a data run would take several hundred iterations. If the processing and restoration of data becomes a routine procedure in the laboratory, tolerances larger than that discussed here would probably be preferable. Tolerances several times larger have produced very satisfactory results.

II. Application

A. ARTIFICIAL DATA

The procedure for implementing the constraint of minimum negativity as given in the preceding paragraphs will first be applied to the deconvolution of the artificial data shown in Fig. 2 of Chapter 12, to permit a better comparison of the effectiveness of this constraint with that of the constraint of finite extent. Figure 1a shows the result of inverse-filtering the

Gaussian peak shown in Fig. 2c of Chapter 12 with six (complex) coefficients retained in its spectrum, and is the same function as that shown in Fig. 4a of Chapter 12. Restoration of 32 (16 complex) coefficients produced the result shown in Fig. 1b. This is slightly better than the result obtained with the constraint of finite extent in which the summation interval included the first negative sidelobes around the peak, which is shown in Fig. 4b of Chapter 12. It is vastly better, however, than the result obtained with the summation interval taken farther away from the peak as shown in Fig. 5 of Chapter 12.

We have found it to be generally true that the constraint of minimum negativity produces results much superior to those obtained with the constraint of finite extent. The minimum-negativity procedure has also been found to be extraordinarily insensitive to noise and other error. This is contrasted with the equations resulting from the constraint of finite extent, for which usually only a narrow band of coefficients may be permitted restoration to achieve a stable solution. The best overall results, however, are obtained with a combination of the two constraints.

Figure 2 shows the constraint of minimum negativity applied to the same deconvolution as that shown in Fig. 1 but with different truncation

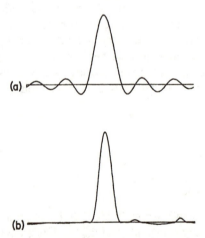

Figure 1 Restoration by inverse filtering of the low-frequency band followed by spectral restoration of the high-frequency band with the constraint of minimum negativity. (a) Inverse-filtered result shown in Fig. 4a of Chapter 12. Six coefficients are retained. (b) Restored function produced by restoring 16 (complex) coefficients to the Fourier spectrum by applying the constraint of minimum negativity.

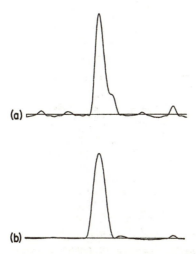

Figure 2 Effect on spectral restoration of choosing a different truncation point for inverse filtering of the peak shown in Fig. 2c of Chapter 12. (a) Restored function produced by restoring 16 coefficients to the spectrum. Seven (complex) coefficients were retained in the inverse-filtered result. The effect of retaining too many error-laden coefficients in the inverse-filtered result is illustrated. (b) Restored function produced by restoring 16 coefficients to the spectrum. Five (complex) coefficients were retained in the inverse-filtered result.

points for the Fourier spectrum. Figure 2a shows restoration to the inverse-filtered estimate with seven (complex) coefficients retained, and illustrates the distortion occurring when too many noise-laden coefficients are retained in the Fourier spectrum. From Fig. 3b of Chapter 12 it is evident that the seventh coefficient contains a large amount of noise error. Figure 2b shows restoration to the inverse-filtered result with only five complex coefficients retained in the Fourier spectrum. It differs little from the restoration with only six coefficients retained in the inverse-filtered estimate shown in Fig. 1. For both cases shown in Fig. 2, 16 complex coefficients were restored.

B. EXPERIMENTAL GRATING SPECTROSCOPY DATA

Figures 3 and 4 illustrate the application of the constraint of minimum negativity for restoring to inverse-filtered infrared spectral lines. A Gaussian function was chosen as an approximation to the impulse response function and 32 (16 unique complex) coefficients were restored in each

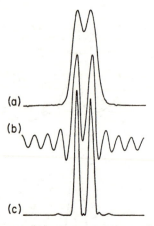

Figure 3 Restoration of Fourier spectrum to the inverse filtering of two noisy infrared peaks using the constraint of minimum negativity. (a) Two merged infrared peaks. (b) Inverse filtering of the infrared peaks with the spectrum truncated after the 10th coefficient. (c) Spectrum restored by minimizing the sum of the squares of the negative regions of the inverse-filtered result. Sixteen (unique complex) coefficients were restored.

case. The plots in Figs. 3a and 3b are the same as the plots in Figs. 6a and 6b of Chapter 12 and permit a comparison of the constraints of minimum negativity and finite extent for experimental data. Figure 3c shows the restoration using the constraint of minimum negativity. It is obviously far superior to the restoration shown in Fig. 6c, which was obtained with the constraint of finite extent. In Fig. 3c the base line is quite straight in the regions outside the immediate vicinity of the peaks and is correct there, because examination of Fig. 3a reveals that no information exists in these regions. If, however, too many high-noise-contaminated coefficients are retained in the inverse-filtered spectrum, false detail will show up in these regions. Examination of the base line away from the peaks in the original data thus provides a crude measure of the error in the restoration, especially that caused by retaining too many error-laden high-frequency coefficients.

The data illustrated in Fig. 4a are methane absorption lines (0.02 cm^{-1} wide) observed with a four-pass Littrow-type diffraction grating spectrometer. For these data also, 256 points were taken. The data were obtained at low pressure, so that Doppler broadening is the major contributor to the

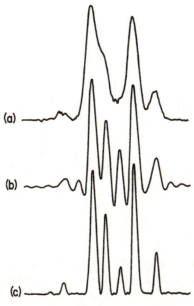

Figure 4 Restoration of Fourier spectrum to the inverse filtering of strongly merged infrared peaks using the constraint of minimum negativity. (a) Noisy infrared data. (b) Inverse filtering of the infrared data with the spectrum truncated after the 15th coefficient. (c) Spectrum restored by minimizing the sum of the square of the negative regions of the inverse-filtered result. Sixteen (unique complex) coefficients were restored.

true width of the lines. The straightforward inverse-filtered estimate with 15 (complex) coefficients retained is shown in Fig. 4b. Figure 4c shows the restored function. The positions and intensities of the restored absorption lines agree well with other independent studies. In data-point units, the original peaks were 19 units wide (full width at half maximum). The restored peaks are about 8–9 units wide, or close to the expected Doppler width. The Gaussian impulse response is approximately 16.5 units wide.

Note that in both Figs. 3 and 4 the negative values of the restored functions are very small.

Figures 5–11 illustrate the restoration process in the presence of a drifting base line. These data are methane absorption lines taken with a four-pass Littrow-type diffraction grating spectrometer. For these data

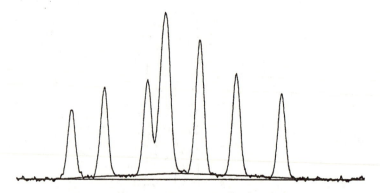

Figure 5 Noisy infrared data (2048 sample points) with a spline fit to the curved base line.

2048 data points were taken. The impulse response function was approximated by a gaussian. The true width of these lines is approximately 0.02 cm^{-1}.

The original data are shown in Fig. 5. A spline fit (De Boor, 1978) has been made to the curved base line. These data are adjusted to a zero base line in accordance with the transmittance relation $T_m = U(x)/U_0(x)$, where T_m is the measured transmittance, $U_0(x)$ the unabsorbed flux, and $U(x)$ the observed flux after absorption by the gas in the sample. The relation between U and U_0 is given by the Bouguer–Lambert law (see

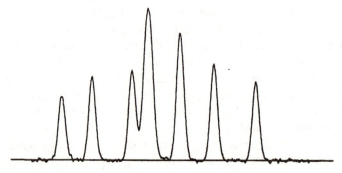

Figure 6 Data of Fig. 5 adjusted to a function with zero base line according to the transmittance relation.

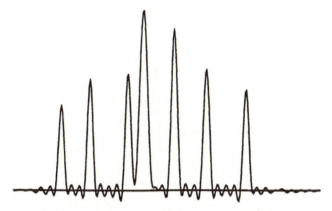

Figure 7 Result of inverse-filtering the corrected data of Fig. 6 with a Gaussian impulse response function having a FWHM of 39 units. The Fourier spectrum was truncated after the 35th (complex) coefficient.

Chapter 2). The corrected data are shown in Fig. 6. Figure 7 illustrates the deconvolution of these corrected data with the Fourier spectrum truncated after the 35th coefficient. The Gaussian impulse response function had a full width at half maximum (FWHM) of 39 (data-point) units. Restoring 32 (16 complex) coefficients to the truncated spectrum by minimizing the sum

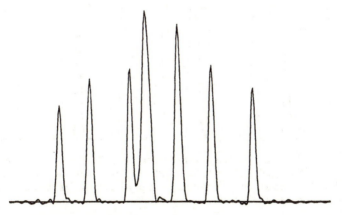

Figure 8 Result of restoring 32 (16 complex) coefficients to the inverse-filtered spectrum of the function shown in Fig. 7 by minimizing the sum of the squares of the negative deviations.

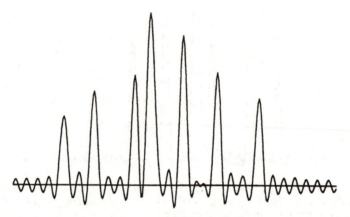

Figure 9   Result of inverse-filtering the corrected data of Fig. 6 with a Gaussian impulse response function having a FWHM of 46 units. The Fourier spectrum was truncated after the 30th (complex) coefficient.

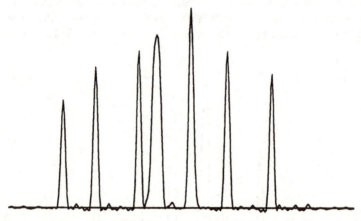

Figure 10   Result of restoring 62 (31 complex) coefficients to the inverse-filtered spectrum of the function shown in Fig. 9 with the constraint of minimum negativity. The broadened spectrum of these narrower peaks necessitates recovering more spectral components to obtain reasonable results.

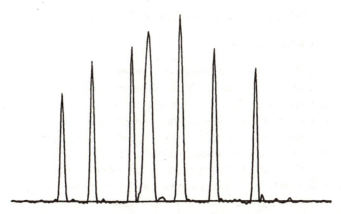

Figure 11 Result of restoring 74 (37 complex) coefficients to the inverse-filtered spectrum of the function shown in Fig. 9 by minimizing the sum of the squares of the negative deviations.

of the squares of the negative values of the function of Fig. 7 produces the improved infrared lines shown in Fig. 8.

Figure 9 shows the result of inverse filtering with a Gaussian impulse response function having a FWHM of 46 units. The Fourier spectrum was truncated after the 30th coefficient. Note that the broader impulse response function should result in narrower restored peaks. Restoring 62 (31 complex) coefficients to the Fourier spectrum of the inverse-filtered result of Fig. 9 by minimizing the sum of the squares of the negative deviations produces the result shown in Fig. 10. Note that these peaks are narrower than those of Fig. 8, which implies a broader Fourier spectrum. This necessitates recovering more spectral components to obtain reasonable results. Restoring 74 (37 complex) coefficients to the Fourier spectrum of the inverse-filtered result of Fig. 9 with the application of the constraint of minimum negativity produces the result shown in Fig. 11. Increased resolution is noted for all the peaks except the fourth one. A more detailed analysis has shown that this peak may be further resolved into three lines. The infrared lines obtained in the detailed restoration are closer to the true Doppler width than those obtained with the impulse response function having a FWHM of 39 units. We may physically interpret the restoration with the impulse response function having a 39-unit FWHM as a removal of only part of the distortion.

C. FOURIER TRANSFORM SPECTROSCOPY DATA

1. Introductory Remarks

The treatment of Fourier transform spectroscopy (FTS) infrared data involves application of essentially the same techniques that apply to the other infrared data. However, sufficient difference exists in practice to warrant a separate discussion. In grating spectroscopy, the data recorded are the infrared lines, with random noise superimposed. The data recorded in Fourier transform spectroscopy form an interferogram. The Fourier transform of the interferogram produces the infrared spectral lines. It is evident, then, that for FTS data the Fourier spectrum of the interferogram and the electromagnetic spectrum are the same. The restoration problem is slightly different for each of these cases. It is seldom possible to obtain the entire nonzero extent of the interferogram experimentally because this generally requires a very large optical path difference. So for FTS data the purpose is to restore the interferogram function by applying constraints to its transform, the infrared spectral lines. For grating spectroscopy, restoration results from the deconvolution operation. In that case it is the inverse-filtered transform that is extended, extrapolated from a function that is truncated because of the increasing noise. In practice, FTS Fourier frequencies are seldom boosted to amplitudes greater than those in the original interferogram unless the lines are considerably broadened.

In the operating mode customarily used, which is to determine the existence, location, and intensity of the spectral lines, the interferometer produces an interferogram that is symmetric about the zero displacement position. If the zero displacement position (the maximum point on the "central fringe") is taken as the origin of the interferogram function, the Fourier transform of this will produce an infrared spectrum that is real and symmetric about its origin. Note that for this case it does not matter whether the forward or inverse Fourier transform is used, because the same result will be obtained. If a one-sided interferogram is given, it is usually symmetrically extended about the origin. An improvement in the efficiency of the required calculations is afforded by a real and symmetric function. See Chapter 17 in Bell (1972) for a detailed discussion of the use of the fast Fourier transform (FFT) in efficient computational schemes of interferometric data.

However, for illustration, only one side of the interferogram and its spectrum will be shown, usually the function of the positive spatial and spectral variable. In other operating modes of the interferometer, asymmetric interferograms are produced that have a complex Fourier transform. Asymmetric interferograms will not be treated in this work. For a more complete discussion of Fourier transform spectroscopy, the reader should consult Bell (1972), Vanasse and Strong (1958), Vanasse and Sakai (1967), Steel (1967), Mertz (1965), the *Aspen International Conference on Fourier Spectroscopy* (Vanasse *et al.*, 1971), and the two volumes of *Spectrometric Techniques* (Vanasse, 1977, 1981). A review of early work, which includes several major contributions of his own, is given by Connes (1969). Another interesting paper on the earlier historical development of Fourier transform spectroscopy is that by Loewenstein (1966).

The incomplete interferogram that is recorded could be represented by the complete interferogram function multiplied by the rectangle function. The transform, of course, is the convolution of the infrared lines with the sinc function, which is the origin of the oscillatory artifacts that alternate positively and negatively about the base line around the peaks. The complete interferogram should yield a set of infrared spectral lines that are all positive, so that the constraint of minimum negativity could appropriately be applied here. If all infrared frequencies outside the band of interest are filtered out before reaching the interferogram, then the constraint of finite extent could also be applied. In practice this constraint is often awkward to apply, however, because the filters never terminate a spectrum abruptly.

For FTS data, artifact removal is a consideration that is as important as resolution improvement for most researchers in this field. Interferogram continuation methods are not as yet widely known in this area. Methods currently in widespread use that are effective in artifact removal involve the multiplication of the interferogram by various window functions, an operation called apodization. A carefully chosen window function can be very effective in suppressing the artifacts. However, the peaks are almost always broadened in the process. This can be understood from the uncertainty principle. A window that reduces the function most strongly closest to the end points will yield a transform for the modified function that must be broader than it was originally. Alternatively we may employ the convolution theorem, which has it that the multiplication of the interferogram by the window function is the same as convolution of the infrared lines with the Fourier transform of the window function. One window function widely used for apodization is the triangular window function, which is very effective in removing negative artifacts. However, because its transform is the sinc-squared function, the small positive sidelobes characteristic of this function show up around sharp peaks in the spectrum.

For most data it is seldom possible to effect a complete restoration of the interferogram. On recovering even a relatively few additional data points, though, a considerable improvement in resolution is usually obtained. However, the artifacts in most cases are still appreciable. It has been found that multiplying the interferogram by an appropriate window function, one that does not go to zero at the end points, will effect, on extending the interferogram, a removal of almost all of the artifacts as well as an improvement in resolution over that of the lines of the original interferogram. However, the ultimate resolution obtained with this window function is not as good as that obtained with the straightforwardly extended interferogram in the absence of the window function. The use of these window functions essentially involves a trade-off between artifact removal and resolution improvement.

For FTS data, then, the general procedure for improvement is first to determine the number of additional points on the interferogram that can be recovered, then to multiply the (unrestored) interferogram by one of a class of window functions appropriate to the number of points to be recovered, and finally to restore the additional points to the interferogram using as many of the applicable constraints as possible. The Fourier transform of the interferogram is almost always computed with the DFT to take advantage of the speed of the FFT algorithm. We may therefore borrow almost all of the DFT formulation of the restoration problem for application to FTS data. Actually, deconvolution with a symmetric impulse response function is the same as multiplying the interferogram by a high-pass filter, because the Fourier transform of a real, symmetric impulse response function is also a real, symmetric function, and would produce no phase shifts. So convolution and inverse filtering would then be only special cases of multiplying the interferogram by a certain class of window functions before its continuation.

2. Simulated Fourier Transform Spectroscopy Data

For a more quantitative analysis of the errors involved, artificial data will be addressed first. To see the effects of truncation and restoration of a single spectral line, we shall first consider a monochromatic electromagnetic source. Figure 12 could represent such a source. Of course, an entirely monochromatic source is physically impossible, but there are many

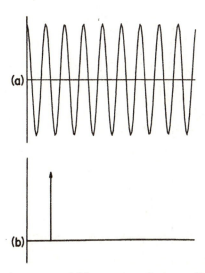

Figure 12 (a) Interferogram and (b) spectrum of a monochromatic source. The interferogram would be a cosine function that extends indefinitely, its Fourier transform the Dirac δ function.

wave trains in nature whose total length is very long compared with its wavelength and would therefore yield a very sharp spectral line. The case that we have illustrated could closely approximate many aspects of such a source. Figure 12a could be a segment of an interferogram of a monochromatic source. Its Fourier transform is shown in Fig. 12b and is a single sharp spectral line (two lines actually, that are symmetric about the origin). Theoretically, the Fourier transform of a monochromatic source of finite intensity is two Dirac δ functions. They would be infinitely high and have zero width. Figure 13 illustrates a more realistic case, that of a finite interferogram. Figure 13a could be the finite interferogram of an approximately monochromatic source. Its Fourier transform, the sinc function, is shown in Fig. 13b. This much-broadened spectral line with the ringing around it closely approximates the lines obtained in practice from the interferograms of approximately monochromatic sources. For improvement of these spectral lines we would like to remove the artifacts as well as increase the resolution. Most presently used methods only remove the artifacts (by the process of apodization). Nearly all of the window functions

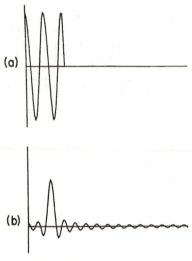

Figure 13 Simulation of a monochromatic source with a finite arm displacement of the interferometer. (a) Truncated cosine function (30 discrete data points). (b) Its Fourier transform, the sinc function, which simulates the infrared spectral line.

employed for apodization have the property of going smoothly to zero at the end point of the interferogram.

One window function widely used is the triangular window function. This function is shown in Fig. 14d. Multiplying the interferogram of Fig. 13a by this window function produces the altered interferogram shown in Fig. 15a. The Fourier transform of this is shown in Fig. 15b. Notice, importantly, that the negative artifacts have been completely removed. The small positive sidelobes remaining are not nearly so prominent as those in the original spectral line. This spectral function is the sinc-squared function, the Fourier transform of the triangle function. Use of the triangular window function has been very effective in reducing artifacts, but a heavy price has been paid. The spectral line has been considerably broadened, as is evident from a comparison of the widths of the lines in Figs. 13b and 15b.

If increasing the resolution is the only consideration, then we may simply extend the interferogram function by the methods discussed earlier. Only the constraint of minimum negativity will be used in the follow-

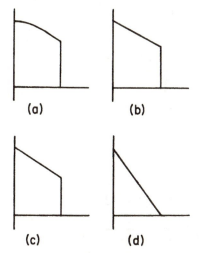

Figure 14   Four window functions used to multiply the interferogram. (a) Gaussian window. (b) Triangular window. (c) Triangular window of greater slope than (b). (d) Triangular window tapering to zero at the end point of the interferogram (used for apodization).

ing illustrations. The straightforward extension of the interferogram of Fig. 13a obtained by minimizing the negative values of the function shown in Fig. 13b produces the restored spectral line shown in Fig. 16b, if the interferogram is extended by the amount shown in Fig. 16a. The extent of the original interferogram is denoted by x_c (for cutoff point) on the plot. A considerable improvement in the resolution of the spectral line has been achieved. However, as is evident from examining the remaining function around the peak of Fig. 16b, the artifacts are still rather large.

To increase the resolution *and* simultaneously reduce the artifacts, the interferogram must be multiplied by the proper window function before extension, as discussed earlier. A variety of window functions have been found that will bring about, to some degree, the simultaneous achievement of both goals. However, they all share the common property that they do *not* taper to zero at the end point of the interferogram. Consider the triangular window function shown in Fig. 14b. Multiplying the interferogram of Fig. 13a by this window function and extending the interferogram to the point where the triangle function would cross the spatial axis yields the result shown in Fig. 17a. The considerably improved spectral line with

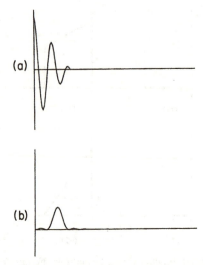

Figure 15 Apodization, or the reduction of artifacts in the spectral line by the multiplication of the interferogram by a window function that tapers to zero at the end point of the interferogram. (a) Cosine interferogram of Fig. 13a multiplied by the triangular window function of Fig. 14d. (b) Resulting spectral line, the sinc-squared function.

negative values almost nonexistent is shown in Fig. 17b. Note that the resolution is improved over that of the original spectral line of Fig. 13b and that the artifacts have been almost completely removed, with only small positive sidelobes remaining. However, note also that the resolution is not quite as good as that of the extended interferogram of Fig. 16b, which has not been multiplied by a window function. This comparison illustrates the trade-off between resolution and artifact removal brought about by the use of a window function. When restoring a given number of points to the interferogram, the researcher should choose the appropriate window function to emphasize whichever of the two aspects he or she deems most important.

Thus far, the discussion has been restricted to triangular window functions. However, it has been discovered that windows of many other functional forms are capable of bringing about improvement in the spectral lines. In this research the author has found that the window of Gaussian shape has produced the best overall results. With the same

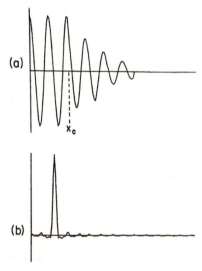

Figure 16 Straightforward extension of the interferogram by minimizing the negative values of the spectrum of Fig. 13b. (a) Interferogram (of 30 data points) extended by 50 data points. (b) Restored spectral line, which exhibits a considerable increase in resolution.

interferogram and extension by the same amount as in the previous example, premultiplication by the Gaussian window function shown in Fig. 14a produced the restored interferogram shown in Fig. 18a. The restored spectral line shown in Fig. 18b has a resolution much improved over that of Fig. 17b, where the triangular window function was used, yet the artifacts are no worse. The researcher should explore the various functional forms of the window function to find the one best suited for his or her particular data.

To see the effects that noise would have on a single spectral line, consider the interferogram of unity amplitude shown in Fig. 19a, where Gaussian noise of root-mean-square (RMS) amplitude 0.1 is added. It is evident that this high noise level considerably distorts the interferogram function. Figure 19b is the resulting spectral line. Note that it is quite smooth, which is not quite what we would expect considering the high noise level in the interferogram. The smoothness comes about because the noise extends over only part of the data field—the part over the finite interferogram. By minimizing the negative values of this function, we

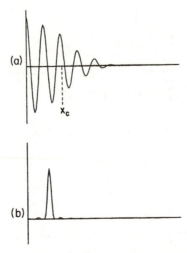

Figure 17 Effectiveness in removing the artifacts from the spectrum of multiplying the interferogram by the proper window function before extending the interferogram by a finite number of points. (a) Cosine interferogram of Fig. 13a premultiplied by the triangular window function of Fig. 14b before extending by 50 data points. (b) Restored spectral line.

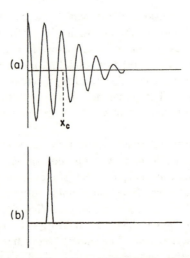

Figure 18 Interferogram of Fig. 13a multiplied by the Gaussian window function of Fig. 14a to effect artifact removal. (a) Interferogram extended by 50 data points. (b) Restored spectral line.

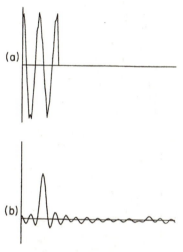

Figure 19   Interferogram of Fig. 13a with Gaussian noise of RMS amplitude 0.1 added. (a) Cosine interferogram of unity amplitude with random noise of RMS amplitude 0.1 superimposed. (b) Single spectral line with the oscillatory artifacts.

obtain the restored interferogram and spectral line of Fig. 20. Note that all cases shown of restoring this interferogram will involve extension by the same number of points. The resolution of this line is very much the same as that of the noise-free case. However, let us remove the debris from the base line to see what remains after reducing the artifacts. Multiplying this interferogram by the triangular window function of Fig. 14b and then extending, we obtain the interferogram shown in Fig. 21a. The restored spectral line is shown in Fig. 21b. Surprisingly, the resolution of the spectral line is almost the same as that of the noise-free case. However, many low-level artifacts have been introduced. Multiplying the interferogram by the Gaussian window function and extending yields the interferogram of Fig. 22a and the restored spectral line of Fig. 22b. The resolution of the spectral line has surprisingly changed very little as compared with the noise-free case. The main effect of the noise seems to be the introduction of many low-amplitude features scattered randomly about the base line.

Considering more complicated spectra, we show in Fig. 23 the (noise-free) interferogram and spectral lines for the superposition of two closely spaced frequencies. This example is shown to demonstrate the ability of the constraint of minimum negativity to separate these two closely spaced

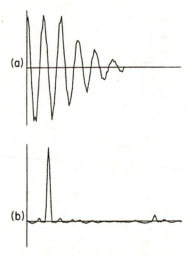

Figure 20 Straightforward extension of the noisy interferogram of Fig. 19a by minimizing the negative values of the distorted spectral line of Fig. 19b. (a) Interferogram extended by 50 data points. (b) Restored spectral line.

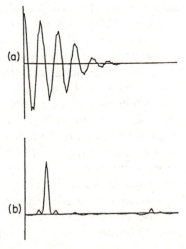

Figure 21 Interferogram of Fig. 19a multiplied by a triangular window function before extension in an attempt to remove the artifacts. (a) Interferogram premultiplied by the triangular window function of Fig. 14b before extending by 50 data points. (b) Restored spectral line.

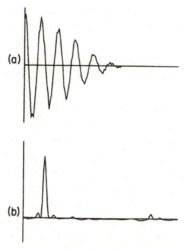

Figure 22 Interferogram of Fig. 19a multiplied by a Gaussian window function before extension in an attempt to remove the artifacts. (a) Interferogram premultiplied by the Gaussian window function of Fig. 14a before extending by 50 data points. (b) Restored spectral line.

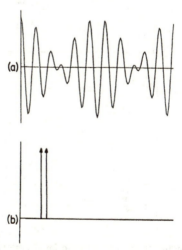

Figure 23 Interferogram and spectrum of two monochromatic sources that differ slightly in wave number. (a) Noise-free interferogram, understood to repeat indefinitely. (b) The two closely spaced spectral lines.

spectral lines. Figure 24a is the finite interferogram. Figure 24b is the spectrum. Note that the lines for this finite interferogram are completely merged. Extending the interferogram by the amount shown in Fig. 25a by minimizing the sum of the squares of the negative values of the merged spectral lines effectively separates these two lines, as Fig. 25b demonstrates.

Considering increasingly more realistic examples, we show in Fig. 26a the interferogram corresponding to the superposition of four monochromatic sources of varying intensity. The largest cosine in the interferogram has an amplitude of 1.6 units. The four sharp spectral lines of this spectrum are shown in Fig. 26b. A physically more realistic case, that of a finite interferogram with Gaussian noise of RMS amplitude 0.1 added, is shown in Fig. 27a. The degraded spectral lines are shown in Fig. 27b, which are now broadened and merged. Minimizing the negative values of this spectrum produces the extended interferogram and restored spectral lines of Fig. 28. As noted earlier, the noise seems to have little effect on the resolution. All the lines are completely resolved. This result also suggests that, even with the noise, there is not much interaction among the peaks

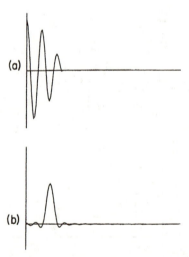

Figure 24 Interferogram of the two monochromatic sources that would be obtained for a finite maximum path difference of the interferometer. (a) Finite interferogram. (b) Recorded spectrum. The two lines are completely merged into one.

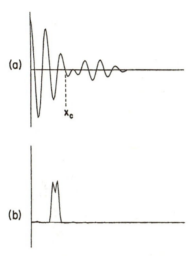

Figure 25 Interferogram of Fig. 24a extended by minimizing the negative values of the spectrum of Fig. 24b. (a) Interferogram extended by 50 data points. (b) Restored spectrum. The lines have been effectively resolved.

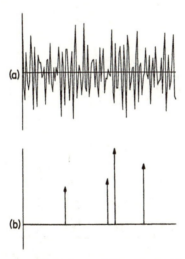

Figure 26 Interferogram and spectrum of the superposition of four monochromatic sources. (a) Noise-free interferogram. The magnitude of the largest of the four cosine functions is 1.6 units. (b) Spectrum composed of four Dirac δ functions.

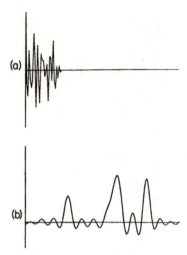

Figure 27 Finite interferogram of the four monochromatic sources of Fig. 26 with Gaussian noise of RMS amplitude 0.1 superimposed and the resulting degraded spectral lines. (a) Interferogram of 30 data points. (b) Merged and distorted spectral lines.

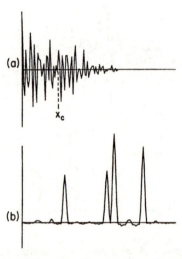

Figure 28 Extended interferogram and restored spectrum produced by minimizing the negative values of the spectrum of Fig. 27b. (a) Interferogram extended by 50 data points. (b) Restored spectral lines.

on restoration. That is, the restored lines are very much like the result that one would obtain by restoring each line separately and then superimposing them. Even though most peaks show little interaction, occasionally a restoration will show wide divergence from this, especially when the noise level is high. Although the resolution of these lines is good, large artifacts still remain in the restoration.

If we multiply the interferogram by the triangular window function of Fig. 14b before extension, we obtain the interferogram and spectral lines shown in Fig. 29. The artifacts have been considerably reduced, but a slight loss of resolution has occurred. The best overall results are obtained by premultiplying the interferogram by the Gaussian window function of Fig. 14a before extension. These results are shown in Fig. 30. The spectral lines of Fig. 30b are better resolved than those of Fig. 29b, yet the artifacts are no worse. The major effect of the noise for all cases seems to be the introduction of randomly located low-amplitude artifacts (over and above those obtained for the noise-free case). Note also that the degradation and restoration operations affect each spectral line the same way, regardless of its wavelength.

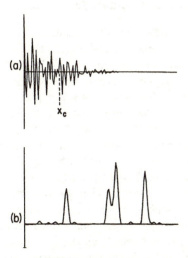

Figure 29 Interferogram of Fig. 27a multiplied by the triangular window function of Fig. 14b before extension to remove the artifacts. (a) Interferogram extended by 50 data points. (b) Restored spectral lines.

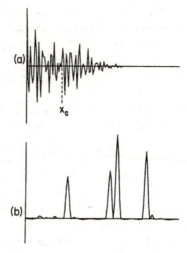

Figure 30 Interferogram of Fig. 27a multiplied by the Gaussian window function of Fig. 14a before extension to remove the artifacts due to a finite interferogram. (a) Interferogram extended by 50 data points. (b) Restored spectral lines.

3. Experimental Fourier Transform Spectroscopy

Finally, experimental FTS data will be addressed. For further discussion and description of the experimental data, and the experimental conditions under which the data were taken, see Toth *et al.* (1981). Given the interferogram, the Fourier transform must be computed to obtain the spectral lines, because the constraints are applied to the spectrum. We may attempt to restore the entire spectrum or isolated sets of spectral lines. The iterative method developed here seems to be adequate to restore a large spectrum consisting of many data points. The method of successive substitutions used is capable of accommodating a large set of equations. Attention here, however, will be restricted to isolated sets of lines. Because we are not sure at what spectral frequencies the information exists for these cases, only the constraint of minimum negativity can be validly applied for improvement. For a small set of spectral lines with only a few data points, it is usually unnecessary to multiply the interferogram function by a window function before extending, because the entire interferogram is recovered in most cases. The reason for this is that real spectral lines have a nonzero width, unlike the Dirac δ-function lines used in the artificial data, and a small number of data points implies a small number of discrete Fourier spectral components.

The experimental data illustrated here are methane absorption lines observed in the 3-μm-wavelength infrared spectral region. The narrow Lorentzian lines due to the molecular transitions are broadened by Doppler effects to produce a final line shape that is approximately Gaussian. See Chapter 1 for a more thorough theoretical investigation of the broadening factors for these lines and a better approximation to the true line shape. A two-sided interferogram is acquired that extends 50 cm on each side of the central fringe. Because 50 cm does not extend to the "washout" point on the interferogram (in practice, the washout point is usually considered to be the point where the RMS amplitude of the noise exceeds the magnitude of the signal), the line shape recorded will be the convolution of the original line shape with the sinc function. Minimizing the negative values of this final spectral function is expected to produce a reasonable approximation to the complete interferogram and to the true line shape.

A very-low-frequency sinusoid was superimposed on these spectral lines owing to channeling. This comes about by reflections from the window surfaces that contain the sample gas. These often result in a spike on the interferogram, which produces a superimposed sinusoid on transforming. Rather than removing the sinusoid from the entire data set, we fitted a smooth curve to the base line of each isolated set of lines treated.

There are three major sources of error in these particular data. There is, of course, the noise. Then there is the inevitable error due to the base-line fit of this strongly varying function. Third, it is impossible to obtain completely isolated lines. Because the data are convolved with the sinc function, the peaks die out slowly, and an "isolated" set of lines never dies out exactly to zero. Some adjustment of the end points of this type of data is almost always necessary to make the data go smoothly to zero there. In spite of these errors, reasonably good restorations were obtained for all three of the data sets treated here.

The simplest case, that of two large infrared lines, is shown in Fig. 31a. A smooth curve was fitted to the base line as shown. A spline-fitting computer program developed by De Boor (1978) was used to obtain this fit very conveniently. After the fit was obtained, the data were adjusted to a flat base line, as shown in Fig. 31b, and the data field was extended by padding with zeros to yield an overall data field of $2^8 = 256$ points. Taking the Fourier transform, we obtained the interferogram function shown in Fig. 32. (Even though it was not obtained directly from the interferometer

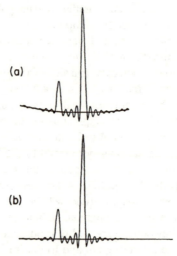

(a)

(b)

Figure 31 Experimentally obtained methane absorption lines in the infrared spectral region. (a) Two isolated methane spectral lines. A smooth curve is fitted to the base line. (b) Replotted lines. All points on the smooth curve were assigned the value of zero, and all values on the spectral lines are adjusted accordingly.

as recorded data, we shall refer to this function as the interferogram function to avoid the introduction of new and possibly confusing terminology.) Note that, as expected, the interferogram function does have a rather sharp cutoff point. The residual low-amplitude interferogram extended beyond the cutoff point is due to the errors discussed in the preceding

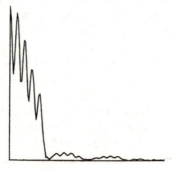

Figure 32 Interferogram function for the set of spectral lines of Fig. 31b.

paragraph. The amplitude of this extended interferogram function provides a crude measure of this error. This residual extended interferogram
function is cleanly truncated so as to yield a smooth noise-free set of
infrared spectral lines to which the constraints may be applied.

Note that the definition of the Fourier spectrum presented here, which
we have given as the Fourier transform of the function or data under
consideration, we may consider the interferogram function to be the
"Fourier spectrum" of the infrared lines under consideration. Although
the term "Fourier spectrum" is usually reserved for the forward transform,
this designation would make little difference on application to the symmetric FTS data treated here. We had earlier referred to the Fourier transform of the experimentally obtained interferogram as the Fourier spectrum.

The infrared spectral lines given by the truncated interferogram function are shown in Fig. 33a. This last step is not absolutely necessary
because the computer program written for restoration will restore a band
of Fourier spectral components for any given spatial function, regardless of

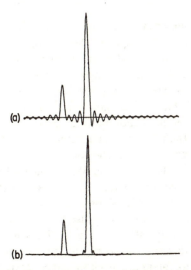

Figure 33 (a) Smooth set of infrared spectral lines produced by truncating the
interferogram function of Fig. 32 just before its sharp dropoff, keeping 27 data
points. (b) Restored infrared spectral lines produced by minimizing the negative
values of a and extending the interferogram by 60 data points.

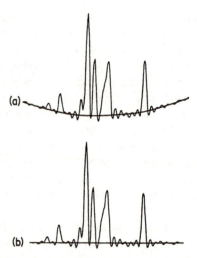

Figure 34 Methane absorption lines in the infrared spectral region. (a) Isolated set of infrared spectral lines. A smooth curve is fitted to the base line. (b) Result of taking the differences from the curved base line drawn and replotting.

whether the spectrum has been truncated or not. So, if we prefer, we may use the program simply to "improve" the spectrum rather than restore the Fourier spectrum of the spatial function after spectral truncation. However, error will occur for an interferogram that is not truncated if the index of the furthermost nonzero point is greater than the highest index in the band of points of the restored interferogram function. Minimizing the negative values of the lines shown in Fig. 33a produces the restored lines shown in Fig. 33b. Both lines show an increase in resolution. The origin of the two small "wings" on each side of the larger peak is unexplained. However, it occurs in the restoration of many lines of this data run.

A more complicated set of spectral lines is shown in Fig. 34a. A spline fit is made to the base line as shown, which allows adjustment to a flat base line, shown in Fig. 34b. Taking the Fourier transform of this yields the interferogram function shown in Fig. 35. Truncating this function after the sharp dropoff that signals the end of the interferogram, we take the Fourier transform to obtain the much smoother set of spectral lines shown in Fig. 36a. Applying the constraint of minimum negativity to these lines yields the restored lines of Fig. 36b. All the lines show an improvement in resolution. Note that the merged peaks in the original data separate into

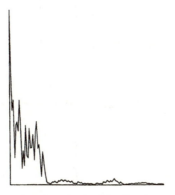

Figure 35 Interferogram function obtained by Fourier-transforming the spectral lines of Fig. 34b.

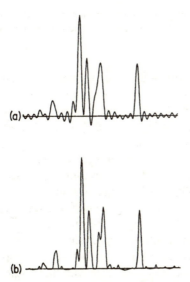

Figure 36 (a) Smooth spectrum produced by truncating the interferogram function of Fig. 35 just before its sharp dropoff, keeping 28 data points. (b) Restored infrared lines resulting from minimizing the negative values of a and extending the interferogram by 60 data points.

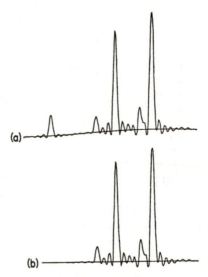

Figure 37 Isolated set of infrared absorption lines in the infrared spectral region (a) with a smooth curve fitted to the base line and (b) adjusted to zero base line.

two distinct lines in the restoration. As mentioned in the discussion of the previous example, two of the Fourier transforms involved here are not absolutely necessary. We could directly apply the constraint of minimum negativity to the data of Fig. 34b if they were relatively error-free.

For the final example, we have the set of spectral lines shown in Fig. 37a. Fitting a smooth curve to the base line and adjusting it to zero gives Fig. 37b. Taking the Fourier transform produces the interferogram function shown in Fig. 38. Choosing a truncation point in the sharp dropoff of this function and transforming produces the smooth set of spectral lines shown in Fig. 39a. Minimizing the negative values of these spectral lines produces the restored infrared spectrum shown in Fig. 39b. Increased resolution is observed. Figure 40 is included to emphasize that a good restoration is not obtained every time with this restoration procedure. By truncating the interferogram function of Fig. 38 and retaining one point less than for the previous figure, the Fourier transform shown in Fig. 40a is produced. Restoration from this function produces the restored lines of Fig. 40b. This restoration is not exactly what is expected from the original data, and is generally inferior to the restoration of Fig. 39b.

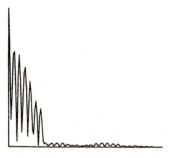

Figure 38 Interferogram function resulting from Fourier-transforming the spectrum of Fig. 37b.

I shall conclude this section with a few remarks on practical restoration with the DFT. To take the fullest advantage of the computational speed afforded by the FFT, the overall data field is usually composed of a number of data points given by 2 raised to some integral power. The data are usually padded with zeros to bring them up to this number. Algorithms

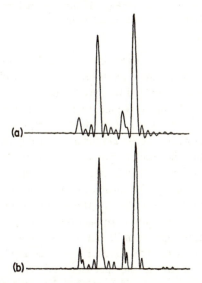

Figure 39 (a) Smooth set of infrared spectral lines produced by the Fourier transform of the truncated interferogram of Fig. 38. (b) Restored infrared spectral lines resulting from minimizing the negative values of a.

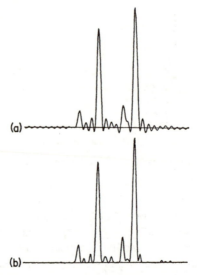

Figure 40 (a) Smooth set of infrared spectral lines resulting from Fourier-transforming the truncated interferogram of Fig. 38 but retaining one datum point less in the interferogram than in Fig. 39. (b) Restored infrared spectral lines resulting from minimizing the negative values of a.

for computing the FFT for bases other than 2, however, are becoming increasingly more popular with improvements in their efficiency.

Often FTS data are obtained that are "tightly packed"; that is, only the minimum number of points needed to represent them is given. If the spectral lines are given, the density of points representing them will be very sparse. The finite interferogram function for this case will extend over all (or nearly all) of the discrete points of the DFT. For the restoration, however, we find that there is little or no interferogram function to extend with this number of points. For further improvement, it is necessary to pad more zeros beyond the end of the interferogram function (or beyond both sides) to double the total number of data points (which brings the number up to the next integral power of 2). On transforming, it will be found that the density of points on the spectral lines has doubled. Applying the constraints will then restore the spectrum and extend the interferogram function. Greater resolution will be obtained for the spectral lines. This implies that more information is now carried by the data points. Many find this result surprising: that the additional data points included actually bring more information with them. Yet this property may be amply verified in practice with both artificial and experimental data. The additional information, of course, is brought in by the constraints. Also, if further extension of the interferogram is needed, more zeros may be simply padded onto it.

III. Concluding Remarks

The number of additional Fourier spectral components to recover is the option of the researcher. The number of iterations to execute with the most general computer program written is also the option of the researcher. A tolerance is presently used to determine the number of iterations performed. However, it is found in practice that only 5 or 10 iterations yield sufficiently accurate results for nearly all experimental data of interest. With the presently used computer program, restoration is to the spatial function, and the improved spatial function and the improved values of the coefficients are both generated with each iteration. If the improved Fourier spectrum is not desired, then additional computational time could be saved by neither reading nor writing the Fourier coefficients. When M data points are treated, the computer memory requirements are seldom more than $7M$ words. If it is not necessary to determine the extended Fourier spectrum, then more than $5M$ words are seldom needed in computer memory.

Furthermore, the iterative algorithm is very versatile in that the four constraints of finite extent, minimum negativity, minimization of values above an upper bound, and recovery of only a finite Fourier spectral band may be simultaneously applied. The formulation could also accommodate other constraints of definite limits. The method of successive substitutions is also able to accommodate a large number of equations, making it possible to solve for a large number of spectral components.

The researcher may want to combine the computer program used for inverse filtering with that used for spectral continuation so as to perform the complete restoration in one step. The truncation frequency of the inverse-filtered spectrum could be automatically determined from the RMS of the noise and the signal, and the amplitude of the spectrum of the impulse response function.

Some of the most valuable recent advances in the field of deconvolution have come from the field of digital image processing. It is therefore appropriate to note in conclusion that all the techniques developed here for spectroscopy may easily be generalized to two dimensions. The three constraints of finite extent, minimum negativity, and recovery of a finite spectral band, for example, are appropriate to image processing. For medical images, some of these constraints would apply. However, ways of applying other constraints appropriate to these data would have to be developed. Also, in two dimensions the memory requirements for the iterative algorithm are considerably less than for the one-dimensional case (in proportion to the number of data points). See Appendix A for the two-dimensional formulation of the iterative equations.

See Appendix B for recent improvements in the numerical efficiency of the iterative equations.

Appendix A

The equations for the coefficients of the sines and cosines in the frequency spectrum to be restored have also been derived for the two-dimensional case. The general expression for these coefficients for a square array is illustrated in the following two equations:

$$
A_{p,q} = -\left(\sum_{x=0}^{N-1} \sum_{y=0}^{N-1} \left\{ u_{x,y} + v_{x,y} - A_{p,q} \cos\left[\frac{2\pi}{N}(px + qy) \right] \right\} \right.
$$
$$
\left. \times \cos\left[\frac{2\pi}{N}(px + qy) \right] \right) \left\{ \sum_{x=0}^{N-1} \sum_{y=0}^{N-1} \cos^2\left[\frac{2\pi}{N}(px + qy) \right] \right\}^{-1} \quad \text{(A.1)}
$$

and

$$
B_{p,q} = -\left(\sum_{x=0}^{N-1} \sum_{y=0}^{N-1} \left\{ u_{x,y} + v_{x,y} - B_{p,q} \sin\left[\frac{2\pi}{N}(px + qy) \right] \right\} \right.
$$
$$
\left. \times \sin\left[\frac{2\pi}{N}(px + qy) \right] \right) \left\{ \sum_{x=0}^{N-1} \sum_{y=0}^{N-1} \sin^2\left[\frac{2\pi}{N}(px + qy) \right] \right\}^{-1} \quad \text{(A.2)}
$$

for $u_{x,y} + v_{x,y} < 0$, where $x, y, p, q = 0, 1, 2, \ldots, N - 1$, and N is the number of rows or columns. The discrete spatial variables are denoted by x and y, and the corresponding spectral variables by p and q. The functions $A_{p,q}$, $B_{p,q}$, $u_{x,y}$, and $v_{x,y}$ are the two-dimensional counterparts of the one-dimensional functions discussed in the text.

If one is looking at the discrete Fourier spectrum of real data with the dc component centered at $(N/2, N/2)$, one wishes to recover a band of

frequencies within the desired spectral range in the first quadrant including the zero frequencies (corresponding to sinusoids oriented along the x
or y axes). In addition, a corresponding band with one negative spectral
variable should be included (exclusive of the zero frequencies). This is
necessary to obtain a complete set of wave forms. This additional spectral
band could either be in the second or fourth quadrant. The choice is
completely arbitrary. No other frequencies are required. The reader will
recall that the complex coefficients in the other two quadrants are the
complex conjugates of those in the two quadrants chosen.

Appendix B

Recent investigations have been made into the computational efficiency
of the restoration methods discussed in Chapters 12 and 13. Rewriting Eqs. (6), (A.1), and (A.2) in different forms to perhaps find a form
that involved fewer computations was explored. Means to bring about a
more rapid convergence of these equations were also investigated. The
two-dimensional equations will be addressed in the following discussion,
although most of the techniques mentioned will apply to the one-
dimensional situation, also.

The summation over the squared sines and cosines, respectively, in the
denominator of Eqs. (A.1) and (A.2) essentially acts as a weighting function applied to each Fourier coefficient. Calculating these summations for
the coefficients restored over several iterations, the author observed that
this weighting function, generally, varied very slowly over the Fourier
frequencies and iterations. This strongly suggested that for the numerical
calculations involved, this function could be replaced by a constant value.

$$1/\sum_x \sum_y \cos^2\left[\frac{2\pi}{N}(px + qy)\right] \approx 1/\sum_x \sum_y \sin^2\left[\frac{2\pi}{N}(px + qy)\right] \quad \text{(B.1)}$$

$$\approx \text{const} = S.$$

Replacing all the weighting functions by a single constant value in the
numerical calculations for several test data sets resulted in little significant
change in the image over the first few iterations. A considerable reduction
in the number of computations results from the use of the constant factor.

To further reduce the computational burden, an attempt was made to
separate the variables. To see how this may be implemented, let us
consider Eq. (A.1), which enforces the minimum negativity constraint.

Note that it may also be written

$$A_{p,q} = -\frac{\Sigma_x \Sigma_y (u_{x,y} + v_{x,y})\cos[(2\pi/N)(px + qy)]}{\Sigma_x \Sigma_y \cos^2[(2\pi/N)(px + qy)]} + A_{p,q} \quad \text{(B.2)}$$

for $u_{x,y} + v_{x,y} < 0$. Replacing the summation by the constant discussed in the preceding paragraph and denoting $u_{x,y} + v_{x,y}$ by $h(x, y)$, we will have

$$A_{p,q} = A_{p,q} - S\sum_x \sum_y h(x, y)\cos\left[\frac{2\pi}{N}(px + qy)\right] \quad \text{(B.3)}$$

for $u_{x,y} + v_{x,y} < 0$. The $A_{p,q}$ on the right-hand side of this equation, as for Eqs. (6), is understood to be the old value of this coefficient. The $A_{p,q}$ on the left-hand side is the new (hopefully improved) value of this coefficient after the current iteration. Finally, the summation over the negative values is equivalent to multiplying $h(x, y)$ by a mask that zeros all but the negative values of this band-limited function. This mask is unity over the negative values and is zero elsewhere. It would be a rect function in one dimension. In two dimensions, though, it could have many varied patterns. Denoting this mask by $m(x, y)$ we now have

$$A_{p,q} = A_{p,q} - S\sum_x \sum_y h(x, y)m(x, y)\cos\left[\frac{2\pi}{N}(px + qy)\right]. \quad \text{(B.4)}$$

The finite extent constraint may easily be included in this formulation by simply letting the nonzero extent of the mask also include all values of the data field outside of the limited extent of the object.

We may include the coefficient for the sine corresponding to any given frequency by writing the coefficients in complex form:

$$A_{p,q} + jB_{p,q} = A_{p,q} - S\sum_x \sum_y h(x, y)m(x, y)\cos\left[\frac{2\pi}{N}(px + qy)\right]$$

$$+jB_{p,q} - jS\sum_x \sum_y h(x, y)m(x, y)\sin\left[\frac{2\pi}{N}(px + qy)\right]$$

$$= A_{p,q} + jB_{p,q} - S\left(\sum_x \sum_y h(x, y)m(x, y)\right.$$

$$\times \left\{\cos\left[\frac{2\pi}{N}(px + qy)\right] + j\sin\left[\frac{2\pi}{N}(px + qy)\right]\right\}\right)$$

$$= A_{p,q} + jB_{p,q} - S\left\{\sum_x \sum_y h(x, y)m(x, y)e^{[j(2\pi/N)(px+qy)]}\right\}.$$

$$\text{(B.5)}$$

The complex exponential separates into factors involving only a single variable. Lumping the constant S with the mask, we now have for the correction term

$$A_{p,q} + jB_{p,q} = A_{p,q} + jB_{p,q} - \sum_x e^{[j(2\pi/N)px]} \sum_y h(x,y)m(x,y)e^{[j(2\pi/N)qy]}.$$

$$(B.6)$$

This separation of the variables allows a vast reduction in the number of calculations. This savings becomes more significant as the number of data points increases. Further, note that the final summation in this equation is in the form of a one-dimensional Fourier transform. This implies that the considerable calculational advantage of the fast Fourier transform (FFT) algorithm may be used here. The entire summation may be performed by repeated application of the one-dimensional FFT. This implies that for any data set that it is practical to apply the FFT, it would be also practical to apply the nonlinear constraints for improvement.

There is no important disadvantage stemming from variable separation and use of the FFT, however. One is limited to the use of the "point simultaneous" procedure in the determination of the coefficients. That is, all of the coefficients must be computed in an iteration before they can be resubstituted back in the iterative equations. With the "point successive" method, the improved coefficient determined by one equation is substituted in the succeeding equation to render additional improvement. With the test data treated in this work, the point simultaneous procedure converged much more slowly than the point successive method.

The application of the point simultaneous procedure to two-dimensional equations yield adequate restorations for many interesting two-dimensional data. However, faster convergence is clearly desirable. Two techniques have been developed by the author that have proved successful in achieving more rapid convergence. One method involves the use of the finite extent constraint applied in the spectral domain. The other procedure involves the use of more or less ad hoc multiplicative factors that weight the spectrum corrections determined from Eq. (B.6), thereby producing spectral values closer to the original.

To understand how the application of the finite extent constraint in the spectral domain produces improved results, a brief review of Fourier theory is necessary. From the first iteration, we would desire the spatial function produced only by the *high frequencies* (beyond cutoff), which cancels the negative values of the given band-limited spatial function (produced by the low-frequency spectral band). Then multiplying this function by the mask would produce the spatial function that is the *negative* of the given (low-frequency) band-limited function multiplied by the mask. The convolution theorem tells us that multiplication by the mask in the spatial domain results in convolution with the Fourier transform of the mask in the frequency domain. This gives us another interpretation for the Fourier spectrum produced by the iteration. It would be the convolution of the Fourier transform of the mask with the (original) low-frequency spectrum. Alternatively, it could be considered as the convolution of the Fourier transform of the mask with the *negative* of the high-frequency spectrum that produces the spatial function that cancels the negative values of the band-limited function.

Interpreting the degradation as a convolution of the desired high-frequency spectrum with the mask transform suggests the use of deconvolution procedures to restore the spectrum. Inverse-filtering would not apply with this type of response function. The constraint of minimum negativity could not appropriately be applied as the spectral components are generally both positive and negative. The constraint of finite extent could appropriately be applied because all values in the low-frequency range could be interpreted as lying outside the extent of the high-frequency range. Because the spectrum we are considering is the Fourier transform of the negative values of the original spatial function, the correct spatial function that lies outside the range of these negative values should produce a Fourier transform that zeros all frequencies lying in the low-frequency spectral range. To implement this constraint numerically, we now apply the mask in the *spectral* domain, the reverse of the usual procedure, to remove those frequencies that lie outside the low-frequency range. (The high-frequency band is saved, of course.) Taking the inverse Fourier transform of this low-frequency band, we multiply the resulting spatial function by the negative image of the original spatial mask, which passes only those values outside the extent of the original negative region and eliminates all others. After taking the Fourier transform of this spatial function, the resulting spectral values are added to the retained high-frequency spectrum to yield the corrected high-frequency band. Finally, the *negative* of this high-frequency band is joined to the original correct low-frequency band to produce the entire improved Fourier spectrum. Taking the inverse Fourier transform of this, we are ready to begin another iteration.

The correction to the spectrum that we discussed in the foregoing may also be considered as a type of "reconvolving factor" because it may be obtained by "reconvolving," or convolving the high-frequency spectrum again with the Fourier transform of the mask that passed only the negative values. This reconvolution was accomplished in the transformed domain because the inverse transform of the high frequencies was multiplied by the mask. The resulting function was then transformed to the frequency domain. These spectral values were then subtracted from the original high-frequency band in an attempt to produce some approximation to the original spectrum before the first convolution. The motivation for applying these corrections, of course, was that as the high-frequency band obtained from the transform of the negative regions is considered a reasonable first approximation to the restored high-frequency spectrum, a reconvolution should yield at least a crude approximation to the changes produced by the first convolution. Adding the negative of these changes to the spectrum should produce a high-frequency band closer, overall, to the original. Reconvolution produces essentially the same corrections as applying the finite extent constraint in the frequency domain, for if one follows the development closely, it is discovered that one set of corrections is the negative of the other. Regardless of the interpretation given these corrections, it has been proved to bring about significant improvement in the resulting images for a variety of test data. However, because two additional FFTs are involved in the generation of these corrections, very little overall reduction in computational time is presently being achieved.

It has also been discovered that a variety of simple multiplicative factors applied to the high-frequency spectrum during an iteration was successful in bringing about more rapid convergence of the equations. The primary motivation for applying these was the observation that the largest spectral components immediately beyond cutoff that were restored in the first few iterations were usually considerably smaller than the original components, but that their signs were nearly all correct. Any simple positive factor that rises slightly above unity for the values immediately beyond cutoff should produce spectral values closer to the original. In practice, a wide variety of factors has been shown to produce improvement. In the test data sets the author used, values from unity to slightly above two could be applied without serious distortion in the image. For the larger factors, however, tapering window functions applied to the Fourier spectrum were needed to reduce the spurious high-frequency components that were occasionally generated. In the test data treated by the author, a window of Lorentzian form that tapered gradually from values of two (for the high-frequency spectrum immediately beyond cutoff) to the higher frequencies produced the best combination of rapid and distortionless restorations. This window was gradually lifted as more of the high frequencies were restored. The factor of two for the lower frequencies was altered only slightly, however, as it was noted that the correction to the spectrum generated on each iteration by the method was always underestimated.

Acknowledgments

There are several people I wish to thank for help in bringing this work to completion. I am deeply indebted to Dr. Robert H. Hunt in ways too numerous to mention. Many enlightening discussions were held with Bob Hunt and Pete Jansson on the subject of Chapters 12 and 13. Their knowledge and expertise were invaluable. I also appreciate their proof-reading the initial drafts. To acknowledge others whose peripheral support was very helpful, I would certainly have to mention Clydeyne Nelson, Ken Ford, Beverly, my wife, as well as many others in the Physics Department at Florida State University.

Some illustrations and passages were borrowed from two papers of mine published in the *Journal of the Optical Society of America* (January 1981 and July 1981).

References

Bell, R. J. (1972). "Introductory Fourier Transform Spectroscopy." Academic Press, New York.

Connes, P. (1969). "Lasers and Light." Freeman, San Francisco.

De Boor, C. (1978). "A Practical Guide to Splines." Springer-Verlag, New York.

Hildebrand, F. B. (1965). "Introduction to Numerical Analysis," pp. 450, 451. McGraw-Hill, New York.

Howard, S. J. (1981). *J. Opt. Soc. Am.* **71**, 819.

Loewenstein, E. V. (1966). *Appl. Opt.* **5**, 845.

Mertz, L. (1965). "Transformations in Optics." Wiley, New York.

Steel, W. H. (1967). "Interferometry." Cambridge University Press, New York.

Toth, R. A., Brown, L. R., Hunt, R. H., and Rothman, L. S. (1981). *Appl. Opt.* **20**, 932.

Vanasse, G. A., ed. (1977). "Spectrometric Techniques," Vol. 1. Academic Press, New York.

Vanasse, G. A., ed. (1981). "Spectrometric Techniques," Vol. 2. Academic Press, New York.

Vanasse, G. A., and Sakai, H. (1967). *Prog. Opt.* **6**, 261. North-Holland, Amsterdam.

Vanasse, G. A., and Strong, J. D. (1958). "Application of Fourier Transforms in Optics: Interferometric Spectroscopy. *In* "Classical Concepts of Optics" (J. D. Strong, ed.), Appendix. Freeman, San Francisco.

Vanasse, G. A., Stair, A. T., Jr., and Baker, D. J., eds. (1971). *Aspen Int. Conf. Fourier Spectrosc.* 1970. AFCRL-71-0019, Spec. Rep. No. 114. A. F. Cambridge Res. Lab., Bedford, Massachusetts.

Chapter 14 | Alternating Projections onto Convex Sets

Robert J. Marks II

Department of Electrical Engineering,
The University of Washington
Seattle, Washington

List of Symbols

$A, B, C, D, A_1, A_2, A_3$	sets of vectors or functions
E	the universal set
$\vec{u}_1, \vec{u}_2, \vec{v}, \vec{z}$	multidimensional vectors
$\vec{u}_A, \vec{u}_B$	vectors in the sets A and B
$u(x), v(x), u_1(x), u_2(x), w(x), z(x)$	functions of a continuous variable
$m(x)$	the middle of a signal
$U(\omega), V(\omega)$	the Fourier transforms of $u(x)$ and $v(x)$
POS	positive
BE	bounded energy
BS	bounded signals

CA	constant area
CP	constant phase
LV	linear variety
P	a projection matrix
P$_S$	projection onto a set **S**
P$_m$	projection onto set m
PI	pseudo-inverse
IM	signals with identical middles
M	number of sets
$\mathscr{O}$	an operator
S	a degradation matrix
T	matrix transposition (used as a superscript)
TL	time limited
Ω	bandwidth
X	duration limit
$d(x)$	a displacement function used to define linear varieties
α	(a) $0 \le \alpha \le 1$ parameterizes the line connecting two vectors; (b) $\alpha > 0$ is used to define a cone in a Hilbert space
λ	relaxation parameter
β, γ	finite real numbers used to define a subspace
(x, y)	coordinates on a two-dimensional plane
$\vec{i}, \vec{o}$	image and object vectors
$\vec{o}^{(n)}$	the result of iteration n on the restoration of a discrete object
$o^{(k)}[n]$	the result of iteration k on restoration of an object, $o[n]$
$o(x)$	an object that is a function of a continuous variable
$o^{(N)}(x)$	the result of iteration N on restoration of an object of a cone
$\mathscr{H}$	a Hilbert space
$\mathscr{I}$	the imaginary part of
$\mathscr{S}$	a subspace (or linear manifold) in a Hilbert space
$\mathscr{P}$	a projection operator
$\mathscr{R}$	the real part of
$\perp \mathscr{S}$	the subspace that is the orthogonal complement of $\mathscr{S}$
$\mathscr{P}_{\mathscr{H}}$	projection operator onto a set A_K
$u_{\mathscr{S}}(x)$	a function in a subspace $\mathscr{S}$
L_2	a Hilbert space with continuous variables
l_2	a Hilbert space with discrete variables
$u[n]$	element n in a sequence u

1	an identity operator
{ }	set
|	in set notation, read "such that"
∩	intersection
∈	is an element of
$\|\cdot\|$	l_2 or L_2 norm
⊥	perpendicular
ρ	the area of a signal

I. Introduction

Alternating projections onto convex sets (POCS)* [1] is a powerful tool for signal and image restoration and synthesis. The desirable properties of a reconstructed signal may be defined by several convex signal sets, which may be further defined by a convex set of constraint parameters. Iteratively projecting onto these convex constraint sets may result in a signal that contains all desired properties. Convex signal sets are frequently encountered in practice and include the sets of band-limited signals, duration-limited signals, signals that are the same (e.g., zero) on some given interval, bounded signals, signals of a given area, and complex signals with a specified phase.

POCS was initially introduced by Bregman [2] and Gubin *et al.* [3] and was later popularized by Youla and Webb [4] and Sezan and Stark [5]. POCS has been applied to such topics as sampling theory [6], signal recovery [7], deconvolution and extrapolation [8], artificial neural networks [9, 10, 11, 12, 13], tomography [1, 14, 15], and time-frequency analysis [16, 17]. A superb overview of POCS with other applications is in the book by Stark [1] and the monograph by Combette [18].

II. Geometrical POCS

Although signal processing applications of POCS use sets of signals, POCS is best visualized viewing the operations on sets of points. In this section, POCS is introduced geometrically.

---

*The *alternating* term is implicit in the POCS paradigm, but traditionally is not included in the acronym.

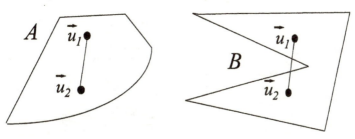

Figure 1 The set A is convex. The set B is not. A set is convex if every line segment with end points in the set is totally subsumed in the set.

A. CONVEX SETS

A set, A, is convex if for every vector $\vec{u}_1 \in A$ and every $\vec{u}_2 \in A$, it follows that $\alpha \vec{u}_1 + (1 - \alpha)\vec{u}_2 \in A$ for all $0 \le \alpha \le 1$. In other words, the line segment connecting $\vec{u}_1$ and $\vec{u}_2$ is totally subsumed in A. If any portion of the cord connecting two points lies outside of the set, the set is not convex. This is illustrated in Fig. 1. Examples of geometrical convex sets include balls, boxes, lines, line segments, cones, and planes.

Closed convex sets are those that contain their boundaries. In two dimensions, for example, the set of points

$$A_1 = \{(x, y)| x^2 + y^2 < 1\}$$

is not closed because the points on the circle $x^2 + y^2 = 1$ are not in the set. The set

$$A = \{(x, y)| x^2 + y^2 \le 1\}$$

is a closed convex set. A is referred to as the *closure* of A_1. Henceforth, all convex sets will be considered closed.

Strictly convex sets are those that do not contain any flat boundaries. A ball is strictly convex whereas a box is not. Neither is a line.[†]

[†]Rigorously, a set A is strictly convex if for any distinct $\vec{u}_1, \vec{u}_2 \in A$, $(\vec{u}_1 + \vec{u}_2)/2$ is an interior point of A. A vector $\vec{u}$ is called an *interior point* of closed set A if $\vec{u} \in A$ and $\vec{u} \notin \text{Closure}[E - A]$, where E is the universal set. In other words, an interior point does not lie on the boundary of a closed convex set.

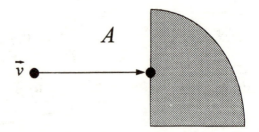

Figure 2   The set A is convex. The projection of $\vec{v}$ onto A is the unique element in A closest to $\vec{v}$.

B.   PROJECTING ONTO A CONVEX SET

The *projection* onto a convex set is illustrated in Fig. 2. For a given $\vec{v} \notin A$, the projection onto A is the unique vector $\vec{u} \in A$ such that the distance between $\vec{u}$ and $\vec{v}$ is minimum. If $\vec{v} \in A$, then the projection onto A is $\vec{v}$. In other words, the projection and the vector are the same.

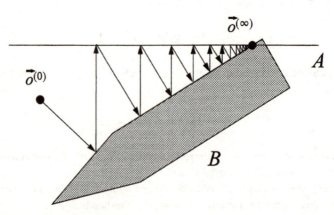

Figure 3   Alternating projection between two or more convex sets with nonempty intersection results in convergence to a fixed point in the intersection. Here, sets A (a line segment) and B are convex. Initializing the iteration at $\vec{o}^{(0)}$ results in convergence to $\vec{o}^{(\infty)} \in A \cap B$.

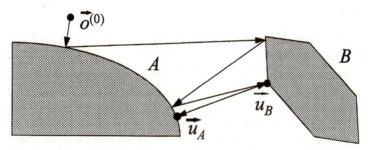

Figure 4   If two convex sets, A and B, do not intersect, POCS converges to a limit cycle, here between the points $\vec{u}_A$ and $\vec{u}_B$. The point $\vec{u}_B \in B$ is the point in B closest to the set A and $\vec{u}_A \in A$ is the point in A closest to the set B.

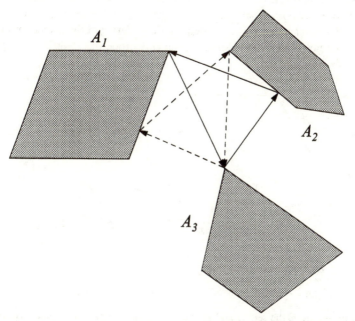

Figure 5   If three or more convex sets do not interest, POCS convergences to greedy limit cycles with no known useful properties. As illustrated here, the limit cycles can differ for different orders of set projection. The limit cycle for the projection order A_1 to A_2 to A_3, shown by the dashed line, differs from that of A_1 to A_3 to A_2.

C. POCS

There are three outcomes in the application of POCS. Each depends on the various ways that the convex sets intersect.

1. The remarkable primary result of POCS is that, given two or more convex sets with nonempty intersection, alternately projecting among the sets will converge to a point included in the intersection [1, 4]. This is illustrated in Fig. 3. The actual point of convergence will depend on the initialization unless the intersection is a single point.

2. If two convex sets do not intersect, convergence is to a *limit cycle* that is a mean square solution to the problem. Specifically, the cycle is between points in each set that are closest in the mean-square sense to the other set [19]. This is illustrated in Fig. 4.

3. Conventional POCS breaks down in the important case where three or more convex sets do not intersect [21]. POCS converges to *greedy limit cycles* that are dependent on the ordering of the projections and do not display any desirable optimality properties. This is illustrated in Fig. 5.

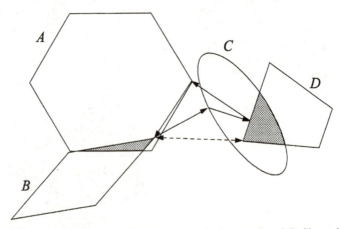

Figure 6 Here, the sets A and B intersect as do the sets C and D. Shown is one of the possible greedy limit cycles resulting from application of POCS. The order of projection is A to B to C to D. The (convex) intersections, $A \cap B$ and $C \cap D$, are shown shaded. Note, though, a greedy limit cycle between the convex sets $A \cap B$ and $C \cap D$ is also possible when using the projection order A to D to C to B. The result is shown by a dashed line. In this case, the limit cycle is akin to that obtained by application of POCS to two nonintersecting convex sets.

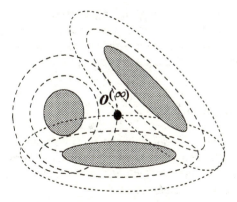

Figure 7 Conventional convex sets can be enlarged (or *fuzzified*) into fuzzy convex sets. Here, the three shaded ellipses correspond to nonintersecting convex sets. Each of the three sets is enlarged to the ellipses shown with the long dashed lines. Enlargement can be done by mathematical morphological dilation of the original convex sets. There is still no intersection among these larger sets, so the sets are enlarged again. Intersection now occurs. The resulting dashed lines are contours, or α-cuts, of fuzzy convex sets. Any group of nonintersecting convex sets can be enlarged sufficiently in this manner so that the enlarged sets (α-cuts) have a finite intersection. Ideally, the minimum enlargement that produces a nonempty intersection is used. A point in that intersection is deemed to be "near" each of the convex sets. In this figure, the point is $o^{(\infty)}$. This point can be obtained by applying conventional POCS to the enlarged sets.

This greedy limit cycle can also occur when some of the convex sets intersect and others do not. This is illustrated in Fig. 6.

Oh and Marks [22] have applied Zadeh's ideas on *fuzzy* convex sets [24] to the case where the convex sets do not intersect. The idea is to find a point that is "near" each of the sets.* Three nonintersecting convex sets are shown in Fig. 7. Each of the convex sets is "enlarged" sufficiently so that the resulting convex sets intersect.† There always exists a degree of

---

*The term "near" is referred to as a *fuzzy linguistic variable*.

†The enlargement of the set is obtained by *morphological dilation* with a convex dilation kernel set. If the kernel is a circle, dilation can be geometrically viewed as the result of rolling the circle on the boundary of the convex set. The larger the diameter of the circle, the greater the dilation. If both A and the dilation kernel set are convex, then so is the dilation. In Fig. 7, enlargement eventually results in the unique intersection point, $o^{(\infty)}$. Description of the details of this process is beyond the scope of this chapter. Details are in the papers by Oh and Marks (22, 23).

enlargement that results in a nonempty intersection of all of the sets.* The smallest enlargement that results in a nonempty intersection is sought. A point in the resulting intersection is then "near" each of the convex sets. Conventional POCS is applied to these sets to find such a point.

III.   Convex Sets of Signals

The geometrical view introduced in the previous section allows powerful interpretation of POCS applied in a signal space, $\mathscr{H}$.[†] For our notation, we will use continuous functions, although the concepts can be easily extended to functions in discrete time.[§]

A.   THE SIGNAL SPACE

An image, $u(x)$, is the $\mathscr{H}$ if**

$$\|u(x)\| < \infty,$$

where the norm of $u(x)$ is

$$\|u(x)\| = \sqrt{\int_{-\infty}^{\infty} |u(x)|^2 \, dx} \, .$$

---

*Proof: Enlarge each set to fill the whole space.

[†]More properly, a Hilbert or L_2 space (20).

[§]That is, from an L_2 to an l_2 space. One advantage of discrete time is the guaranteed strong convergence of POCS. Denote the nth element of a sequence u by $u[n]$ and let $o^{(k)}[n]$ be the kth POCS iteration. Let $o[n]$ be the point of convergence and $u[n]$ is any point (including the origin) in l_2, then

$$\lim_{k \to \infty} \|u[n] - o^{(k)}[n]\|^2 = \|u[n] - o[n]\|^2$$

for all $u[n]$. The square of the l_2 norm of a sequence $u[n]$ is

$$\|u[n]\|^2 = \sum_{n=-\infty}^{\infty} |u[n]|^2.$$

In continuous time, only weak convergence can generally be assured [4]. Specifically,

$$\lim_{n \to \infty} \int_{-\infty}^{\infty} u(x) o^{(n)}(x) \, dx = \int_{-\infty}^{\infty} u(x) o(x) \, dx$$

for all $u(x)$ is L_2. Strong convergence assures weak convergence.

**Signals in a Hilbert space are also required to be Lebesgue measurable, although this will not be of concern in our treatment.

The *energy* of $u(x)$ is

$$E = \|u(x)\|^2. \tag{1}$$

The signal space thus consists of all finite energy signals. Geometrically, $\vec{x} = u(x)$ can be visualized as a point in the signal space a distance of $\|u(x)\|$ from the origin, $u(x) \equiv 0$. Similarly, the (mean square) distance between two points $u(x)$ and $v(x)$ is simply $\|u(x) - v(x)\|$.

Two objects, $w(x)$ and $z(x)$, are said to be *orthogonal* if

$$\int_{-\infty}^{\infty} w(x)z^*(x)\,dx = 0, \tag{2}$$

where the superscript asterisk denotes complex conjugation.

Some examples of signal sets follow.

1. Convex Sets

Analogous to the definition given for sets of vectors, a set of signals, A, is convex if, for $0 \le \alpha \le 1$, the signal

$$\alpha u_1(x) + (1 - \alpha)u_2(x) \in A \tag{3}$$

when $u_1(x), u_2(x) \in A$.

2. Subspaces

Subspaces (also called linear manifolds) can be visualized as hyperplanes in the signal space. A set of signals, $\mathscr{S}$, is a subspace if

$$\beta u_1(x) + \gamma u_2(x) \in \mathscr{S} \tag{4}$$

when $u_1(x), u_2(x) \in \mathscr{S}$, and β, γ are any finite real numbers. Comparing Eq. (4) with Eq. (3) reveals that, in the same sense that lines and planes are geometrically convex, so are subspaces convex in signal space. The origin, $u(x) \equiv 0$, is an element of all subspaces.[‡]

3. Linear Varieties

A *linear variety* [20] is any hyperplane in the signal space. It need not go through the origin. All subspaces are linear varieties. A linear variety, $\mathscr{L}$, can always be defined as

$$\mathscr{L} = \{u(x) = u_{\mathscr{S}}(x) + d(x) | u_{\mathscr{S}}(x) \in \mathscr{S}\},$$

[‡]If $u(x) \in \mathscr{S}$, then, with $u_1(x) = u(x) = -u_2(x)$ and $\beta = \gamma = 1$, we have, from Eq. (4), that $0 = u(x) - u(x) \in \mathscr{S}$.

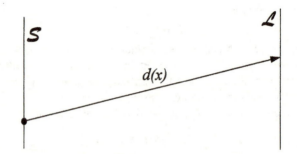

Figure 8   The subspace, $\mathscr{S}$, is shown here as a line. The function $d(x)$, assumed not to lie in the subspace, $\mathscr{S}$, is a displacement vector. The tail of the vector, $d(x)$, is at the origin $[u(x) \equiv 0]$. The set of all points in $\mathscr{S}$ added to $d(x)$ forms the linear variety, $\mathscr{L}$.

where $\mathscr{S}$ is a subspace and $d(x) \notin \mathscr{S}$ is a displacement vector. The formation of a linear variety from a subspace is illustrated in Fig. 8.

4. Cones

A *cone* with its vertex at the origin is a convex set. If $u(x) \in$ cone, then $\alpha u(x) \in$ cone for all $\alpha > 0$. Orthants* and line segments drawn from the origin to infinity in any direction are examples of cones.

B. SOME COMMONLY USED CONVEX SETS OF SIGNALS

A number of commonly used signal classes are convex. In this section, examples of these sets and their projection operators are given. Projection operators will be denoted by a $\mathscr{P}$. The notation

$$u(x) = \mathscr{P}_K v(x)$$

is read "$u(x)$ is the projection of $v(x)$ onto the convex set A_K." Note, if $v(x) \in A_K$, then $\mathscr{P}_K v(x) = v(x)$. Also, projection operators are *idempotent* in that

$$\mathscr{P}_K^2 = \mathscr{P}_K.$$

In other words, once one projects onto a convex set, an additional projection onto the same convex set results in no change. Most projections

---

*In two dimensions, an orthant is a quadrant; in three, an octant.

are intuitively straightforward. In many instances, the projection is obtained simply by forcing the signal to conform with the constraint in the most obvious way.

In the definition of some sets and projections, the Fourier transform of a signal is used. It is defined by

$$U(\omega) = \int_{-\infty}^{\infty} u(x)e^{-j\omega x}\,dx.$$

The inverse Fourier transform is

$$u(x) = \frac{1}{2\pi} \int_{-\infty}^{\infty} U(\omega)e^{j\omega x}\,d\omega.$$

1. Band-Limited Signal

The set of band-limited signals with bandwidth Ω is

$$A_{\mathrm{BL}} = \{u(x)|U(\omega) = 0 \qquad \text{for} \quad |\omega| > \Omega\}. \tag{5}$$

The set, A_{BL}, is a subspace because $\beta u_1(x) + \gamma u_2(x)$ is band limited when both $u_1(x)$ and $u_2(x)$ are band limited. To project an arbitrary signal, $v(x)$, onto A_{BL}, the Fourier transform, $V(\omega)$, is first evaluated:

$$u(x) = \mathscr{P}_{\mathrm{BL}}v(x)$$

$$= \frac{1}{2\pi} \int_{-\Omega}^{\Omega} V(\omega)e^{j\omega x}\,d\omega. \tag{6}$$

In other words, $u(x)$ is obtained by passing $v(x)$ through a low-pass filter.

2. Duration Limited

The convex set of duration-limited signals is, for a given $X > 0$,

$$A_{\mathrm{TL}} = \{u(x)|u(x) = 0 \qquad \text{for} \quad |x| > X\}. \tag{7}$$

Because the weighted sum of any two duration-limited signals is duration limited, the set A_{TL} is recognized as a subspace. To project an arbitrary $v(x)$ onto this set, the portion of $v(x)$ for $|x| > X$ is simply set to zero.

$$u(x) = \mathscr{P}_{\mathrm{TL}}v(x)$$

$$= \begin{cases} v(x), & |x| \le X, \\ 0, & |x| > X. \end{cases} \tag{8}$$

3. Real Transform Positivity

The set of signals with real and positive Fourier transform is

$$A_{POS} = \{u(x)|\mathscr{R}U(\omega) \geq 0\}, \tag{9}$$

with $\mathscr{R}$ denotes "the real part of." The set A_{POS} is a cone. To project onto this set, the Fourier transform of $v(x)$ is first evaluated. The projection follows as

$$u(x) = \mathscr{P}_{POS}v(x)$$

$$= \frac{1}{2\pi} \int_{-\infty}^{\infty} j\mathscr{I}V(\omega)[\text{POS } \mathscr{R}V(\omega)]e^{j\omega x} \, d\omega. \tag{10}$$

where POS performs a positive value operation and $\mathscr{I}$ denotes the "imaginary part of." In other words, the Fourier transform is made to be real and positive and is then inverse transformed to find the projection onto A_{POS}.

4. Constant Area

Over a given interval, $2a$, the set of signals with given area, ρ, is

$$A_{CA} = \left\{ u(x) | \int_{-a}^{a} u(x) \, dx = \rho \right\}.$$

This is a linear variety. The subspace that is displaced to form A_{CA} is that corresponding to $\rho = 0$. For $|x| \leq a$, a displacement vector is $d(x) = \rho/(2a)$.

5. Bounded Energy

The set of signals with bounded energy, for a given bound η, is

$$A_{BE} = \{u(x)|E \leq \eta\},$$

where E is defined in Eq. (1). The set A_{BE} is a ball. The projection follows as

$$u(x) = \mathscr{P}_{BE}v(x)$$

$$= \begin{cases} v(x), & \|v(x)\|^2 \leq \eta, \\ \dfrac{\sqrt{\eta} v(x)}{\|v(x)\|}, & \|v(x)\|^2 > \eta. \end{cases}$$

6. Constant Phase

The convex set of signals whose Fourier transform has a specific phase, $\varphi(\omega)$, is

$$A_{CP} = \{u(x)|\theta(\omega) = \varphi(\omega)\},$$

where arg $U(\omega) = \theta(\omega)$.*

7. Bounded Signals

For a given real signal, $a(x)$, the set of bounded signals

$$A_{BS} = \{u(x)|\mathscr{R}u(x) \le a(x)\}$$

is convex. If $a(x) \equiv$ constant, then A_{BS} is a box. As in other cases, the choice of projection is obvious. The signal is set to $\mathscr{R}u(x) = a(x)$ if $\mathscr{R}v(x)$ is too large. Otherwise, the signal remains as is.

$$u(x) = \mathscr{P}_{BS}v(x)$$

$$= \begin{cases} v(x); & \mathscr{R}v(x) \le a(x) \\ a(x) + j\mathscr{I}v(x); & \mathscr{R}v(x) > a(x). \end{cases}$$

8. Signals With Identical Middles

Let $a(x)$ be a given signal on the interval $|x| \le X$. Define

$$A_{IM} = \{u(x)|u(x) = a(x) \quad \text{for} \quad |x| \le X; \quad \text{anything for} \quad |x| > X\}. \tag{11}$$

This set is a linear variety. The parallel subspace corresponds to $a(x) = 0$ and the displacement vector is $d(x) = a(x)$. To project onto A_{IM}, one merely inserts the desired signal, $a(x)$, into the appropriate interval.

$$u(x) = \mathscr{P}_{IM}v(x)$$

$$= \begin{cases} a(x), & |x| \le X, \\ v(x), & |x| > X. \end{cases}$$

*In other words, the transform can be written in polar form as $U(\omega) = A(\omega)e^{j\theta(\omega)}$.

IV. Examples

A number of commonly used reconstruction and synthesis algorithms are special cases of POCS. In this section, we look at some specific examples.

A. VON NEUMANN'S ALTERNATING PROJECTION THEOREM

When all of the convex sets are linear varieties, POCS is equivalent to Von Neumann's alternating projection theorem (25).

B. THE PAPOULIS–GERCHBERG ALGORITHM

The Papoulis–Gerchberg algorithm is a method to restore band-limited signals when only a portion of the signal is known. It is a special case of POCS.

Given an analytic (entire) function on the complex plane, knowledge of the function within an arbitrarily small interval is sufficient to perform an *analytic continuation* of the function to the entire complex plane. The value of the function and all of its derivatives can be evaluated at some point interior to the interval and extension performed using a Taylor series. In practice, noise and measurement uncertainty prohibit evaluation of all derivatives. If the measurement of only the first three derivatives can be done with some degree of certainty, for example, only a cubic polynomial could be fitted to the point.

A Taylor series expansion at a point does not use all of the known values of the function within the interval. Slepian and Pollak [26, 27] were the first to explore analytically the possibility of reconstructing a band-limited signal* using all of the known portion of the signal. Using prolate spheroidal wave function analysis, Slepian was able to show that the restoration problem is ill-posed.

Papoulis [28–30] and later Gerchberg [31] formulated a more straight-forward, intuitive, and simple technique [27] for the same problem considered by Slepian. Assume we are given a portion of a band-limited object, $o(x)$:

$$i(x) = \begin{cases} o(x), & |x| \leq X, \\ 0, & |x| > X. \end{cases}$$

*All bandlimited signals are analytic (entire) everywhere on the finite complex plane.

We further assume we know the bandwidth, Ω, of $o(x)$. The Papoulis–Gerchberg algorithm simply alternatingly imposes the requirements that the signal (a) is band-limited and (b) matches the known portion of the signal. It consists of the following steps.

1. Initiate iteration $N = 0$ and $o^{(N)}(x) = i(x)$.
2. Pass $o^{(N)}(x)$ through a low-pass filter with bandwidth Ω.[†]
3. Set the result of the filtered signal to zero in the interval $|x| \leq X$.
4. Add the known portion of the signal, $i(x)$, over the interval $|x| \leq X$.
5. The new signal, $o^{(N+1)}(t)$, is no longer band limited. Therefore, set $N = N + 1$ and go to step 2. Repeat the process until desired convergence.

This is illustrated in Fig. 9. The numbers in Fig. 9 correspond to the numbered steps above. In the absence of noise, the Papoulis–Gerchberg algorithm has been shown to converge (32) in the sense that

$$\lim_{N \to \infty} \|o(x) - o^{(N)}(x)\| = 0.$$

Restoration of a band-limited signal knowing only a finite portion of the signal and the signal's bandwidth is ill-posed. In other words, a small bounded perturbation on $i(x)$ cannot guarantee a bounded error on the restoration. However, (1) additional, possibly nonlinear, constraints can be straightforwardly added to the iteration to improve the problem's posedness; (2) numerical results are typically good "near" where the signal is known—the restoration is known to be band limited and therefore smooth; and (3) the problem described is one of *extrapolation*. The Papoulis–Gerchberg algorithm can also be applied to interpolation, i.e., finding $o(x)$ from $o(x) - i(x)$; interpolation is well posed (32).

Youla [33] was the first to recognize the Papoulis–Gerchberg algorithm as a special case of POCS between a subspace and a linear variety. There are two convex sets:

1. The set of all signals equal to $o(x)$ on the interval $|x| \leq X$[‡]:

$$A_{\mathrm{IM}} = \{u(x) | u(x) = i(x), \quad |x| \leq X\}.$$

[†]That is, Fourier transform $o^{(N)}(x)$, set the transform equal to zero for $|\omega| > \Omega$ and inverse transform.

[‡]Equivalent to the definition in Eq. 7 with middle $m(x) = i(x)$.

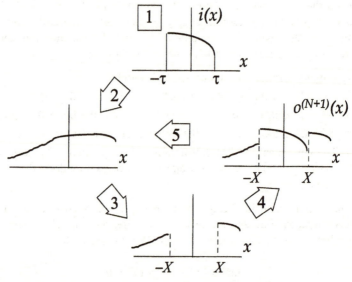

Figure 9 Illustration of the Papoulis–Gerchberg algorithm. Because no signal that is time limited can be band limited, the first estimate, $o^{(N)}(x) = i(x)$ $(N = 0)$, is incorrect. It is made band limited by the process of low-pass filtering. The bandwidth of the filter is Ω. The new signal no longer matches $o(x)$ in the middle. Thus, the signal is set to zero on the interval $|x| \le X$ and the known portion of the signal is added. The result, $o^{(N+1)}(x)$, is no longer band limited. The sharp discontinuities at the edges prohibit it from being so. Therefore, it is made band limited by the process of low-pass filtering. The process is repeated until the desired accuracy is achieved.

2. The set of all band-limited signals, A_{BL}, with a bandwidth Ω or less. This set is defined in Eq. (5).

The geometry of the Papoulis–Gerchberg algorithm in a signal space is illustrated in Fig. 10. Shown is the subspace, A_{BL}, consisting of all band-limited signals with a bandwidth not exceeding Ω. Also shown is the subspace $A_{\mathrm{IM}=0}$ consisting of all signals that are identically zero in the interval $|x| \le X$. Consider the set

$$A_{\perp} = \{u(x) | u(x) = 0, \qquad |x| > X\}.$$

The subspace $A_{\perp}$ is orthogonal* to $A_{\mathrm{IM}=0}$ because, for every $w(x) \in$

*$A_{\perp}$ is said to be the *orthogonal complement* of $A_{\mathrm{IM}=0}$. Note that, if $\mathscr{P}_{\mathrm{IM}}$ projects onto $A_{\mathrm{IM}=0}$ then $1 - \mathscr{P}_{\mathrm{IM}}$ projects onto $A_{\perp}$ where 1 is an identity operator.

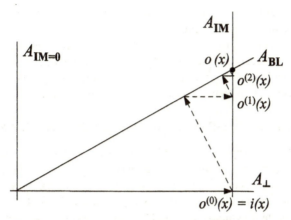

Figure 10   The Papoulis–Gerchberg algorithm in signal space. The two convex sets are (1) a (subspace) set of band-limited signals and (2) the linear variety of all signals with $i(x)$ in the middle.

$A_{IM=0}$ and $z(x) \in A_{\perp}$, Eq. (2) is true.[†] As always, the origin $[u(x) \equiv 0]$ is common to all of the subspaces.

The known portion of the signal, $i(x)$, clearly lies on the subspace, $A_{\perp}$. The signal to be recovered, $o(x)$, lies on the A_{BL} subspace. As illustrated in Fig. 10, the given signal, $i(x)$, can be visualized as the projection of $o(x)$ onto the subspace $A_{\perp}$. The linear variety, A_{IM}, in Fig. 10 is the displacement of the subspace, $A_{IM=0}$, by the orthogonal vector, $i(x)$.

In the signal space setting of Fig. 10 the Papoulis–Gerchberg algorithm can be described. The signal to be restored, $o(x)$, lies on the intersection of band-limited signals (A_{BL}) and the set of signals equal to $i(x)$ in the middle (A_{IM}). We know only that the signal to be reconstructed looks like $i(x)$ in the middle and lies somewhere in the space of band-limited signals.

Beginning with the initialization, $o^{(0)}(x) = i(x)$, we perform a low-pass filter operation. This is equivalent to projecting $i(x)$ onto the subspace of band-limited signals. The next step, as illustrated in Fig. 9, is to throw away the middle of the signal. This is the equivalent in Fig. 10 of projecting onto the $A_{IM=0}$ subspace. To this signal, we vectorally add $i(x)$ to obtain $o^{(1)}(x)$. The process is repeated. The signal $o^{(1)}(x)$ is projected onto the set of band-limited signals, projected onto $A_{IM=0}$, and then added to $i(x)$ to

[†]Indeed, $w(x)z^*(x) = 0$.

form $o^{(2)}(x)$, etc. Clearly, the iteration is working its way along the set A_{IM} toward the desired result, $o(x)$.

The alternating projections in the Papoulis–Gerchberg algorithm are performed between the subspace A_{BL} and the linear variety A_{IM}. Inspection of Section IV.A reveals the Papoulis–Gerchberg algorithm as a special case of Von Neumann's alternating projection theorem.

C. HOWARD'S MINIMUM-NEGATIVITY-CONSTRAINT ALGORITHM

Howard [34, 35] proposed a procedure for extrapolation of a interferometric signal known in a specified interval when the spectrum of the signal was known (or desired) to be real and nonnegative. The technique was applied to experimental inteferometric data and performs quite well.

As shown by Cheung et al. (36), Howard's procedure was a special case of POCS. Iteration is between the cone of signals with non-negative Fourier transforms defined by A_{POS} in Eq. (9) and the set of signals with identical middles, A_{IM}, as defined in Eq. (11). As with the Papoulis–Gerchberg algorithm, the middle is equal to the known portion of the signal.

The geometrical interpretation of Howard's minimum-negativity-constraint algorithm is similar to that of the Papoulis–Gerchberg algorithm pictured in Fig. 10, except that the subspace A_{BL} is replaced by a cone corresponding to A_{POS}.

D. RESTORATION OF LINEAR DEGRADATION

In this section, POCS is applied to a linear degradation of a discrete time signal. Denote the degradation operator by the matrix $\mathbf{S}$ and the degradation by

$$\vec{i} = \mathbf{S}\vec{o}. \tag{12}$$

If the degradation matrix, $\mathbf{S}$, is not full rank, a popular estimate of $\vec{o}$ is the *pseudo-inverse**

$$\vec{o}_{PI} = [\mathbf{S}^T\mathbf{S}]^{-1}\mathbf{S}^T\vec{i}.$$

The *projection matrix,*[†] $\mathbf{P_S}$, projects onto the column space of $\mathbf{S}$.

$$\mathbf{P_S} = \mathbf{S}[\mathbf{S}^T\mathbf{S}]^{-1}\mathbf{S}^T \tag{13}$$

---

*The solution, $\vec{o}_{PI}$, is also referred to as the *minimum norm solution.*

[†]As is necessary for a projection operator, the projection matrix is idempotent: $\mathbf{P_S^2} = \mathbf{P_S}$.

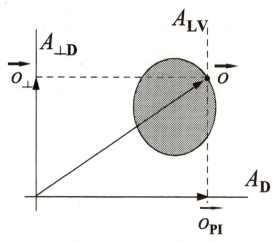

Figure 11 Illustration of restoration of a linear degradation. The signal to be reconstructed, $\vec{o}$, lies both on the linear variety $A_{\perp S}$ and on the convex set $A_\mathscr{C}$.

Consider the signal space illustrated in Fig. 11. The pseudo-inverse solution of Eq. (12) lies on the subspace A_S onto which $\mathbf{P}_S$ projects. Denote the orthogonal complement of A_S by $A_{\perp S}$. The projection operator onto this space is simply

$$\mathbf{P}_{\perp S} = \mathbf{1} - \mathbf{P}_S,$$

where $\mathbf{1}$ is an identity matrix. If the projection onto the orthogonal complement is

$$\vec{o}_\perp = \mathbf{P}_{\perp S}\vec{o},$$

we are assured that

$$\vec{o} = \vec{o}_{PI} + \vec{o}_\perp . \qquad (14)$$

Define the linear variety

$$A_{LV} = \left\{ \vec{u} \mid \vec{u} = \vec{\imath} + \mathbf{P}_{\perp S}\vec{v} \right\},$$

where $\vec{v}$ is any vector in the space.

The pseudo-inverse solution can be improved using POCS if $\vec{o}$ is also known to satisfy one or more convex constraints. In other words

$$\vec{o} \in A_m,$$

where $\{A_m | 1 \leq m \leq M\}$ is the set of constraint sets and M is the number of constraints. A set of corresponding projection operators can be defined:

$$\{\mathscr{P}_m | 1 \leq m \leq M\}. \tag{15}$$

In other words, $\mathscr{P}_m$ projects an arbitrary signal onto A_m.* Because the desired solution, $\vec{o}$, is known to lie in each of the constraint sets, it follows that

$$\mathscr{P}_m \vec{o} = \vec{o}, \qquad 1 \leq m \leq M, \tag{16}$$

and

$$\vec{o} = \mathscr{P}_M \mathscr{P}_{M-1} \cdots \mathscr{P}_2 \mathscr{P}_1 \vec{o}$$
$$= \mathscr{P}_{\mathscr{C}} \vec{o}, \tag{17}$$

where

$$\mathscr{P}_{\mathscr{C}} = \mathscr{P}_M \mathscr{P}_{M-1} \cdots \mathscr{P}_2 \mathscr{P}_1. \tag{18}$$

Correspondingly, define the (nonempty) convex set

$$A_{\mathscr{C}} = A_M \cap A_{M-1} \cap \cdots \cap A_2 \cap A_1,$$

where $\cap$ denotes intersection. Note that $\mathscr{P}_{\mathscr{C}}$ does not project onto $A_{\mathscr{C}}$. As illustrated in Fig. 11, the desired signal to be reconstructed, $\vec{o}$, is known to lie both in the convex set $A_{\mathscr{C}}$ and the set A_{LV}. A point in the intersection of these sets can be found using POCS. We have the POCS iteration

$$\vec{o}_{(N+1)} = \mathscr{P}_{\mathrm{LV}} \mathscr{P}_M \mathscr{P}_{M-1} \cdots \mathscr{P}_2 \mathscr{P}_1 \vec{o}_{(N)}$$
$$= \mathscr{P}_{\mathrm{LV}} \mathscr{P}_{\mathscr{C}} \vec{o}_{(N)}$$
$$= \vec{i} + (\mathbf{I} - \mathbf{P}_{\mathrm{S}}) \mathscr{P}_{\mathscr{C}} \vec{o}_{(N)} \tag{19}$$

The iteration is guaranteed to converge to a point on a solution set,

$$A_{\mathrm{solution}} = A_{\mathscr{C}} \cap A_{\mathrm{LV}}.$$

The solution will, in general, be nearer to $\vec{o}$ in the mean-square sense than was the pseudo-inverse. If A_{solution} consists of a single point, the solution will be unique.

The restoration in Eq. (19) is a generalization of (1) the Papoulis–Gerchberg algorithm for $\mathscr{P}_{\mathscr{C}} = \mathscr{P}_{\mathrm{BL}}$ in Eq. (6) and (2) Howard's algorithm for $\mathscr{P}_{\mathscr{C}} = \mathscr{P}_{\mathrm{POS}}$ in Eq. (10).

*When A_m is a subspace, the projection operator is a matrix. That is, $\mathscr{P}_m = \mathbf{P}_m$.

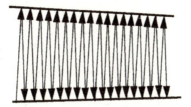

Figure 12 A geometrical example of slowly converging POCS. The intersection of the two linear varieties, far to the right, is the ultimate fixed point of the iteration.

V. Notes

In certain instances, POCS can converge painfully slowly. An example is shown in Fig. 12. One technique to accelerate convergence is relaxing the projection operation by using a relaxed projection with parameter λ [4]:

$$\mathscr{P}_{\text{relaxed}} = \lambda\mathscr{P} + (I - \lambda)I. \qquad (20)$$

An operator, $\mathscr{O}$, is said to be *contractive* if, for all $\vec{w}$ and $\vec{z}$,

$$\|\mathscr{O}\vec{w} - \mathscr{O}\vec{z}\| < \|\vec{w} - \vec{z}\|. \qquad (21)$$

In other words, operating on the two vectors places them closer together. This is illustrated in Fig. 13. A useful property of contractive operators (37) is, for any initialization, the iteration

$$\vec{o}^{(N+1)} = \mathscr{O}\vec{o}^{(N)} \qquad (22)$$

converges to a unique fixed point

$$\vec{o}^{(\infty)} = \mathscr{O}\vec{o}^{(\infty)}.$$

A POCS projection is, however, not contractive.[†] Projection operators, though, are *nonexpansive*. The $\mathscr{O}$ operator is nonexpansive if

$$\|\mathscr{O}\vec{w} - \mathscr{O}\vec{z}\| \leq \|\vec{w} - \vec{z}\|.$$

For nonexpansive operators, the iteration in Eq. (22) can converge to a number of fixed points. A relaxed nonexpansive operator, as in Eq. (20),

---

[†]If, for example, both $\vec{w}$ and $\vec{z}$ are in the convex set, and $\mathscr{O}$ is a projection operator, then $\|\mathscr{O}\vec{w} - \mathscr{O}\vec{z}\| = \|\vec{w} - \vec{z}\|$ and Eq. (21) is violated.

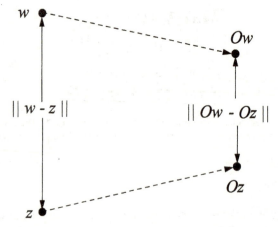

Figure 13 A geometrical example of slowly converging POCS. The intersection of the two linear varieties, far to the right, is the ultimate fixed point of the iteration.

however, is contractive (19). Applications of contractive operators do not have the elegant geometrical interpretation of POCS.

VI. Conclusions

Restoration of degraded signals can, in many cases, be posed as a special case of alternating projection onto convex sets, or POCS. The object to be restored is known to lie in two or more convex constraint sets. Restoration can be achieved by projecting alternately on each of the sets. If the sets have a nonempty intersection, then the projection will approach a fixed point lying in the intersection of the sets. If there are two sets that do not intersect, POCS will converge to a minimum mean-square error solution. If there are three or more sets with empty intersection, POCS yields results that are not generally useful. Fuzzy POCS, however, can be used to obtain a result that is "close" to each of the constraint sets.

POCS is particularly useful in ill-posed deconvolution problems. The problem is regularized by imposing possibly nonlinear convex constraints on the solution set. Using the projection onto to the column space of the convolution kernel as one of the constraints, POCS can be used, in many cases, to craft a desired result.

References

1. H. Stark, editor, "Image Recovery: Theory and Application." Academic Press, Orlando, Florida, 1987.
2. L. M. Bregman, Finding the common point of convex sets by the method of successive projections. *Dokl. Akad. Nauk. SSSR*, **162** (No. 3), 487–490 (1965).
3. L. G. Gubin, B. T. Polyak, and E. V. Raik, The method of projections for finding the common point if convex sets. *USSR Comput. Math. Math. Phys. (Engl. Transl.)* **7** (No. 6), 1–24 (1967).
4. D. C. Youla and H. Webb, Image restoration by method of convex set projections: Part I—Theory. *IEEE Transactions on Medical Imaging* **MI-1**, 81–94 (1982).
5. M. I. Sezan and H. Stark, Image restoration by method of convex set projections: Part II—Applications and Numerical Results. *IEEE Transactions on Medical Imaging* **MI-1**, 95–101 (1982).
6. S. J. Yen and H. Stark, Iterative and one-step reconstruction from nonuniform samples by convex projections. *J. Opt. Soc. Am. A* **7**, 491–499 (1990).
7. Hui Peng and H. Stark, Signal recovery with similarity constraints. *J. Opt. Soc. Am. A* **6** (No. 6), 844–851 (1989).
8. R. J. Marks II and D. K. Smith, Gerchberg-type linear deconvolution and extrapolation algorithms. *In* "Transformations in Optical Signal Processing" (W. T. Rhodes, J. R. Fienup, and B. E. A. Saleh, eds.), SPIE 373, pp. 161–178 (1984).
9. M. Ibrahim Sezan, H. Stark, and Shu-Jen Yeh, Projection method formulations of Hopfield-type associative memory neural networks. *Appl. Opt.* **29** (No. 17) 2616–2622 (1990).
10. Shu-jeh Yeh and H. Stark, Learning in neural nets using projection methods. *Optical Computing and Processing* **1** (No. 1), 47–59 (1991).
11. R. J. Marks II, A class of continuous level associative memory neural nets. *Appl. Opt.* **26**, 2005–2009 (1987).
12. R. J. Marks II, S. Oh, and L. E. Atlas, Alternating projection neural networks. *IEEE Trans. Circuits Syst.* **36**, 846–857 (1989).
13. S. Oh, R. J. Marks II, and D. Sarr, Homogeneous alternating projection neural networks. *Neurocomputing* **3**, 69–95 (1991).
14. R. J. Marks II, (ed.) "Advanced Topics in Shannon Sampling and Interpolation Theory." Springer-Verlag, Berlin, 1993.
15. S. Oh, C. Ramon, M. G. Meyer, and R. J. Marks II, Resolution enhancement of biomagnetic images using the method of alternating projections. *IEEE Trans. Biomed. Eng.* **40** (No. 4), 323–328 (1993).
16. S. Oh, R. J. Marks II, L. E. Atlas, and J. W. Pitton, Kernel synthesis for generalized time-frequency distributions using the method of projection onto convex sets. *SPIE Proc. Int. Soc. Opt. Eng.* **1348**, 197–207, (1990).

17. S. Oh, R. J. Marks II, and L. E. Atlas, Kernel synthesis for generalized time-frequency distributions using the method of alternating projections onto convex sets. *IEEE Transactions on Signal Processing* **42** (No. 7), 1653–1661 (1994).

18. P. L. Combettes, The Foundation of Set Theoretic Estimation. *Proc. IEEE* **81**, 182–208 (1993).

19. M. H. Goldburg and R. J. Marks II, Signal synthesis in the presence of an inconsistent set of constraints. *IEEE Trans. Circuits Syst.* **CAS-32** 647–663 (1985).

20. D. G. Luenberger, "Optimization by Vector Space Methods." Wiley, New York, 1969.

21. D. C. Youla and V. Velasco, Extensions of a result on the synthesis of signals in the presence of inconsistent constraints. *IEEE Trans. Circuits Syst.* **CAS-33** (No. 4), 465–468 (1986).

22. S. Oh and R. J. Marks II, Alternating projections onto fuzzy convex sets. *Proceedings of the Second IEEE International Conference on Fuzzy Systems*, (*FUZZ-IEEE '93*), San Francisco, March 1993 **1**, 148–155 (1993).

23. R. J. Marks II, L. Laybourn, S. Lee, and S. Oh, Fuzzy and extra-crisp alternating projection onto convex sets (POCS). *Proceedings of the International Conference on Fuzzy Systems* (*FUZZ-IEEE*), Yokohama, Japan, March, 20–24, 1995, pp. 427–435 (1995).

24. L. A. Zadeh, Fuzzy sets. *Information and Control* **8**, 338–353 (1965); reprinted in J. C. Bezdek, (ed.) "Fuzzy Models for Pattern Recognition" IEEE Press, 1992.

25. J. Von Neumann, "The Geometry of Orthogonal Spaces." Princeton Univ. Press, Princeton, New Jersey, 1950.

26. D. Slepian and H. O. Pollak. Prolate spheroidal wave function Fourier analysis and uncertainty I. *Bell Syst. Tech. J.* **40**, 43–63 (1961).

27. R. J. Marks II, "Introduction to Shannon Sampling and Interpolation Theory." Springer-Verlag, Berlin, 1991.

28. A. Papoulis, A new method of image restoration. *Joint Services Technical Activity Report* **39** (1973–1974).

29. A. Papoulis, A new algorithm in spectral analysis and bandlimited signal extrapolation. *IEEE Trans. Circuits Syst.* **CAS-22**, 735–742 (1975).

30. A. Papoulis, "Signal Analysis." McGraw-Hill, New York, 1977.

31. R. W. Gerchberg, Super-resolution through error energy reduction. *Opt. Acta* **21**, 709–720 (1974).

32. R. J. Marks II. "Introduction to Shannon Sampling and Interpolation Theory." Springer-Verlag, Berlin, 1991.

33. D. C. Youla, Generalized image restoration by method of alternating orthogonal projections. *IEEE Trans. Circuits Syst.* **CAS-25**, 694–702 (1978).

34. S. J. Howard, Fast algorithm for implementing the minimum-negativity constraint for Fourier spectrum extrapolation. *Appl. Opt.* **25**, 1670–1675 (1986).

35. S. J. Howard, Continuation of discrete Fourier spectra using minimum-negativity constraint. *J. Opt. Soc. Am.* **71**, 819–824 (1981).

36. K. F. Cheung, R. J. Marks II, and L. E. Atlas, Convergence of Howard's minimum negativity constraint extrapolation algorithm. *J. Opt. Soc. Am. A* **5**, 2008–2009 (1988).

37. A. W. Naylor and G. R. Sell, "Linear Operator Theory in Engineering and Science." Springer-Verlag, New York, 1982.

Index